ALLE ZEIT WACH
1842

Victor A. Drits   Cyril Tchoubar

X-Ray Diffraction by Disordered Lamellar Structures

Theory and Applications to Microdivided Silicates and Carbons

With the Collaboration of
G. Besson, A. S. Bookin, F. Rousseaux
B. A. Sakharov and D. Tchoubar

Foreword by André Guinier

Springer-Verlag Berlin Heidelberg New York
London Paris Tokyo Hong Kong Barcelona

Professor Victor A. Drits
Geological Institute
Academy of Sciences
7 Pyzhevsky perspekt
109017 Moscow, USSR

Professor Cyril Tchoubar
Université d'Orléans
Laboratoire de Cristallographie
(associé au CNRS)
Rue de Chartres
F-45067 Orléans Cedex, France

Translated from French by:
R. Setton, National Center of Scientific Research (CNRS), C.R.S.O.C.I., Orléans, France

Library of Congress Cataloging-in-Publication Data
Cyril Tchoubar. X-ray diffraction by disordered lamellar structures: theory and applications to microdivided silicates and carbons / Victor A. Drits, Cyril Tchoubar [sic] with the collaboration of G. Besson ... [et al.]; [translated from the French by R. Setton]. Includes bibliographical references.

ISBN-13: 978-3-642-74804-2 e-ISBN-13: 978-3-642-74802-8
DOI: 10.1007/978-3-642-74802-8

1. X-ray crystallography. 2. X-rays – Diffraction. I. Tchoubar, Cyril, 1932– . II. Title. III. Title: Disordered lamellar structures. QD945.D75 1990 548'.83 – dc 20

Softcover reprint of the hardcover 1st edition 1990

Typesetting: K+V Fotosatz GmbH, Beerfelden
2132/3145 543210 – Printed on acid-free paper

Foreword

Saying that X-ray diffraction reveals the atomic structure of solids would be commonplace if one were not to add that two types of problems are involved, each requiring its own method of approach.

It is possible to obtain, from a "proper", isolated crystal of adequate size, a diffraction pattern yielding the atomic structure unambiguously, even if the number of atoms in the lattice is large. The problem can be considered as solved, as witnessed by the importance of its results in chemistry, biology, etc.

A different matter is that of imperfect crystals, in which the periodicity of the atomic positions is only partial, or even approximate. The difficulty is further compounded by the fact that, in general, only microcrystalline and poorly oriented powders are on hand. The experimental data are then insufficient for a complete determination of the structure at the atomic level since X-ray diffraction is not equivalent to a microscope with a resolution of the order of the atomic diameters, enabling us to "see" the structure. What could then be the maximum information to be drawn from the experimental determinations?

The answer to this question is provided by the authors for lamellar solids, a very numerous class of imperfect crystals, of which clays are the best known example. The method consists in calculating the diffraction pattern of a model drawn up using all the available information on the sample, then in adjusting arbitrarily chosen parameters to obtain the best fit with the experimental data. The corresponding structure is thus a possible structure, it may even be the likely one, but it has not been proved to be *the* actual true structure.

From the many examples fully discussed in this work, a few capital ideas emerge:

– one can no longer be content, as was done a few decades ago, with approximate or even merely qualitative results; calculations on models are possible and the computerized comparison of these results with the experimental data can bring out exact values of unknown parameters;

– it is imperative that the determination of the intensity of the diffracted wave be accurate and free from systematic errors, since results obtained otherwise could correspond to anything. Modern X-ray sources, new detectors, and the monochromatization of the incident beam permit excellent measurements, at least if one is willing to take the trouble to obtain them. The insistence of the authors on this most important point, a prerequisite to the increase in our knowledge of disordered structures, is all to their credit.

The richness and progress in the knowledge of the structure of layered materials brings to our mind the merit and insight of one of the pioneers in this field, namely

Jacques Méring. A gifted theoretician, he successfully computed the diffraction by model structures, while simultaneously impressing his co-workers with the need for excellence in the quality of the measurements. Many of the authors in this book were his students. Their extensive achievements are a fitting tribute to the memory of Jacques Méring.

Paris, August 1990 André Guinier

Preface

It is well known that many physical and physico-chemical properties of solids are directly related to certain defects in their crystal structure: the structure of an actual crystal is always somewhat different from its idealized representation as determined by the symmetry group of the crystal and by the atomic motif, on the scale of the unit cell.

Divergences from the idealized structure may vary greatly. They may be related, within the actual crystal, to isolated point defects due to voids or, in contrast, to interstitial atoms, or to the isomorphic replacement of certain atoms within the motif by atoms of a different nature. Structural defects may also gather in zones: for example, the hearts of screw or wedge dislocations are the origin of whole areas of linear defects, whereas stacking or twinning faults, or polysomatism, induce the formation of planar defects. Furthermore, many crystals are also subject to internal microtensions which perturb the perfect periodicity of the crystal structure along one or more directions in space. In the same way, among all natural or synthetic compounds, some are found to have a so-called interstratified structure with a more or less regular stacking of layers differing in chemical composition as well as in crystal structure. In some cases, these interstratified compounds are made up of two-dimensional lattices non-commersurable with each other. Lastly, one may also find distortions or structural elements associated in such a way that the structure of the crystal is modulated and, possibly, non-commensurable as well.

The crystallographer seeking to characterize such structures will have to determine parameters specific to different types of defects, as well as the proportions in which they occur, their exact location and their possible interactions. Two methods of analysis are used to determine these structural characteristics:

– on one hand, spectroscopic methods – IR, Mössbauer, NMR, EXAFS, etc. – which allow the determination of local structures related to the order-disorder at short distances; these methods are suitable for the study of point defects, of deformed interatomic bindings, of the nature of sites where isomorphic replacements occur and for the study of their distribution at short distances, etc.;

– on the other hand, diffractometric methods – of X-rays, electrons or neutrons – which, with the help of statistical parameters, allow the characterization of the distribution of point defects, of the nature of linear and planar defects, their localization, and the determination of their interactions within the solid.

The diffractometric technique most commonly used in determining these characteristics is X-ray diffraction (XRD). This is due, first of all, to the relative ease of application of the method and to the moderate cost of the equipment; in the second place, compounds thus analyzed generally do not need to be put under experimental conditions likely to modify them, such as vacuum. However, the crystallographer's task becomes particularly tricky when the object is microdivided – a case frequently encountered – for it will then only be possible to exploit data from a polycrystalline system. Thus, over a long period of time, most interpretations of complicated diffraction effects produced by microdivided systems with a partially disordered crystal structure were based on an intuitive approach; it is only within the last few decades that methods of interpretation of diagrams have excluded the excessively large emphasis formerly placed on intuition.

Our aim, in the present book, is to describe the methods best suited to the determination, by XRD, of deviations from real and idealized average structures, the basic idea being that the only methods permitting the determination, at one and the same time, of the nature, quantity, position and interaction of structural defects in a given solid are indirect or modelization methods. These are generally used in several steps: a first study examines all the structural models containing various types of defects compatible with the specific crystallo-chemical family to which a given solid belongs. The second step is the calculation of the 'synthetic' diagrams corresponding to each of the structural models examined and the determination of the effect produced by the variation of the parameters characteristic of each model on the intensity distribution of the diffracted waves. Finally, retaining only the model or models producing synthetic diagrams closest to the experimental spectrum, the characteristic parameters of this model or models are modified step by step to obtain the best agreement between the experimental and calculated spectra. The concordance of the two spectra justifies the attribution of the parameters of the best possible model, considered to be representative of the real structure, to the solid under analysis.

We have purposely limited our presentation to *microdivided systems with a layered structure*, a choice that may seem restrictive. However, layered systems include large numbers of mineral families and synthetic compounds and it will be seen that the methods of analysis described in this book can often be applied to structural studies of some non-layered crystals, since their diffraction patterns are easier to interpret when viewed as diffraction phenomena originating from a layered system.

The book is divided into three parts of unequal length and importance. The first part, consisting of five chapters, is devoted to theoretical and mathematical developments which describe the diffraction from microdivided systems of layered structures with different kinds of structural defects. The first, introductory, chapter presents the entire set of defects likely to exist in lamellar systems. The second chapter is a description of the particulars of powder XRD patterns in the absence of structural defects but in the presence, on one hand, of extremely small domains of interferential coherence – limited, at most, to a few tens of unit cells – and, on the other hand, to preferential mutual orientation of crystallites in the powder, a case often occurring in layered systems forming highly anisotropic shapes of crystallites. The third chapter describes the modifications in the diffraction caused by the presence of position faults among stackings of identical layers, one of the most commonly occurring defects in microcrystalline-layered systems. The specific modifications of the XRD patterns are described for randomly distributed faults without mutual interaction in the following cases: faults due to arbitrary rotations or translations of layers in their own plane; faults due to partially defined rotations or translations (such as $n \times 120°$ rotations, with n integral); fluctuations about a mean value of the relative position of first-neighbor layers, etc. The third chapter also introduces the matrix formalism best adapted to computer calculation of the synthetic powder patterns. The fourth chapter deals with the description of a second, most important category of structural defects, i.e. the more or less regular interstratification of different types of layers. Various probabilistic parameters are defined which allow the description and characterization of any interstratified system, whatever the type of sequence of layers or distances, these parameters taking into account a greater or lesser range of interaction – which some authors call "Reichweite" – as well as the possible presence of a partial segregation among layers of each type. The concluding chapter of this part introduces the matrix formalism which will allows – with the help of the synthetic diagrams – the determination of the probabilistic parameters characteristic of a given interstratification.

The second part of our book (Chapter 6) is devoted to the description of the experimental conditions necessary to obtain and record diagrams adequate for the modelization methods. This chapter purposely avoids too detailed a description of XRD techniques since these have already been given in a number of excellent works. We simply mention the precautions that must be taken to prevent distortions in these diagrams caused by inappropriate experimental conditions since these distortions could be as large as those due to the presence of structural defects in the solid.

The last three chapters of the book give different applications of the modelization methods. Chapter 7 deals with the problems met with more or less graphitized carbons and goes on to show just how accurate characterization is likely to be when the structure of the layer itself is simple. Chapter 8 presents characterizations of actual structures of some microcrystallized layered silicates, both in regard to layer sequence – i.e. stacking faults – and to internal structure of the layers and interlayer space organization, particularly where the distribution of water molecules is concerned, as is the case with certain swelling clays. Lastly, Chapter 9 gives numerous examples of the application of modelization methods to the characterization of varied layer sequences in complex interlayered minerals, including minerals with a high proportion of lacunar defects which can be described using a formalism adapted to interstratified systems.

We have tried to make our book interesting to scientists in many different fields, such as the chemistry of solids, geology, mineralogy, the physics of materials and the science of soils, and mathematical developments therefore always begin at a fairly elementary level. Similarly, every practical application of the modalization methods is preceded by a detailed account of the characteristics of the mineral group studied.

The seven French and Soviet scientists who jointly contributed to this book are attached to two groups which have specialized, for several decades, in the study of structural defects in microdivided systems. For about 10 years now, they have been able to collaborate thanks to the support granted by the scientific sections of the department of international relations of the Soviet Academy of Sciences, the French C.N.R.S., the ministries of Education and of Foreign Affairs, to all of whom we particularly wish to extend our thanks.

We should also like to recall that all the co-authors of this book were fervent admirers and sometimes former students or collaborators of the late Jacques Méring, Director of Research at the C.N.R.S. and one of the main pioneers in the field of modelization methods described in the present book.

Professor André Guinier, member of the French Académie des Sciences and undisputed world specialist in all subjects concerning the structure of matter, did us the honor of reading our book and writing a Preface for it, and we wish to thank him for his invaluable advice concerning the contents of the present work. Some of us had the good fortune to work under his direction but, to all of us, Professor Guinier's achievements were the basis of our knowledge of and enthusiasm for crystallography.

We also wish to thank Ralph Setton, Director of Research at the C.N.R.S. and himself a specialist on carbons, who faithfully translated this book without changing its initial spirit in any way and helped us, all along, with his remarks on its subject matter.

Lastly, we would like to thank Madame Marcelle Chauvette, of the Laboratory of Crystallography at the University of Orléans, who took on the thankless, tricky and tiresome task of typing the final text of this book.

Moscow, Orléans
August 1990

Victor A. Drits, Cyril Tchoubar

Contents

Foreword V

Preface VII

Chapter 1
Overall Description of Imperfect Lamellar Crystals 1

1.1 Some Reminders on the Specific Characteristics of Crystals with a Triperiodic Structure 1
1.2 Range of Validity of the Direct Methods of Structural Analysis .. 1
1.2.1 Crystals with Point Defects 2
1.2.2 Crystals with Planar Defects 3
1.3 Indirect Structural Analysis of Partially Disordered Lamellar Systems. Principles of Their Modelization 4
1.4 Determination of the Structural Characteristics of the Layers 5
1.5 General Characteristics of Triperiodic Layer Stackings 8
1.5.1 Characteristic Translations of Layer Stackings 8
1.5.2 Polytypic Modifications 9
1.5.3 Triperiodic Lamellar Structures with Layers Containing Isomorphic Substitutions or with Different Types of Layers 10
1.6 Principal Characteristics of Lamellar Structures with Stacking Faults 10
1.6.1 Translation Stacking Faults 11
1.6.2 Rotation Stacking Faults 13
1.6.3 Stacking Faults Due to Enantiomorphism 13
1.6.4 Well-Defined Stacking Faults 17
1.6.5 Random Stacking Faults 18
1.6.6 Stacking Faults Due to Fluctuations in the Position of the Layers. Disorders of the First and Second Types 18
1.6.7 Particles, Crystallites and Interferential Coherence Domains 21
1.7 Principal Characteristics of Interstratified Minerals 22
1.7.1 Interstratified System Characterization by the Stacking Mode of the Layers 23
1.7.2 Order-Disorder in the Sequence of Layers of Different Types 24
1.8 Commensurate and Incommensurate Structures in Interstratified Systems 27
References 31

Chapter 2
Theory of the Diffraction Phenomenon Produced by Powders of Microcrystals with a Lamellar Structure ... 33

2.1 Diffraction from an Isolated Layer of Finite Extent ... 33
2.2 Diffraction from a Defect-Free Stack of Identical Layers ... 37
2.2.1 General Description of the Diffraction ... 37
2.2.2 Effect of the Thinness of the Interferential Coherence Domains on the Intensity Distribution. Apparent Irrationality of the $00l$ Reflections ... 40
2.3 Diffraction by a Powder of Particles with Totally Random Orientation ... 43
2.3.1 General Expression for the Intensity of the Wave Diffracted by an Isotropic Powder ... 43
2.3.2 The Tangent Cylinder Approximation ... 46
2.3.3 Physical Significance of and General Expression for T(U) ... 47
2.3.4 Computation of T(U) for Rectangular Interferential Coherence Domains ... 52
2.4 Diffraction from a Powder of Partially Oriented Particles ... 54
2.4.1 Definition of the Spatial Distribution of the Particles in a Powder 55
2.4.2 Diffraction from a Partially Oriented Powder in a Symmetrical $\theta - 2\theta$ Transmission Mounting ... 57
2.4.3 Diffraction from a Partially Oriented Powder in an Asymmetrical Transmission Mounting or in a Reflection Arrangement ... 62
2.4.4 Diffraction from a Partially Oriented Powder in the Particular Case of the (00) Rod ... 64
References ... 66

Chapter 3
Diffraction from Lamellar Crystals with Stacking Faults ... 69

3.1 General Expression for the Diffraction Produced by Stacks of Layers with Position Defects ... 70
3.1.1 Mathematical Description of the Diffraction ... 70
3.1.2 The Matrix Formalism ... 71
3.2 Diffraction Produced by Stacks Containing Rotation or Translation Faults Without Mutual Interaction ... 77
3.2.1 Effects of Random Rotation or Translation Stacking Defects on the Diffraction ... 77
3.2.2 Effect of Well-Defined Translation Defects on the Diffraction ... 81
3.2.3 Effect of Well-Defined Rotation Defects on the Diffraction ... 83
3.3 Diffraction Produced by Stacks with Defects Due to Fluctuations in the Positions of the Layers ... 87
3.3.1 Position Fluctuations Leading to a Disorder of the First Type ... 89
3.3.2 Position Fluctuations Leading to a Disorder of the Second Type . 90

3.3.3 Determination of the Mean Standard Deviation of the Fluctuations Affecting the Interlayer Distances by Direct Profile Analysis of the 00*l* Reflections 94
3.3.4 Comparison of the Effects of Random Defects and of Position Fluctuations on the Diffraction 96
3.3.5 Comparison of the Physical Significances Attached to the Concepts of Random Defects and of Position Fluctuation Defects 98
References 101

Chapter 4
Statistical Models and Parameters Used to Describe Interstratified Lamellar Systems 103

4.1 General Parameters Characterizing the Stacking of Different Layers in Interstratified Structures 103
4.2 Interstratified Structures with S = 0 106
4.3 Interstratified Structures with S = 1 107
4.3.1 Determination of the Independent Parameters Characterizing Two-Component Structures 107
4.3.2 Classification of Two-Component Structures as a Function of the Degree of Order in the Sequence of Layers 109
4.3.3 Interstratified Structures with Three Types of Layers 112
4.4 Interstratified Structures with S = 2 114
4.4.1 Relationships Between the Proportions of Different Types of Layers and the Conditional Probabilities 114
4.4.2 Choice of the Independent Parameters 116
4.4.3 Classification of Structures with S = 2 as a Function of the Degree of Order in the Sequence of Layers 118
4.4.4 Interstratified Structures with S = 2 and g Types of Layers 122
4.5 Interstratified Structures with S = 3 122
4.6 Degree of Homogeneity for Powders of Thin Particles with Markovian Interstratification (Quasi-Homogeneous System) 124
4.7 Parameters for the Characterization of Homogeneous Interstratified Systems 127
4.7.1 Homogeneous Two-Component (A and B) Systems with S = 0 ... 129
4.7.2 Homogeneous Two-Component Systems with S ≠ 0 and Restrictive Conditions for the Sequence of Layers 131
References 132

Chapter 5
Diffraction Methods Adapted to the Structural Analysis of Interstratified Systems 135

5.1 Direct Methods of Structural Analysis 136
5.1.1 The Method of D'yakonov 136

5.1.2 Computation of the Function $\varphi'(z)$ ... 138
5.1.3 Comparison of the Mac Ewan and D'yakonov Direct Methods of Structural Analysis ... 139
5.2 Indirect Methods of Structural Analysis Based on the Computation of the Intensities of Basal Reflections ... 140
5.2.1 Calculation of an Interference Function Using a Single Structure Factor ... 140
5.2.2 Methods of Intensity Calculation Using Different Structure Factors ... 142
5.3 Diffraction by Systems with g Types of Layers, with a Specific Translation r Between the Adjacent i-Type and j-Type Layers, for any Given Value of S ... 148
5.3.1 Expressions for the Matrices [W], [Φ], and [Q], when $S = 0$ or 1 150
5.3.2 Expressions for [W], [Φ], and [Q] when $S = 2$ ... 150
5.3.3 Expressions for [W], [Φ], and [Q] when $S = 3$ ... 153
5.3.4 The Matrices [W], [Φ] and [Q] in the Case of Interstratified Systems with g Components, for any Given Value of S ... 155
5.4 Intensity of the Wave Diffracted by Systems with g Types of Layers, for any Value of S and R ... 156
5.4.1 Matrix Formalism for Systems with Identical Layers in the Same Azimuthal Orientation, with Translational Defects and an Interaction Parameter $R \geq 1$ ... 156
5.4.2 Matrix Formalism for Interstratified Structures with any Number of Translations Without Mutual Interaction ($R = 0$) ... 157
5.4.3 Matrix Formalism in the General Case of Interstratified Systems . 159
5.5 General Remarks ... 163
References ... 163

Chapter 6
Experimental Techniques Adapted to the Study of Microdivided Lamellar Systems ... 165

6.1 Survey of the Techniques Most Frequently Used in Powder Diffractometry ... 165
6.1.1 The Powder Diagram ... 165
6.1.2 The Debye-Scherrer-Hull Mountings ... 166
6.1.3 Use of a Recording Counter and of a Monochromator ... 168
6.1.4 Advantages and Drawbacks of the Reflection and Transmission Mountings ... 170
6.2 Adaptation of Transmission Techniques to the Study of Microdivided Lamellar Systems ... 171
6.2.1 The X-Ray Source ... 171
6.2.2 The Monochromator ... 172
6.2.3 Particular Features of the Specimen ... 174
6.2.4 The Goniometer ... 174

6.2.5 The Detector and Counting Equipment 175
6.3 Perturbing Factors which can be Minimized 177
6.3.1 Choice of the Slit-Widths in the Path of the X-Ray Beam 177
6.3.2 Optimal Slit Height 178
6.3.3 Choice of Sample Thickness 179
6.4 Principal Corrections on the Diffraction Patterns 181
6.4.1 Correction of Effects Due to Polarization of the X-ray Beams 181
6.4.2 Correction of Effects Due to Sample Absorption 181
6.4.3 Correction for the Nonlinear Response of the Localization Detector 185
6.5 Perturbing Factors Introduced in the Computation of the Theoretical Diffractograms 187
6.5.1 Lorentz Factor 187
6.5.2 Orientation Function for the Particles in a Powder 190
6.6 Determination of the Absolute Intensity Scale 194
6.6.1 Definition of the Absolute Scale 194
6.6.2 Determination of the Absolute Scale 195
6.6.3 Examples and Applications 195
References 198

Chapter 7
Structural Characteristics of Carbons 199

7.1 General Characteristics of Carbon Materials 199
7.1.1 General Description 199
7.1.2 Basic Features of the Graphitization Process 200
7.2 Structural Characteristics of the Graphitization Process 201
7.2.1 Structural Study of the Carbon Layers 202
7.2.2 Examples of Structural Evolution in the Carbon Layer as a Function of the Thermal Treatment 207
7.3 Organization of the Stacks 220
7.3.1 Structure of the Stacks in the two Graphite Polytypes 220
7.3.2 Structure of the Stacks in a Carbon Undergoing Graphitization 221
References 230

Chapter 8
The Modelization Method in the Determination of the Structural Characteristics of Some Layer Silicates: Internal Structure of the Layers, Nature and Distribution of the Stacking Faults 233

8.1 Structural Defects in Kaolinite 233
8.1.1 Common Features of the Layers in Kaolin Minerals 235
8.1.2 Common Features of the 1:1 Layers in Dickite and Nacrite 235
8.1.3 Characteristics of the 1:1 Layer in Kaolinite 237
8.1.4 Comparison of the Kaolinite and Dickite Unit Cells 238

8.1.5 Models for the Stacking Faults in Kaolinite 240
8.1.6 Comparison Between Calculated and Experimental XRD Patterns 249
8.2 Distribution of the Cations in the *cis* and *trans* Octahedral Sites of Dioctahedral Smectites 254
8.2.1 Preparation Techniques for Smectite Samples Used in the Diffractometric Determination of the Distribution of Cations in Octahedral Sites 255
8.2.2 Determination of the Octahedral Cation Distribution in K-Smectites by Oblique Texture Electron Diffraction 256
8.2.3 Determination of the Distribution of Octahedral Cations in K^+-Nontronites by XRD 262
8.2.4 Analysis of the XRD Powder Patterns from Dioctahedral Cs-Smectites 267
8.3 Determination of the Distribution of Cations and Water Molecules in the Interlamellar Spaces of Dioctahedral Smectites . 272
8.3.1 Experimental Conditions 272
8.3.2 Analysis of the Profile of the 00*l* Reflections from a Two-Water-Layer Na-Beidellite (Sample E_2) 273
8.3.3 Qualitative Description of the (02, 11), (20, 13) and (04, 22) Bands Given by Two-Water-Layer Na^+-Beidellite 275
8.3.4 Determination of the x, y Coordinates of the Sites Occupied by Water Molecules 277
8.3.5 The Different Possible Stackings of Layers in Two-Water-Layer Na-Beidellite 280
8.3.6 Determination of the Structural Characteristics of the Two-Water-Layer Na^+-Beidellite by Fitting the Calculated Pattern to the Experimental XRD Data 280
8.3.7 Structural Characteristics of One-Water-Layer Na^+-Beidellite. Comparison with the Two-Water-Layer Hydrate 281
8.4 Structural Defects in Glauconites 284
8.4.1 Structure of the Glauconites 284
8.4.2 Choice of the Samples, Experimental Conditions and Description of the Experimental Diffractograms 285
8.4.3 Determination of the Unit Cell Parameters and of the Atomic Coordinates 286
8.4.4 Structural Models for Glauconites Devoid of Stacking Defects ... 289
8.4.5 Models with Well-Defined $\pm 120°$ Rotational Stacking Faults 293
8.4.6 Structural Models with Enantiomorphic Layers 294
8.4.7 Structural Model with n 60° Rotational Stacking Faults and $R = 0$ 296
8.4.8 Structural Model with n 60° Rotational Stacking Faults and $R = 1$ 298
8.4.9 Determination of the Structural Parameters Characteristic of Glauconites 298
References 300

Chapter 9
Determination of the Structural Characteristics of Mixed-Layer Minerals 305

9.1 The Method of D'yakonov ... 305
9.1.1 Practical Example of the Use of the D'yakonov Method ... 306
9.1.2 Appraisal of the D'yakonov Method ... 310
9.2 General Guidelines for the Use of Modelization of X-Ray Diffractograms in the Study of Mixed-Layer Minerals ... 315
9.2.1 Determination of the Nature of the Layer Types ... 316
9.2.2 Chemical Composition and Structure of the Layers and of the Interlayer Spaces ... 318
9.2.3 Choice of the Origin for the z Ordinates of Atoms in the Scattering Units ... 318
9.3 Calculation of the Reference X-Ray Diffractogram for Quasi-Homogeneous Interstratified Minerals ... 320
9.3.1 Specific Features of the X-Ray Diffractograms Given by Interstratified i-m Systems with S = 0 or 1 ... 322
9.3.2 Characteristics of the X-Ray Diffractograms Given by i-m Systems with S = 2 and $W_i > W_m$ ... 325
9.3.3 Specific Features of the X-Ray Diffractograms from Interstratified i-m Structures with S = 3 ... 330
9.3.4 Comparison of the Specific Features of the Diagrams Given by Systems with S = 0, 1, 2, 3 ... 331
9.4 Parameters Other than W and S which Influence the Profile of the Calculated X-Ray Diagrams of Two-Component Interstratified Systems ... 332
9.4.1 Influence of the Thickness of the Scattering Units ... 332
9.4.2 Influence of the Thickness of the Interferential Coherence Domains 333
9.4.3 Physical Mixtures of Quasi-Homogeneous Mixed-Layer Systems . 335
9.4.4 Homogeneous Mixed-Layer Models ... 341
9.5 Quantitative Determination of the Structural Characteristics of Interstratified Dioctahedral Mica-Smectite Minerals ... 341
9.5.1 Two-Component Interstratified Minerals: Celadonite-Nontronite .. 343
9.5.2 Three-Component Interstratified Minerals: Leucophyllite-Montmorillonite-"Vermiculite" with S = 1 ... 344
9.5.3 Two-Component Interstratified Minerals: Leucophyllite-Montmorillonite, with S = 3 ... 346
9.6 Semi-Quantitative Determinations of the Structural Characteristics of Interstratified Minerals ... 346
9.6.1 Two-Component Interstratified Minerals: Illite-Montmorillonite .. 347
9.6.2 Interstratified Minerals with Kaolinite 1:1 Layers ... 348
9.6.3 Study of Hydrated Talcs ... 355
References ... 358

Author Index ... 361

Subject Index ... 365

CHAPTER 1

Overall Description of Imperfect Lamellar Crystals

1.1 Some Reminders on the Specific Characteristics of Crystals with a Triperiodic Structure

Periodicity is the characteristic common to all crystallized objects; it is determined by the position in space of all the elements which constitute the crystal: atoms, ions, molecules, etc. The distribution of these elements in the volume of the crystal constitutes the crystal structure.

Because of this treble periodicity, the crystal contains a large number of equivalent points around which are found the same atoms, at the same distances and orientations. The set of these equivalent points constitutes the triperiodic *lattice* of the crystal.

The crystal lattice is usually characterized by a single parallelepiped called the *unit cell*, which conforms with the symmetry of the lattice and which is defined by three fundamental and noncoplanar vectors, $\vec{a}_1$, $\vec{a}_2$, $\vec{a}_3$, with the same origin, each one parallel and equal in length to one of the edges of the parallelepiped. One could also say that the crystal lattice is, in a way, a geometric representation akin to an operator which returns all the equivalent points and characterizes the periodicity of position of the atoms in the whole crystal. Consequently, a three-dimensional crystal structure can be described by merely specifying the arrangement of the elements: atoms, ions, molecules, etc. within a unit cell with specific shape and dimensions. Such an arrangement, often called the **basis,** will here be called the **atomic motif** or **atomic pattern.** The position of all the atoms in the bulk of the crystal is obtained by translating such a unit cell along the vectors corresponding to all the linear combinations $p_1\vec{a}_1+p_2\vec{a}_2+p_3\vec{a}_3$ of the fundamental vectors (p_1, p_2, p_3 = integers).

1.2 Range of Validity of the Direct Methods of Structural Analysis

When the crystallized object to be analyzed is a perfect crystal, the problem of its structural analysis can be reduced to the determination of the unit cell and of the atomic pattern. The data obtained by X-ray or neutron diffraction from single crystals are used to solve this problem. At present, this entails using a four-circle goniometer and recording the integrated intensity of the reflections produced by single crystals; the diffracted beams are deemed to reach maximum intensity at the center of the nodes of the three-dimensional reciprocal lattice. Thanks to the present development of computational facilities, it is now easy to get back the positions of the atoms and the effective factors of atomic scattering by using one of the direct methods of structural analysis, such as that of the Patterson series. Refinements of the structural characteristics are then obtained, e.g., by using least squares.

When the crystal contains no – or hardly any – structural defects, this determination of the atomic pattern constitutes the characterization of the true crystal structure. This, however, will not be the case when the crystal structure is marred by the presence of structural defects, which can be of various types.

1.2.1 Crystals with Point Defects

We shall first consider the case of crystals with isomorphic substitutions. In general, these structures do not show true triperiodicity in the real sense of the word since, as a rule, there is no long-range order in the distribution of isomorphic substitutions, as homologous points of the structure may be occupied by different kinds of atoms. An isomorphic substitution can be considered as a point defect which results, for instance, in the replacement of a cation of a certain size by another cation of a different size and/or with a different charge. As a consequence, the atoms in an actual crystal with this type of defect are displaced with respect to the position which they would occupy in the perfect lattice. However, since the displacements are small, it is possible to describe the structure of a crystal with isomorphic substitutions in terms of a mean idealized lattice not very different from the true lattice. As with perfectly ordered crystals, the X-ray diffractograms of isomorphically substituted mineral species also show individualized reflections centered on the nodes of the reciprocal periodic lattice. Hence, the direct analytical methods of structural analysis can still be used to determine the composition and structure of the unit cell but the characteristics thus determined correspond to **mean values** obtained by averaging, over the whole volume of the crystal, the chemical composition and the positions of the atoms of all the slightly different unit cells. In other words, the use of direct methods of structural analysis in the case of crystals containing point defects created by isomorphic substitutions artificially introduces an ideal periodicity. This is due to the fact that we are using integrated intensities which we fix at the center of the nodes of a perfect reciprocal lattice without taking into account the small variations shown by these intensities within the volume of the nodes. It is evident that only mean characteristics can thus be obtained and that direct methods of analysis will not, for instance, lead to a determination of short-range order in species with isomorphic substitutions.

To get at the actual structure, it is necessary to use other means, particularly spectroscopic methods which are specifically sensitive to local order-disorder phenomena.

To show the limits of the classical methods of structural analysis, we can take muscovite $2M_1$ as an example. The monoclinic cell of this mica is made up of two layers and its structural formula is

$$2\{K_2Al_4[Si_6Al_2]O_{20}(OH)_4\} .$$

Refinement of the structure obtained by diffraction methods yields a random placement of the Si^{4+} and Al^{3+} cations in two symmetrical independent tetrahedral sites of the unit cell (Rothbauer 1971). However, high-resolution NMR (Sanz and Serratosa 1983) has brought out the fact that the distribution of cations is not totally random: e.g., as required by Lovenstein's rule, the first neighbors of each Al^{3+} cations are exclusively Si^{4+} cations.

Lastly, it should be noted that even if the direct methods of structural analysis do not generally give full information on the actual structure of minerals with point defects, they are nevertheless extremely useful, since the crystallochemistry of these systems, which are so often found with isomorphic substitutions in their natural state, is based on the knowledge of their mean atomic motif. Furthermore, the mean characteristics provided by the direct methods are necessary for the interpretation of the spectrometric data. However, it should always be remembered that the determination of mean characteristics will correspond to a physical reality only if the structure is sufficiently homogeneous for different parts of the crystal to have identical mean values of the structural characteristics.

1.2.2 Crystals with Planar Defects

Another very common category of faults in natural or synthetic crystals corresponds to the presence of planar defects such as stacking faults in lamellar systems. Each stacking fault introduces a sudden modification of the relative position of the atoms situated on either side of the defect as compared to their mutual relative position in a fault-free crystal.

With imperfect structures rich in these planar defects, the direct methods of analysis are pointless, since the very notion of an atomic pattern and of a mean cell, as well as of their determination, no longer correspond to any physical reality. Getting at the actual structural characteristics of these crystals requires the elaboration of specific methods, whose potentialities and descriptions constitute the basis of the present work.

1.3 Indirect Structural Analysis of Partially Disordered Lamellar Systems. Principles of Their Modelization

The only diffraction method[1] which permits the structural analysis of partially disordered systems is based on the simulation of diffractograms for specified structural models. This method of modelization must take into account all the factors (including those due to the apparatus) likely to modify the distribution of the diffracted beam intensity and allow a direct comparison between the experimental and the synthetic computed diagrams. It is evident that solving the problem of obtaining the computed diagram must depend on the main structural features of the objects under study. The present work will be limited to microcrystals with an imperfect lamellar or pseudo-lamellar structure.

A first approach will consist in applying the notion of layers to the case of crystals containing domains sufficiently large that there exists a quasi-ideal biperiodicity which can be described by a 2 D-lattice with two fundamental vectors $\vec{a}_1$ and $\vec{a}_2$. This concept of layers has its counterpart in physical reality when bonding between atoms within a layer is much stronger than that between atoms on neighboring layers. In some cases, nonlamellar structures are found to contain planar defects: it is then convenient to describe the structure using as layers flat fragments parallel to planar defects.

There are usually several ways of stacking layers. Since they all lead to practically equivalent interlamellar spaces, the resulting sequences do not much differ from each other as regards the energy of each system. This possibility of coexistence of several types of stackings of identical layers lies at the basis of the formation of polytypes. It is also one of the main reasons for the loss of true periodicity, with consequent imperfection of the crystal structure.

If layers with different structure and chemical composition can give rise to nearly equal interlayer interactions, they can coexist within one and the same crystal, whether the parameters of their two-dimensional cell be close to, identical with or fairly different from each other. Any crystal in which different types of layers are to be found, in whatever proportion, stacked in a more or less ordered sequence, belongs to the category of *interstratified or mixed layer systems*. Interstratification can also be obtained artificially by letting atoms or molecules penetrate within the interlayer space of a lamellar crystal. The structure is then described as being that of an *intercalation compound* and the penetration of "foreign" atoms or molecules in the lamellar matrix is the *intercalation process or reaction*.

In the next chapters we shall therefore present a mathematical formalism applicable to the structural characteristics of the layers and to their stacking. It must be capable of dealing with the eventual presence of defects likely to perturb the periodicity and must be applicable to systems with a single or with different types of layers.

---
[1] Except high resolution electron microscopy.

In this connection, we should note that the development of the mathematical formalism adapted to the study of the structural characteristics of lamellar systems did not originate only in the necessity of correctly describing this abundant category of minerals, but also in the fact that it is relatively easier to work out an indirect method of analysis in layered structures. Since these systems generally show good biperiodicity in the plane of the layers, the structural analysis is practically reduced to the one-dimensional problem which consists in determining the laws governing stacking and interactions perpendicular to the planes. Note also that the mathematical formalism used to compute the "synthetic" diffractograms corresponding to models with a given imperfect structure will not by itself provide the full solution of the problems: we still need to start from models which, as can be judged by general crystallochemical criteria, are a priori compatible with (and whose features are close or analogous to) those of the species considered. This aspect of the problem, which is just as important as the mathematical formalism itself, consists in first determining all the distinctive structural features to be found in a given type of layer, then in determining the principal stacking faults which can exist in a given category of objects. One should also consider, for interstratified systems, all the possible modes of association of the different kinds of layers likely to appear in the genus of the mineral under study. Lastly, an appropriate statistical description is needed to describe the imperfect structure in terms of the probability of finding each type of defect and its law of recurrence in the structure.

1.4 Determination of the Structural Characteristics of the Layers

One of the basic problems to be solved is the determination of the structural characteristics of the layers which make up the mineral; these are needed to detail the crystallochemistry of the system and to explain the features of the diffraction, itself dependent on the mode of stacking and on the intrinsic structure of the layers. Since the latter generally show good in-plane periodicity, the problem is reduced to the determination of the atomic pattern in the unit cell of the layer.

The first step of the procedure uses, as an initial approximation for the calculation to follow, the refined atomic coordinates of a triperiodic mineral species containing the same type of layers as the faulty structure. If these coordinates are lacking one can, in certain cases, determine the atomic motif by regression algorithms (Appelo 1978; Bailey 1966, 1980, 1984; Drits 1971, 1975; Bookin and Smoliar 1985) which yield fairly precise values of the atomic coordinates from the parameters of the unit-cell and the chemical composition. It is necessary to take into account the different possible cation and anion distributions among the compatible structural sites and then correct the atomic coordinates for each possible ion distribution.

In the course of the next step in the comparison between the synthetic and experimental diffractograms, a refinement of the atomic coordinates is performed. This touching-up is necessary, since any modification in the relative position of the layers with respect to that in the perfect triperiodic crystal may bring about slight changes in the position of the atoms, as compared to those in the perfect crystal.

A few examples will illustrate these points.

1. Many lamellar silicates are built on *2:1-layer stackings.* In each 2:1 layer, two plane sheets of tetrahedra (TO_4) mark off an octahedral sheet between them (Fig. 1.1). Each tetrahedron is connected by three of its apices to neighboring tetrahedra, thus forming a plane of surface oxygens. Within this plane, the oxygen atoms form hexagonal cavities which are regular when the tetrahedral sheets themselves show hexagonal symmetry (see Fig. 1.2a). However, depending on the nature and concentration of the cations in the tetrahedral and octahedral sites and in the interlamellar space, these cavities may be deformed, particularly as a result of a slight rotation of the tetrahedra around an axis perpendicular to the plane of the surface oxygens and passing through the fourth *apical oxygen* of each tetrahedron (see Fig. 1.2b). Hence, one of the problems connected with the study of the structure of lamellar silicates is the determination, among other characteristics of 2:1-layer stackings, of the angle of rotation of the tetrahedra.

2. The second example deals with variations in the type of the octahedral crystallographic sites occupied by cations in the 2:1-layer silicates. Among these

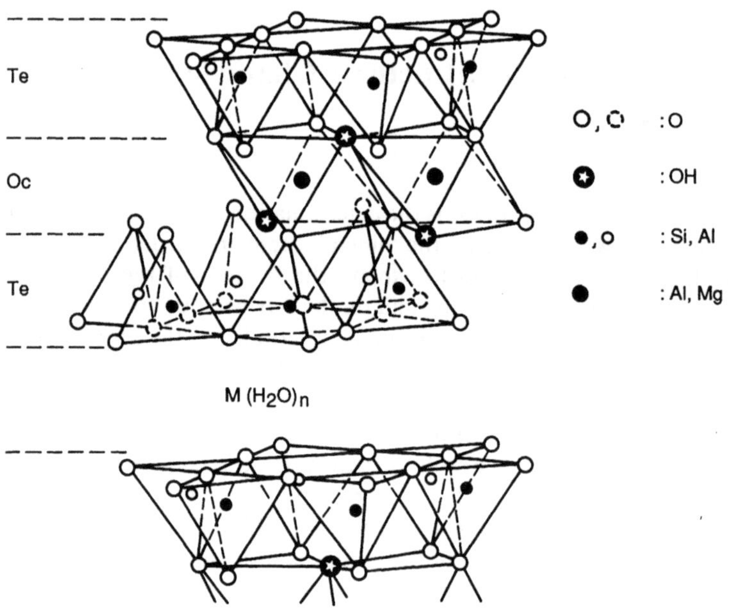

Fig. 1.1. Isometric projection of a 2:1-layer stacking. (Grim 1968). M[H_2O] stands for the hydrated interlamellar cation. The abbreviations *Te* and *Oc* respectively identify tetrahedral and octahedral sheets

Fig. 1.2. Plane of surface oxygens in lamellar silicates. **a** when the tetrahedral sheet has hexagonal symmetry; **b** when the tetrahedra have undergone rotation around an axis perpendicular to the sheet and passing through their apical oxygens

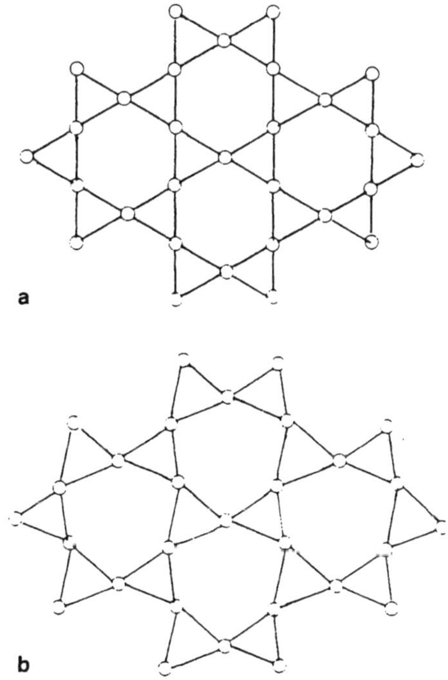

layers, the category in which cations occupy only two octahedral cavities out of three is called *dioctahedral.* The following possibilities can then be distinguished, each leading to different types of layers:

a) Layers in which only the *cis-octahedra* are occupied; the latter are such that their apex bearing an OH group are on the same edge (see Fig. 1.3). In this case, the third octahedral cavity is unoccupied: it is a *trans-octahedron* in which the two OH groups are diametrically opposite. These layers, with vacant *trans*-octahedra, are called *T-layers.*
b) Layers in which the cations occupy the *trans*-octahedra and the *cis*-octahedra C_2 while the *cis*-octahedra C_1 are vacant: these layers are called *C_1-layers.* Similarly, one could have *C_2-layers*, enantiomorphous to C_1-layers and in which the C_2 cavities are unoccupied.
c) Layers in which the three types of octahedral cavities T, C_1, and C_2, are *statistically* occupied, with an equal proportion, namely 2/3: these are called *M-layers.* It is clear that, in this case, the vacant cavities will be different points of the layer and any atomic motif which will be defined can only be a mean motif averaged over the whole extent of the layer.

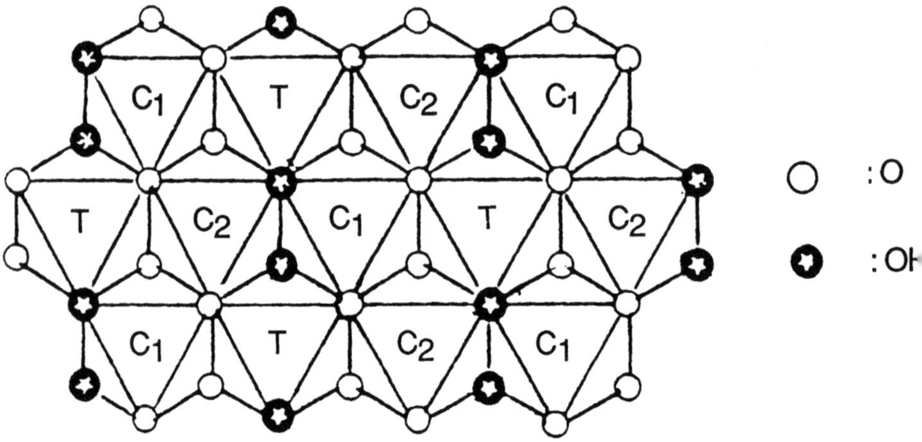

Fig. 1.3. Idealized octahedral sheet 2:1-layer silicate, showing the different types of octahedral sites

1.5 General Characteristics of Triperiodic Layer Stackings

1.5.1 Characteristic Translations of Layer Stackings

If the internal structure of the layers is known, it will only be necessary to determine the mutual arrangement of successive layers for this triperiodic stacking.

Let us consider a case in which all layers have the same chemical composition, internal structure, and *azimuthal orientation (AO)*. The latter implies that, in all the layers, certain elements of symmetry such as the binary axes or the mirror planes are parallel to each other. This stacking is then characterized by translations, i.e., by vectors $\vec{r}_0$ connecting two equivalent (or homologous) points in two adjacent layers. If the translations connecting two successive pairs of layers are equal to $\vec{r}_0$, the crystal structure is described by a lattice with periodicity equal to a single layer. The vector $\vec{r}_0$ is then equal to the parameter c and can be written as

$$\vec{r}_0 = p_0\vec{a} + q_0\vec{b} + d_0\vec{u}_2 \ , \tag{1.1}$$

in which $p_0\vec{a}$, $q_0\vec{b}$ and $d_0\vec{u}_2$ are respectively the projection of $\vec{r}_0$ on the axes $\vec{a}$, $\vec{b}$ and on the normal to the layers. The unit vector along this last direction is $\vec{u}_2$ and d_0 is the projection on the normal to the layer of the distance between two equivalent points on two adjacent layers, expressed in appropriate units (such as nm or Å) while p_0 and q_0 are expressed as fractions of the unit-cell parameters.

The vector $\vec{r}_0$ is generally chosen so that p_0, q_0 and d_0 – which depend on the chemical composition of the layers, on their internal structure, and on that of the interlayer space – have the smallest possible values.

1.5.2 Polytypic Modifications

We have already mentioned that the adjacent layers of lamellar minerals can form several types of associations while keeping practically identical structures and chemical bonding of their interlayer space. This possibility lies at the basis of the existence of *polytypes*, which generally contain the same layers in different mutual arrangements. Thus, the mica known[2] as muscovite 1M is made up of stacks of 2:1-layers with identical *AO*, whereas muscovite $2M_1$ has these same 2:1-layers alternately turned through $\pm 120°$, which induces a two-layer periodicity; lastly, muscovite 3T shows a three-layer periodicity due to a rotation of 120° between any two successive layers (Smith and Yoder 1956; Zvyagin 1967; Bailey 1980).

A characteristic feature of regular polytypes is their homogeneity. This is shown by the fact that all the layers "see" the same local environment: for instance, in muscovite $2M_1$ whatever the *AO* of a given layer, it will always have two nearest neighbors with the same AO, but turned by an angle of 120° relative to the initial layer.

An n-layer polytype exists not only when the *AO* is modified from one layer to the next, but also when there is a regular sequence of different translations along any crystallographic direction parallel to the (001) plane. Thus, graphite presents two polytypic forms, 2H and 3R, which differ in the mode of succession of a translation vector $\vec{\Delta}$ between adjacent layers. A single layer of graphite is a monoatomic sheet of carbon atoms and the translation vector $\vec{\Delta}$ is defined as $\vec{\Delta} = \vec{a}_1/3 + 2\vec{a}_2/3$ (see Fig. 1.4) where $\vec{a}_1$, $\vec{a}_2$ are the vectors of the hexagonal cell. In 2H graphite, the individual layers are alternatively translated by $\pm\vec{\Delta}$, thus forming a noncompact, ABAB... stacking with two-layer periodicity. In 3R rhombohedral graphite, each layer is translated by $\vec{\Delta}$ with respect to the previous one in the stack; this corresponds to an ABCABC... stacking and three-layer periodicity. Hence, the different polytype can be described by the relative translations of the layers as well as by their successive azimuthal orientations.

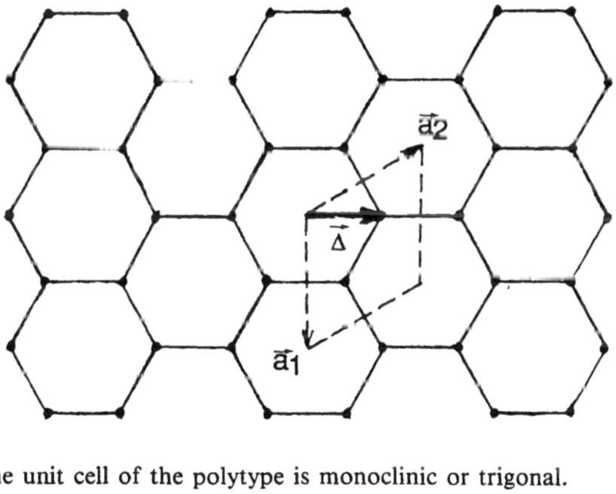

Fig. 1.4. Position of the carbon atoms in a single layer of graphite and definition of the translation vector $\vec{\Delta}$. (After Gay and Gasparoux 1965)

[2] The symbols M or T indicate that the unit cell of the polytype is monoclinic or trigonal.

Note that the unit-cell parameter c of a n-layer periodic structure is still given by Eq. (1.1) but with d_0 replaced by $d = n\,d_0$.

It is then evident that we must have

$$\vec{c} = \sum_{i=1}^{n} \vec{r}_i \,,$$

where $\vec{r}_i$ is the translation between two adjacent layers forming the i-th pair.

1.5.3 Triperiodic Lamellar Structures with Layers Containing Isomorphic Substitutions or with Different Types of Layers

A given mineral family includes a large number of species, each triperiodic but differing structurally from each other by the differences in their isomorphic substitutions. This category of minerals includes, among others, those in which the layers have different chemical compositions but identical or similar stacking sequences. This is, for instance, the case for the $2M_1$, mica polytypes in which the layers have different chemical compositions but follow identical rules of succession

- muscovite $2\{K_2(Al_4)(Si_6Al_2)O_{20}(OH)_4\}$
- phengite $2\{K_2(Al_3Mg)(Si_7Al)O_{20}(OH)_4\}$
- phlogopites-biotites $2\{K_2(FeMg)_6(Si_6Al_2)O_{20}(OH)_4\}$
- etc.

Other lamellar crystals have triperiodic structures in which different kinds of layers are stacked according to a specific regular scheme, as in most graphite intercalation compounds with a well-defined *stage* n, which have stacks of n carbon layers between each two successive layers of intercalated species. Thus KC_8 is a 1st stage compound with regularly alternating negatively charged layers of C atoms and positive K^+ ions (Lagrange et al. 1978). Similarly, chlorites are formed by the regular alternation of 2:1-layers and of brucite-type layers (see Fig. 1.11 a). These systems also give rise to polytypic modifications which, again, differ from each other by translations and/or the *AO* of successive layers.

1.6 Principal Characteristics of Lamellar Structures with Stacking Faults

Creating structural models containing different types of defects requires a thorough knowledge of the features of ordered lamellar crystals described in Section 1.5. This is especially true for stacking faults, since well-ordered multi-layered structures can provide much-needed information on

- the different mutual arrangements possible for successive layers;
- the different translations which can exist between adjacent layers;
- the influence of the various possible mutual arrangements on the internal structure of the layers and the structure of the interlayer space.

Indeed, one can expect lamellar structures with stacking faults to present rather the same mutual arrangements as regular polytypes except for a slight disorder in the successions. Hence, using as basis the known and refined structures of regular polytypes prevents taking into consideration structural models with stacking faults unlikely to be acceptable from the crystallochemical point of view.

In the present paragraph, we shall consider only structures with stacking faults affecting identical layers.

1.6.1 Translation Stacking Faults

Imperfect lamellar systems shall be considered to have stacking faults due to translations when successive layers have the same *AO* but are connected by several translations distributed in the stack with different degrees of order-disorder. For instance (Fig. 1.5) in the single-layer polytype of triclinic talc, one can obtain the same mutual position of the two tetrahedral sheets on either side of the interlayer spaces for three different possible translations $\vec{r}_0$, $\vec{r}_0+\vec{b}/3$ and $\vec{r}_0-\vec{b}/3$, where $\vec{r}_0$ is the translation in the ordered structure. If the succession of these translations is random, the probabilities of finding one of these vectors between a given pair of layers are respectively W_0, W_1 and W_2; these are also the *relative occupancy* of the translation vectors $\vec{r}_0$, $\vec{r}_1$ and $\vec{r}_2$ and we can write:

$$W_0+W_1+W_2=1 \ .$$

Other sequences are possible for these translations. Thus, if the successive pairs of adjacent layers be numbered 1, 2, 3 . . ., the translation vectors $\vec{r}_1$ and $\vec{r}_2$ would be associated with probabilities W'_1 and W'_2, which could have different values depending on the parity of the number associated with the particular pair. In other words, the presence of a translation vector $\vec{r}_i$ between two adjacent layers would depend on the previous translation and there would be an interaction between first-neighbor translations.

More generally, if the presence of a given type of translation is determined by R previous translations between successive layers, the value of R is a measure of the interactions between translations and of the short range order in the sequence of translations. It is characteristic of each structure with stacking faults. In the example given above, if $\vec{r}_0$, $\vec{r}_1$, and $\vec{r}_2$ occur at random (lack of correlation between the translations), the structure would have $R=0$, while the second type of structure would be characterized by $R=1$.

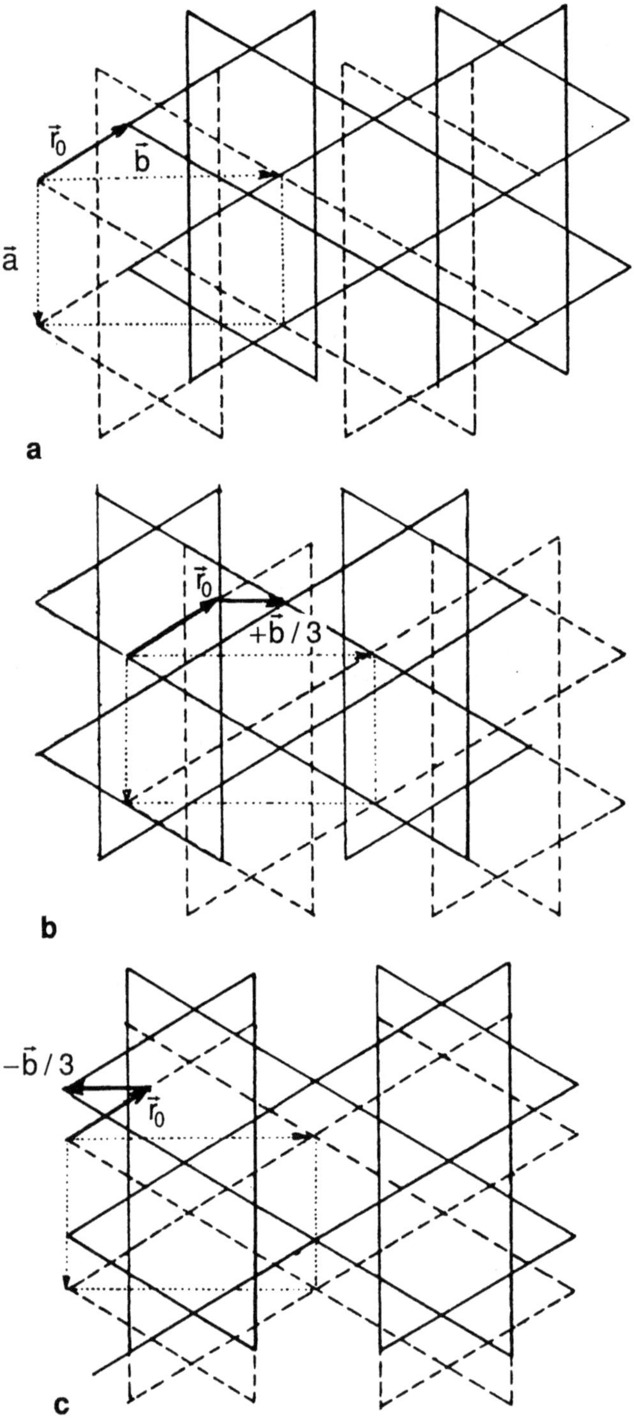
$\vec{r}_0$
$\vec{b}$
$\vec{a}$
a
$\vec{r}_0$
$+\vec{b}/3$
b
$-\vec{b}/3$
$\vec{r}_0$
c

1.6.2 Rotation Stacking Faults

Loss of periodicity in the mutual arrangement of the layers can also happen through irregular rotation of the layers in their own plane. This type of defect arises whenever the regular sequence of azimuthal orientations is modified. Thus a fragment of the $2M_1$ polytype is formed in 1M muscovite when a layer is rotated by 120° in its own plane (Amouric et al. 1978), while muscovite $2M_1$ may locally be transformed into the 1M polytype when successive layers have the same *AO*.

Note that the mutual relation between adjacent layers with given *AO* is also characterized by the translation vector $\vec{r}$ connecting two homologous points in these layers. When there are rotation stacking faults, it is necessary to differentiate between two categories of structures:

- those in which the translation is unique and equal to $\vec{r}$ for two adjacent layers with given *AO*;
- those in which different translation vectors $\vec{r}_i$ can exist for two adjacent layers with given *AO*. Such structures can also contain translation defects.

1.6.3 Stacking Faults Due to Enantiomorphism

Many lamellar minerals, with the silicates among them, are made up of layers having hexagonal or pseudo-hexagonal lattice symmetry. Yet it appears that the elaboration of models to describe stackings with rotation faults leads to particular difficulties when the rotational angles are equal to n 60° (with n = integer).

Let us first imagine that the layers are all identical, with a true hexagonal 2D-lattice; for convenience, we shall replace the true unit cell with a centered rectangular unit cell with parameters a and b such that $b = a\sqrt{3}$. If one of the two adjacent layers forming a pair is rotated by n 60°, it is still possible to describe the pair by a single hexagonal lattice common to both layers. Furthermore, whenever the *AO* of the two layers differ by 60° or 120°, they are enantiomorphs and there is a glide plane in the lattice (Fig. 1.6).

In the actual structures, the condition $b = a\sqrt{3}$ is not always strictly true; the lattice is then only pseudo-hexagonal and $b/a = \sqrt{p}$, with $p \neq 3$. As a start, we shall, however, retain the value of the angle γ (between the $\vec{a}$ and $\vec{b}$ axes) as $\gamma = 90°$.

Consider now a pair of layers with a difference of 120° in their *AO*. Each has a pseudo-hexagonal lattice, in this case with $p < 3$ (shown in Fig. 1.7), but they can no longer be included in a common unit cell since the parameter a (vector $\overrightarrow{OC'}$)

Fig. 1.5. Idealized projection, on the (001) plane, of the surface oxygen atoms of two adjacent layers in talc. (After Drits et al. 1975). The oxygen atoms on the upper and lower layers are the apices of the triangles drawn respectively with full or with dashed lines. The unit cell is shown by the dotted lines. The translations between the two layers are $\vec{r}_0, \vec{r}_0 + \vec{b}/3$ and $\vec{r}_0 - \vec{b}/3$ respectively in **a**, **b** and **c**

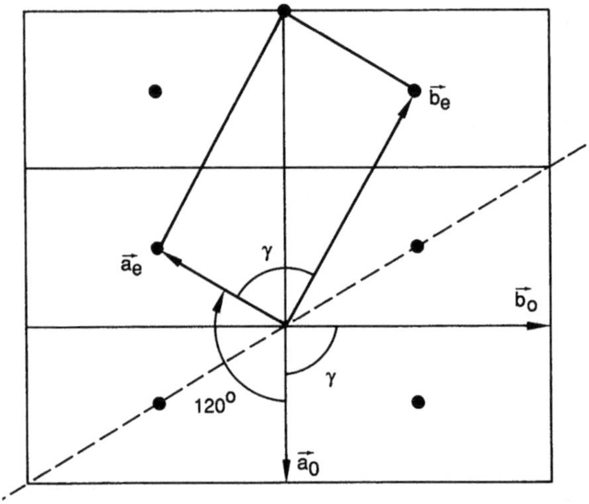

Fig. 1.6. Projection on the same plane of the enantiomorphous unit cells a_0, b_0 and a_e, b_e, ($b = a\sqrt{3}$ and $\gamma = 90°$ for both) corresponding to two adjacent layers rotated in their plane by 120°. The *dashed line* indicates the glide plane which transforms one lattice into the other

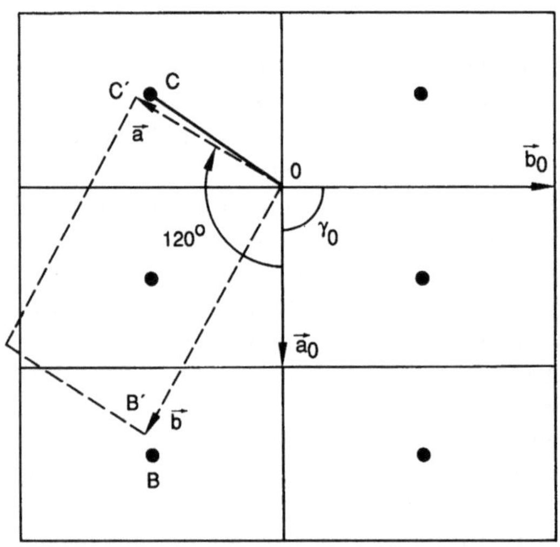

Fig. 1.7. Rotation by 120° of one of the two adjacent layers in a pair when the parameters of the centered unit cells are such that $b_0 < a_0\sqrt{3}$, with $\gamma = 90°$. *Dashed lines* indicate the unit cell of the faulty layer

of the cell after rotation does not coincide, in length and direction, with the vector $\overrightarrow{OC}$ which connects the origin with the nearest equivalent node of the nonrotated layer. As a consequence, structures which contain layers rotated by n 60° and with p ≠ 3 are characterized by the existence of noncoinciding sublattices which correspond to the individual layers out of alignment.

Let us now suppose that the faulty layer be turned with respect to the reference layer by an angle such that the pair of layers can be described using the unit cell of the reference layer. Such a description, based on a single lattice, is only possible when the lattice of the faulty layer is somewhat deformed. As seen in Fig. 1.8, this cell is not rectangular, the fundamental vectors $\vec{a}_1 = \overrightarrow{OC}$ and $\vec{b}_1 = \overrightarrow{OB}$ are not respectively equal to $\vec{a}_0$ and $\vec{b}_0$ and the faulty layer can no longer be obtained by a simple rotation of the reference layer since the angle between the two Ox axes is not equal to that between the two Oy axes. Hence, the characterization of a stack with this type of fault cannot call upon the strict concept of rotation. Moreover, it is obvious that the unit cell of the faulty layer cannot be rectangular and that its parameters must be different from those of the reference layer.

From these considerations, it is clear that there must be differences in the finer characteristics of faultly and faultless layers and that the presence of a rotation fault creates deformations in the lattice of the faulty layers: thus, a C2/m initial symmetry for the faultless layer must go down at least to $C\bar{1}$ for the faulty layer. Furthermore, it is not a priori certain that one should use a specific unit cell to describe the faultless layers and one could legitimately ask whether the deformation in the faulty layers would have no effect on the adjacent layers or whether, on the contrary, there is sufficient interaction between all the layers that they would tend towards an identical structural configuration.

If this situation is real, the same unit cell must be applicable to all the layers and it should be possible to describe the structure of the stack starting from a single biperiodic lattice. When $p \neq 3$, this leads to the following conditions: on one hand, the unit cell of the reference layers can no longer be rectangular; on the other hand, the stack should not be viewed as made up of layers with different azimuthal orientations but of enantiomorphs. The parameters must be the same for the unit cell of the reference and of the enantiomorph, so that $a_0 = a_e$, $b_0 = b_e$ and $\gamma_0 = \gamma_e$, which entails (see Fig. 1.9):

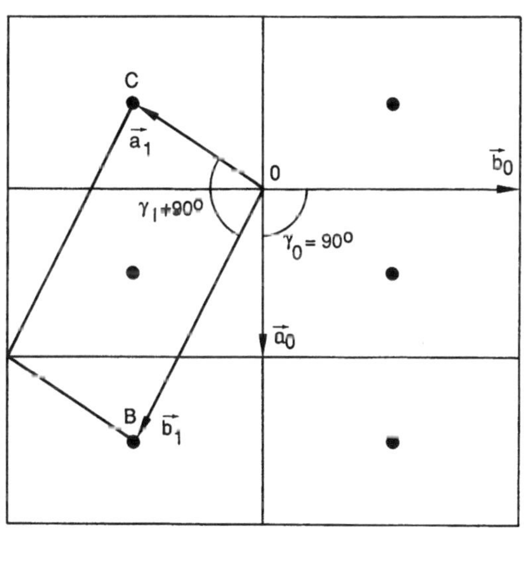

Fig. 1.8. Rotation of a layer by an angle different from 120°, with $b_0 < a_0\sqrt{3}$ and with a centered unit cell (a_1, b_1, γ_1) deformed so as to permit the description of the pair of adjacent layers by the same lattice

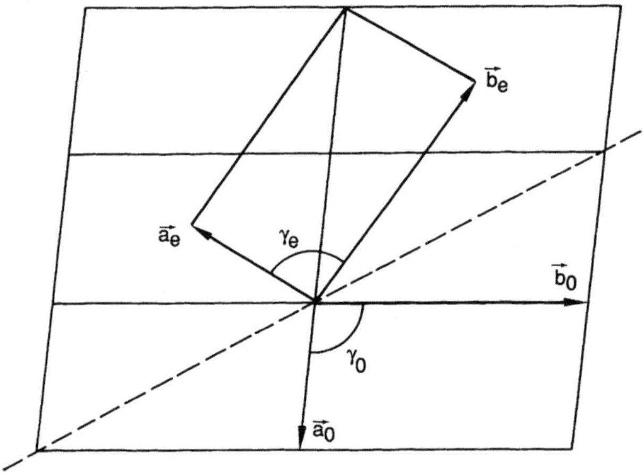

Fig. 1.9. The centered oblique unit cells (a_0, b_0, γ_0) and (a_e, b_e, γ_e) of two enantiomorphous layers, in which $b_0 < a_0\sqrt{3}$ and $\gamma_0 = \gamma_e > 90°$. The *dashed line* indicates the glide plane connecting the two lattices

$$(2\vec{a}_0)^2 = (\vec{a}_e + \vec{b}_e)^2 = a_e^2 + b_e^2 - 2a_e b_e \cos\gamma_e$$

and

$$4a_0^2 = a_0^2 + b_0^2 - 2a_0 b_0 \cos\gamma_0 ,$$

where

$$\cos\gamma_0 = \cos\gamma_e = \frac{p-3}{2\sqrt{p}} . \tag{1.2}$$

This shows that $\gamma > 90°$ when $p < 3$ and $\gamma < 90°$ when $p > 3$. In both cases a common orthogonal cell exists for the enantiomorphous layers.

The preceding developments can be summarized as follows:

1. **When** $p = 3$ **and** $\gamma = 90°$, rotation by n 60° of the layers in the stack still permits the elaboration of a single biperiodic lattice. Moreover, this type of stack with a regular sequence of alternate rotations by ±120° would be indistinguishable from one with a regular alternation of enantiomorphs and both structures would have a glide plane parallel to the $\vec{a}$ axis of the unit cell.
2. **When** $p \neq 3$ **and** $\gamma = 90°$, three limit cases can exist:

a) The stacking faults correspond to *rotations of the layers by exact multiples of* 60°, with each layer retaining its individual characteristics. The stack cannot be described by a single unit-cell and the structure is that of a set of noncoinciding sublattices, one for each group of layers with the same *AO*. Since each of these sublattices has its own counterpart in reciprocal space, the interpretation of the diffractograms can become quite complicated.

b) The stacking faults correspond to *rotations of layers through angles not corresponding to exact multiples of* 60° and such that the crystal can be described using a single biperiodic lattice identical to that of a fault-free layer. The deformation in each faulty layer depends on the value of p, but the whole stack can be described using a 2D-lattice with a rectangular unit cell. The difficulty in modeling such a system lies in the description of the deformation of the faulty layers; however, when p is close to 3, this deformation can be neglected.

c) The presence of rotation faults results in a *general deformation of the entire structure* so that all the layers have the same structural characteristics, can be described by one and the same 2D oblique lattice.

3. **When** $p \neq 3$ **and** $\gamma \neq 90°$, a stack containing faults by enantiomorphism of the layers can be described by a single lattice only if p and γ for the individual layers are connected by Eq. (1.2).

All these different types of structures, which contain layers in which $p \neq 3$, are interesting from the crystallochemical point of view. They are concrete examples of the fact that certain crystallographic directions, which would be equivalent in an ideal layer structure, are no longer so when, for instance, the distributions of isomorphous cations along these directions are not identical.

1.6.4 Well-Defined Stacking Faults

In the preceding paragraphs, we described different types of stacking faults (translations, rotations, enantiomorphisms) whose common feature is that the presence of any one of them in a pair of adjacent layers determines unequivocally the relative position of these layers. Thus, for translation faults, we assumed that the translation vector is of specific length and direction; again, for rotation faults, the angle and direction of rotation were assumed to be quite unambiguous. Hence, all the stacking faults described so far may be classified as *well-defined stacking faults.*

It should be pointed out that a lamellar system may contain a set of well-defined faults, bound to each other by simple relationships, and which can be described as a group. There may, for instance, be rotations of 60°, 120°, ... 300°: they constitute a set of five rotation faults, each well-defined and appearing between nearest-neighbor layers with a certain probability. The set of these rotations[3] is known by the generic name *"n 60° rotations"* (with n an integer and $1 \leqslant n \leqslant 5$). Again, a set of translation faults could have the translation vectors described by a single equation

$$\vec{t} = m\vec{a}/v + n\vec{b}/w$$

where v and w would be values specific to a certain type of defect while the integers m and n could take different values, each leading to a specific well-defined translation fault.

[3] Prior to the present work, such a set of faults was called "partially defined" in some papers (Tchoubar 1984, 1986) because of the indeterminacy in n and/or m.

1.6.5 Random Stacking Faults

In contrast to the preceding cases, there may be other rotation or translation faults with constitute *random stacking faults*; they can occur, for instance, when a translation vector has a random amplitude or direction or when there is a random amplitude of rotation of the layers in their own plane. Since there cannot be any connection among random faults, they cannot be regrouped into a describable set and two adjacent layers with a random fault between them may have any mutual arrangement.

1.6.6 Stacking Faults Due to Fluctuations in the Position of the Layers. Disorders of the First and Second Types

Most crystals are subject to slight deformations which perturb the regular sequence of their constituent atoms. Hence, their description must include the fact that the atoms do not occupy positions strictly defined with respect to the regular lattice nodes but are more or less displaced from their ideal positions. These deformations are distributed throughout the bulk of the crystal rather than localized in certain regions of the structure, as are random or well-defined rotation or translation stacking faults.

There is a clear distinction between two types of deformations called *disorders of the first and of the second type* (Guinier 1964). Consider, to this effect, a one-dimensional array of atoms (see Fig. 1.10). If this periodic structure is perfect, the atoms are separated by strictly equal distances and are at the nodes of a regular 1D lattice characterized by the parameter a (Fig. 1.10a).

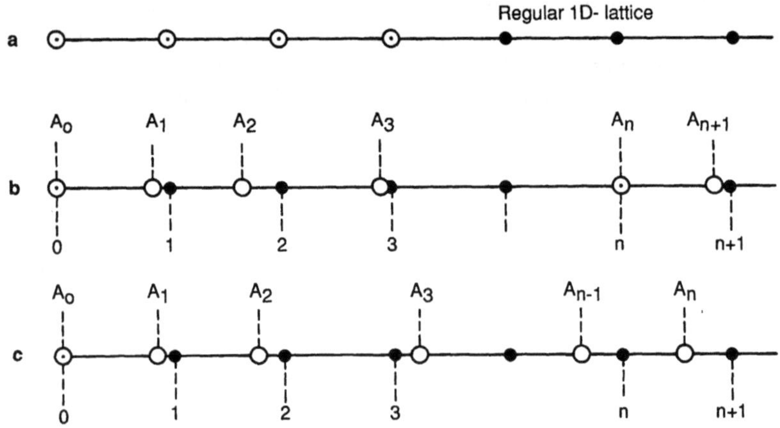

Fig. 1.10. a An ideal one-dimensional structure; the structures shown in **b** and **c** correspond to disorders of the first and second type respectively (see text) (After Guinier 1964)

Let there now be slight fluctuations in the positions of the atoms in the row. To any pair of first-neighbor atoms separated by the distance x_1 we can associate a variation ε_1 with respect to the ideal value a. Although always much smaller than a, ε_1 may vary from one pair of atoms to the other and the variation of x_1 may be described by a statistical distribution function $\gamma_1(x)$ of the distances between nearest neighbors. In the general case of nth-nearest neighbors, we have the general relation

$$x_n = na + \varepsilon_n \ ,$$

where ε_n is the difference between the "ideal" distance na and the true distance x_n.

Here too, ε_n may vary for each pair of nth-nearest neighbors and the variation of x_n may be described by a statistical distribution function $\gamma_n(x)$.

Depending on the nature of the interactions between the atoms and the physical reason for the fluctuations in the interatomic distances, two cases may occur:

1. The distance between atoms m and (m+1) depends on the distances between the pairs (m−1) and m, (m−2) and (m−1)..., or, in other words, the interatomic distances are correlated. The law of distribution of the variations is then correlated with the distribution laws $\gamma_1, \gamma_2, \ldots$, which describe the distances between first-, second- ..., neighbors. More particularly, the correlation may be such that a single law describes both short-range and long-range order so that $\gamma_n(x)$ is the same for $n = 1, 2, 3, \ldots$ and *the law which describes long-range order is independent of the size of the crystal.* This kind of imperfection is a *disorder of the first type* (Guinier 1964, see Fig. 1.10b). From the physical point of view, crystalline defects of the first type correspond to a deformation of the crystal such that the bulk of the structure yields in an attempt to cancel statistically the consequences of the presence of the defects. This, for instance, is the case when there is a random distribution of isomorphous substitutions which causes *statistical* deformations of the positions of the atoms. Thermal motion will lead to a similar result (this deformation is dynamic).
2. In the second type of disorder, the distance between first-nearest neighbors may still fluctuate around the mean value "a" while following a statistical law of distribution $\gamma(x)$, but this distance is now *independent of the distance between other pairs of neighbors.* In other words there is no correlation between the distance of successive adjacent atoms. If one of the atoms, say A_0, be chosen as the origin and the following atoms be identified as $A_1, A_2, \ldots A_n, \ldots$, we can describe the distance with respect to the origin A_0 of the successive atoms, $A_1, A_2, A_3, \ldots$ (see Fig. 1.10c)

–for two 1st nearest neighbors: $A_0A_1 = x_1 = a + \varepsilon_1$, with a probability of existence of x equal to $\gamma(x)\,dx$ where the distribution function $\gamma(x)$ is normalized through

$$\int_0^\infty \gamma(x)\,dx = 1 \ ,$$

and where the mean value of $\gamma(x)$ is given by

$$\int_0^\infty x\gamma(x)\,dx = a \ ;$$

– for two 2nd-nearest neighbors: $A_0A_2 = x_2 = 2a + \varepsilon_2$ with a probability of existence equal to $\gamma_2(x)$, where

$$\gamma_2(x) = \int_0^\infty \gamma(u)\gamma(x-u)\,du \ ,$$

in which u is one of the *real* positions of atom A_1. By definition, this integral is the convolution (or *Faltung*) of the function $\gamma(x)$ by itself so that

$$\gamma_2(x) = \gamma(x) * \gamma(x) \ ;$$

– for two 3rd-nearest neighbors $A_0A_3 = x_3 = 3a + \varepsilon_3$

and
$$\gamma_3(x) = \gamma_2(x) * \gamma(x) = \gamma(x) * \gamma(x) * \gamma(x) \ ;$$

– for two nth-nearest neighbors $A_0A_n = x_n = na + \varepsilon_n$

and
$$\gamma_n(x) = \underbrace{\gamma(x) * \gamma(x) * \ldots * \gamma(x)}_{\text{n times}} \ .$$

The statistical law of distribution of distance for nth-nearest neighbors is therefore the $(n-1)$th self-convolution of the law of distribution for 1st-nearest neighbors and it has been shown (Guinier 1964) that $\gamma_n(x)$ will be normalized to 1 for all values of n. This means that the difference between the actual distance separating an atom and its nth-nearest neighbor and the mean value n an increase with n. Hence, as from a certain value of n, atom A_n can be at a position such that it may be at or before the midpoint between the nodes $(n-1)$ and n, as well as at or after the midpoint of the nodes n and $(n+1)$: this correspond to the disappearance of long-range order, and this type of structural defect has been called a *disorder of the second type* (Guinier 1964).

True lamellar species also exhibit similar fluctuations due, for instance, to variations in the chemical composition of the layers and/or in the occupancy of the interlamellar space. They can appear as variations in the translation vector or in the azimuthal orientation angle and/or in the distance separating two adjacent layers, as found in partly graphitized carbons (Maire and Méring 1970), in some microcrystallized muscovites (Kodama et al. 1971) or in hydrated or anhydrous sodium beidellite (Ben Brahim et al. 1983).

1.6.7 Particles, Crystallites and Interferential Coherence Domains

Powdery crystalline solids are made up of aggregates or of individual particles which can be separated by means of differences in their size, or in their density, etc. When these powders are analyzed by diffraction of X-rays, electrons, or neutrons, the reflections are broadened to a varying extent and this fact has been used in attempts at getting back to the mean dimensions of the structural entities responsible for the diffraction, using for instance the Scherrer relation

$$\Delta(2\theta) \approx \lambda/\mathrm{L}\cos\theta \ ,$$

in which $\Delta(2\theta)$ is the angular width at half-maximum intensity of an hkl reflection, λ is the wavelength of the incident radiation, θ is the Bragg angle for the (hkl) family of reflecting planes and L is the extension of the structural entity in the direction normal to the (hkl) planes.

If the powder is made up of monocrystalline particles or, as is more often the case, of particles in which the position of the atoms can be described using a single mean lattice, the mean dimensions of the structural entities, as determined from the reflections, are equal to the dimensions of the particles themselves: the atoms in the whole volume of each particle diffract the incident radiation coherently and the volume constitutes the *interferential coherence domain.*

However, in powdered solids with structural defects, this notion of coherence domain is not so obvious. For one thing, the particles are often made up from a number of small single crystals (or crystallites) with a slight mutual misorientation, with each crystallite slightly turned or translated with respect to its neighbors, the whole forming a *mosaic crystal.* Within each crystallite, the distribution of atoms can be described by a single regular lattice and it is the mean volume of all the crystallites which constitutes the domain of coherence as determined, for instance, by the Scherrer formula. In this case, the dimension of the structural entities drawn from the diffraction data is much smaller than the dimensions of the particles observed, for instance, by electron microscopy. Note that the smaller the scattering angle, the lesser the perturbations in the coherence of the scattered waves due to mosaicity. In other words, if the size of the domain of coherence is calculated from reflections indexed $\mathrm{n}h, \mathrm{n}k, \mathrm{n}l$, low values of n lead to domains of coherence larger than those calculated from data corresponding to larger values of n. The large domains calculated for low value of n are then greater than the mean volume of a crystallite of the mosaic crystal, and a coherence domain corresponding to the mean volume of the crystallite will be obtained only from values of n greater than a certain lower limit beyond which the calculated size of the coherence domain will remain constant. The presence in lamellar crystals of distortions in layers, singly or in groups, as well as the presence of dislocations, will have the same average effect of limitation on the interferential coherence. Here too, changing the value of n in $\mathrm{n}h, \mathrm{n}k, \mathrm{n}l$ reflections will lead to different values of the apparent dimensions of the domains. In the case of lamellar species, these variations in the apparent domains of coherence were first described by Bouraoui and Méring

(1964) for partially graphitized carbon using an interpretation already given some years before (Méring and Longuet-Escard 1954). The same concept of apparent modification of coherence domains was later used by Plançon and Tchoubar (1975) in their structural analysis of partially disordered kaolinites; since then, it has formed the basis of various studies bearing on the shape of the *hk* diffraction bands in microcrystalline layer silicates (cf. Chap. 8).

There is another aspect of the concept of interferential coherence domains specifically related to the presence of a disorder of the second type. As seen above, particles with this type of defect do not have spatially defined domain boundaries, as in a mosaic crystal, but, as one gets further away from any given point, the atoms occupy positions which differ more and more from the positions which they would occupy in a regular lattice. As a consequence, the phase differences between the waves scattered in a given direction by successive atoms are no longer constant and the loss of long-range order results in a loss of interferential coherence. X-ray data will therefore yield sizes of domains which do not correspond to actual borders in particles but to limit dimensions beyond which the atomic arrangement can no longer be described by a periodic lattice.

To conclude this paragraph, let us compare the size of the structural entities as given by X-ray diffractometry (XRD), by classical bright-field transmission electron microscopy (TEM), or by scanning TEM (STEM). As determined by these last two techniques, the dimensions of the "particles" are often much larger than the dimensions of the coherence domains given by XRD. Electron microscopy will yield the dimension of the coherence domains only when *dark field techniques* are used (Hirsch et al. 1965; Gard 1971). Yet, even with this technique, the visualization of the interferential coherence domains is only possible if they are physically bounded (mosaic domains, distortions, etc.). However, coherence domains will not even be visible by dark-field electron microscopy when they correspond to the presence of structural disorders of the second type. In this case, there is an apparent contradiction between the results obtained using XRD and TEM or STEM.

1.7 Principal Characteristics of Interstratified Minerals

We have already mentioned (Sect. 1.3) that certain crystalline systems are built up from two-dimensional fragments having different structures and compositions. These systems are not only met in lamellar structures (Drits and Sakharov 1976; Hérold 1979; Reynolds 1980; Solin 1982) but also in chain (Veblen et al. 1977; Drits 1987) and in channel structures (Terner and Busek 1979, 1981). Interstratified lamellar compounds, in which successive layers and interlayer spaces have different structures, or compositions, or both, are found very frequently. The interlamellar spaces of many microcrystalline lamellar compounds often constitute zones where structural modifications can occur without influencing the internal structure of the layers themselves. Because of this characteristic, and depending on the physicochemical parameters of the medium in which they happen to be, certain minerals successively transform into other species. This evolution occurs by a selective modification, within the crystal, of the composition and of the structure of certain interlamellar spaces rather than by sudden changes which would, for instance, transform all the interlamellar spaces into those of another species. The result is the formation of a series of interstratified phases which reflect the successive stages of the transformations.

Another important factor at the basis of the existence of interstratified systems is the presence of certain peculiarities of the surfaces on either side of the interlamellar spaces. Thus, the geometry of surface oxygens in 2 : 1 lamellar silicates favors the formation of hydrogen bonds with the surface hydroxyl groups of brucitic or kaolinitic type of layers. Hence, many interstratified silicates are constituted by stackings of 2 : 1 and 1 : 1 or 2 : 1 and brucitic layers.

The great variety shown by interstratified systems is due to the fact that they are made up of different types of layers whose relative abundance and position can vary within a wide range from one mineral to the other. In most of these compounds, there are two types of layers, but interstratified systems are known which contain three or more components. Furthermore, it is often convenient to describe some minerals with a single type of layer as interstratified systems when some layers are faulty owing to well-defined rotation defects: all the layers with a given *AO* are considered as being of one kind.

1.7.1 Interstratified System Characterization by the Stacking Mode of the Layers

Three groups of structures can be distinguished among the ways in which different types of layers can follow each other in interstratified systems:

1. Structures in which a given sequence is always observed, with true periodicity in the direction perpendicular to the layers. This type of interstratification is found, for instance, in rectorite (Reynolds 1980) in which mica-type layers (A) and montmorillonite-type layers (B) form a regular ABAB... sequence. Similarly, all the homogeneous stages of the graphite intercalation compounds belong to the category of regular interstratified structures.
2. Irregular interstratified structures in which the different kinds of layers are stacked following an order-disorder law. The specific characteristic of this category which, incidentally, is the one found most frequently, is that the species are always formed from layers with markedly different thicknesses. The irregular sequence and the presence of different interlayer distances yield diffractograms which could, at first sight, seem surprising: the basal reflections form a succession which does not correspond to a series of rational and integral values of l, so that the *apparent* interplanar spacings do not correspond to the various basal plane distances existing in the mineral species. This irrational series is therefore a good criterion for the presence of interstratification. Note, however, that lack of rationality in the $00l$ series may also be due to overthin

particles or oversmall coherence domains (cf. Sect. 2.2.2). Furthermore, this qualitative criterion is not applicable to some interstratified mineral species with irregular sequences owing to the presence of layers with thickness which are integral multiples of each other. This is, for instance, the case in the "penninite-lizardite" group in which chlorite- and serpentine-type layers, respectively 1.4 and 0.7 nm, alternate irregularly (see Chap. 9). The powder diagrams of these minerals yield basal reflections corresponding to a series of rational values of l strongly reminiscent of the basal reflections given by chlorites when layers of this species abound in the interstratified compound mineral. In this case, verification of the interstratified character of the mineral requires a precise analysis of the position, intensity, and *profile* of the basal reflections.
3. Lastly, a third group of interstratified systems is formed by species in which the layers have the same thickness and similar, but not identical, internal structure. For instance, in certain dioctahedral lamellar silicates, T-type and C-type 2:1 layers (cf. Sect. 1.4) are superposed; these species show all the characteristics of interstratified systems, since no symmetry operation returns one type of layer from the other.

From the preceding paragraphs, it is evident that the determination of the structural characteristics of interstratified species (nature of the layers, concentration of each component, mode of stacking, etc.) from the data provided by the diffractograms is a fairly complex problem. As in the study of crystals with stacking faults, one of the most successful methods to be used is the modelization of the structure and the elaboration of a synthetic diffractogram to be compared with the experimental results. The specific parameters needed in each model of interstratified structure and the mathematical formalism used in the computation of the theoretical diagram are presented in Chapters 4 and 5 respectively.

1.7.2 Order-Disorder in the Sequence of Layers of Different Types

As was done above for the characterization of stacking faults (cf. Sect. 1.6.2), we can use a parameter which is a measure of the range of interaction between layers to describe the order-disorder in the sequence of layers in an interstratified structure. This factor, which was called *Reichweite* by Jagodzinski (1949), is equal to the number of preceding layers upon which depends the probability of finding a layer of a given type at any place in the stack. Let S be this number of layers:

1. When the alternation of layer types is random, the probability of finding any particular type of layer at a given point in the stack does not depend on the nature of the surrounding layers and $S = 0$. It is said of these structures that they are **totally disordered as regards the sequence of layers.** Their characterization only requires a knowledge of the number of different kinds of layers and of the probability W_i of occurrence associated with each type i.

2. When the probability of finding a given type of layer depends on the nature of the previous layer, then S = 1. The simplest example of this type of structure is that of an interstratified system containing two types of layers, namely A and B, with a sequence ABABA ... In this case, the probability of finding a layer A (or B) is predetermined by the necessity of having a previous layer of the opposite kind.

Suppose now that there be an interstratified structure with components A and B, in proportions $W_A = 0.75$ and $W_B = 0.25$, with a constraint excluding the occurrence of a pair BB. Such a structure would also have S = 1, since the constraint imposes that after a B-layer there be a A-layer (even though, in the present case, the reverse condition is not true). There is, however, another way of presenting the same stacking sequence, which consists in looking upon it as made up of the structural elements C and D, where C is a pair of layers AB and D is an A layer. The proportions must now be such that $W_C : W_D = 1/3 : 2/3$. The same stack, which with A and B layers and the BB pair exclusion constraint corresponded to S = 1, can now be described as made up of the elements C and D with S = 0 since the stacking sequence can be random while still including the exclusion of BB pairs.

3. If the probability that the nth layer of a stack be of a given type depends on the nature of the two previous layers, (i.e., those numbered $n-1$ and $n-2$), one can speak of second-nearest neighbor interaction and S = 2. The simplest such interstratified structure is the sequence AABAAB ... in which the probability of finding a B layer is contingent upon the occurrence of an AA pair immediately before it. It is also possible to envisage a more general case consisting of a stack of A and B layers in proportion such that $W_A : W_B = 0.75 : 0.25$ with a constraint excluding both the pairs BB and the triplets BAB. Here too, the factor describing the range of interactions would be S = 2, since a B layer can only be found after an AA pair. Still, it is again simpler to describe such a stacking as a random sequence of AAB triplets and A layers, with S = 0.

Clearly, the examples of interstratified structure presented above are but a few among a large variety of possibilities which will be more fully discussed in Chapter 4. Presently, let us however note that describing a stack by the range of interaction implies that the model for interstratification follows *Markovian statistics* according to which the *probabilities* of finding, in a fixed direction, a layer of a given type depends only on the nature of its nearest preceding neighbors. If the particles in a powder are fairly large, using Markovian statistics ensures that, in each crystallite, the occurrence of layers of a given type, say 'i', will be approximately equal to the mean occurrence W_i of this type of layer in the powder; the latter can then be said to be *homogeneous as regards the composition of the particles*. On the contrary, if the system only contains particles whose thickness is that of 10 to 15 layers, one of the consequences of the Markovian model is that the occurrence of each type of layer may not be the same for all the particles (cf. Chap. 6.6) and one speaks of a *quasi-homogeneous powder* (Drits et al. 1984). Even if most particles

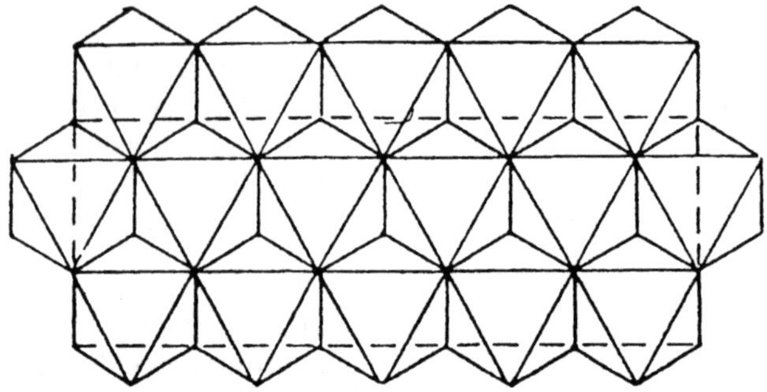

a

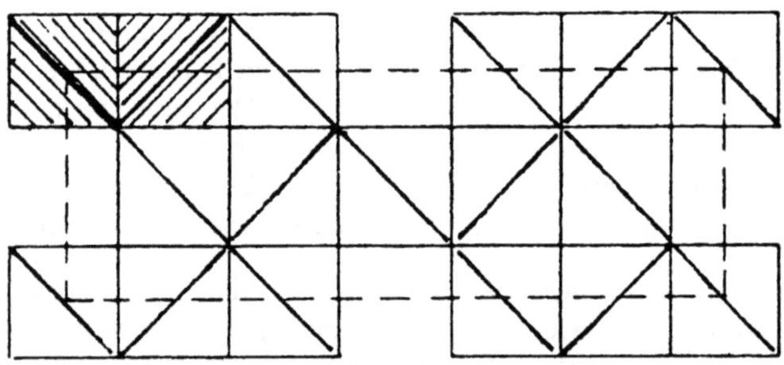

b

Fig. 1.11. Diagrammatic representation of layers of brucite (**a**) and mackinovite (**b**). *Dashed lines* indicate the multiple cell common to both types of layers

have values of W_i equal or close to the mean value of W_i in the powder, one cannot exclude, in such a system, the existence of particles containing a single type of layer. However, when drawing up a model for the powder, we must not a priori discard systems in which the proportion of each type of layer is the same in all particles (i.e., when the powder is again described as being *homogeneous*) and we shall therefore give, in Chapter 4, the characteristics of quasi-homogeneous and homogeneous powdered interstratified systems.

Lastly, we should note that the consequences derived from the use of Markovian statistics for the description of powders of ultra-thin particles are the same as those yielded by the description of the interactions characterized by the factor R which describes stacking faults. As will be seen in Chapter 6, we shall once again distinguish between homogeneous and quasi-homogeneous powders, but these definitions will refer to the occurrence of each type of stacking fault in the particles.

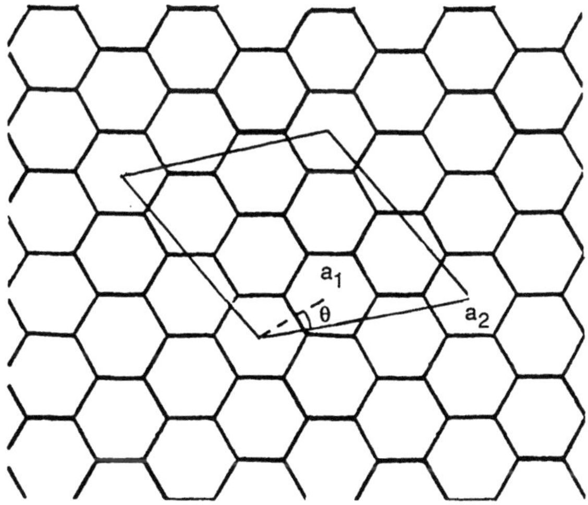

Fig. 1.12. Hexagonal unit cell ($\sqrt{7}\times\sqrt{7}$) R 19.11°. The angle of orientation of the unit cell with respect to the [100] direction of the graphite lattice is $\theta = 19.11°$ and $a_2 = a_1\sqrt{7}$

1.8 Commensurate and Incommensurate Structures in Interstratified Systems

It has already been mentioned (cf. Sect. 1.7) that one of the factors responsible for the ease of formation of interstratified structures is the existence of layers[4] with fairly similar geometrical characteristics of their basal surfaces in spite of differences in their nature. Because of this, layers with different composition or internal structure can form stacks leading to interstratified structures.

In certain interstratified systems, there are rational relationships among the parameters, the shapes, and the orientation of the 2D-cells of the different layers whose lattices are then said to be *commensurate*. This will occur, for instance, when the hexagonal unit-cells of two different kinds of layers are such that $a_2 = a_1\sqrt{3}$, and there is an angle of 30° between the 2D-lattices. An example is chlorite (Bailey 1980) in which talc-like ($a_2 = 0.52$ nm) and brucite-like ($a_1 = 0.30$ nm) layers alternate regularly. Another more complex example is given by tochilinite (Organova et al. 1972) in which there is a regular alternation of brucitic layers and 'sulfide' type layers of mackinovite. Figure 1.11 is a schematic description of the correspondence between the square unit-cell of the mackinovite lattice (Fig. 1.11 b) and the hexagonal or orthohexagonal unit-cell of brucite (Fig. 1.11 a).

[4] The word 'layer' is used here in a most general sense; it can describe the lamellar entity from which the host material is built, or the mono or polyatomic sheet of the intercalated substance (as in the intercalation compounds), or the different types of layers found in natural interstratified minerals.

Other commensurate structures are found in the graphite intercalation compounds; for instance, the first stage $C_{14}SbCl_5$ has molecules of antimony chloride forming a plane monomolecular and hexagonal sheet with a unit cell parameter $a_2 = 0.651$ nm (Mélin and Hérold 1975). This value corresponds to a unit cell noted $(\sqrt{7} \times \sqrt{7})$ R 19.11°, where it is indicated that the hexagonal unit cell of the chloride is rotated by 19.11° with respect to the unit cell of graphite, as shown in Fig. 1.12, and that the relation between the two parameters is $a_2 = a_1\sqrt{7}$, where the parameter a_1 of the unit cell of graphite is 0.246 nm.

The concept associated with a crystalline state of matter in which two lattices coexist (as in the interstratified minerals), each with its own periodicity and without any rational relation among their parameters, is known as *incommensurability* or *noncommensurability*. The lack of rationality may be between the 2D lattices of the layers, or along their perpendicular direction.

One can distinguish two different types of incommensurate structures:

1. A first type is that in which each layer has its own lattice, totally independent of the others. Thus, Fig. 1.13 shows the selected area electron diffraction (S.A.D.) given by a valeriite single crystal consisting of sulfide-like and brucite-like layers. Only two examples of this category are found in the layer silicates: the first is composed of a random succession of nonswelling trioctahedral and swelling dioctahedral types of chloritic layers (Novikov et al. 1973), while the other corresponds to a silicate in which layers of dioctahedral smectite and trioctahedral chlorite alternate regularly (Blatter et al. 1973). Similarly, there are ordered and disordered incommensurate structures in the series of manganese oxides (Chukrov et al. 1980, 1982, 1983). Thus, asbolite is built from a regular alternation of MnO_2 and $(Ni, Co)(OH)_2$ sheets in which the difference in the ionic radii of the small Mn^{4+} and the larger Ni^{2+} cations is the cause of the incommensurability of the two layers, so that two different lattices must be used to describe the periodicity. Note that one of the two layers in this mineral is no longer continuous in two dimensions but consists of isolated island-like layer fragments to enable the formation of the necessary hydrogen bonds (Chukhrov et al. 1982).

   In the same way, many cases of incommensurability corresponding to sequences of different and quasi-independent lattices are met in the intercalation compounds of graphite, as for instance in $C_{n(6.6)}FeCl_3$ where n is an integer corresponding to the stage of the compound (Rousseaux et al. 1982).
2. In the second type of incommensurate structures, the lattices undergo deformations to reach a common compromise; this results in the formation of a modulation wave which may affect either the position of the atoms or that of the ions, or some other local property such as the spin density or the charge density which is then characterized by a period incommensurate with that of the lattice. The system is no longer invariant with respect to the fundamental lattice translations, as would be the case if the wave were absent. These structures exhibit a new type of 'disorder'. From the microscopic point of view, the origin of the modulation wave depends on the nature of the compounds, but all have

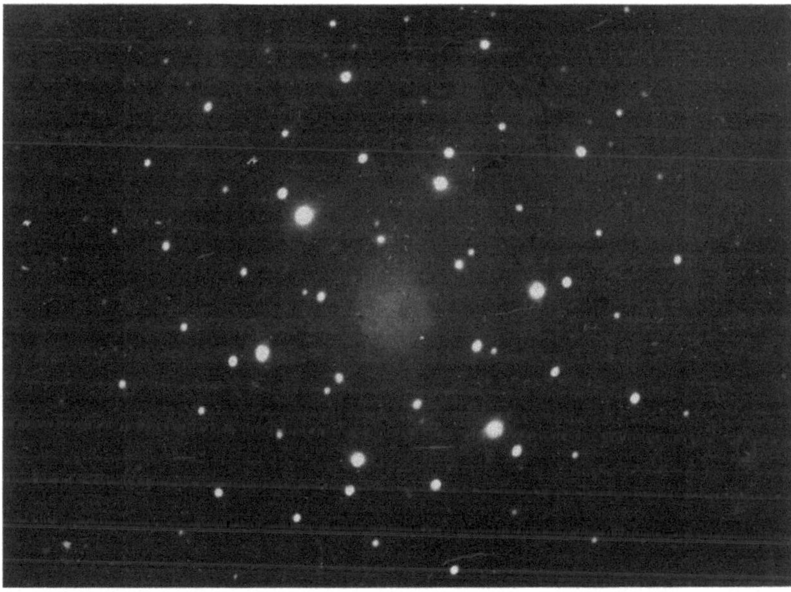

Fig. 1.13. S.A.D. pattern from valeriite single crystal having incommensurate sulfide-like and brucite-like layers. (Organova et al. 1973)

one common feature: the existence of an antagonism between many types of forces which tends to promote a different order. This conflict among forces, 12 years ago called *frustrations* by P.W. Anderson (Nobel Prize, 1981), has been popularized by G. Toulouse since 1977. For instance, this conflict exists in the magnetic interactions between first and second neighbors of certain magnetic metals. Again, the interactions among atoms of an adsorbed sheet may favor incommensurability while the interactions between adsorbed atoms and the substrate would tend to promote a commensurate structure, leading to a competition among elastic forces with different ranges.

The diffraction pattern of these stuctures contains two sets of reflections. The first – or **principal** – set is also present in the diffractogram of the nonmodulated phase constituting the basic structure; it is distributed on a periodic lattice $\vec{a}_1^*, \vec{a}_2^*, \vec{a}_3^*$, and each reflection is at the center of a second set of regularly placed reflections – known as **satellites** – with their own periodicity characterized by the wave vector $\vec{k}^*$ of the modulation, with

$$\vec{k}^* = k_1\vec{a}_1^* + k_2\vec{a}_2^* + k_3\vec{a}_3^* \ .$$

The presence of incommensurability requires that at least one of the k_i be irrational. The integrated intensity of the satellites is generally weaker than that of the principal reflections, so that it is easy to distinguish between the principal (or 'mean') lattice and the satellites. In Fig. 1.13, we show, as an example, the room-

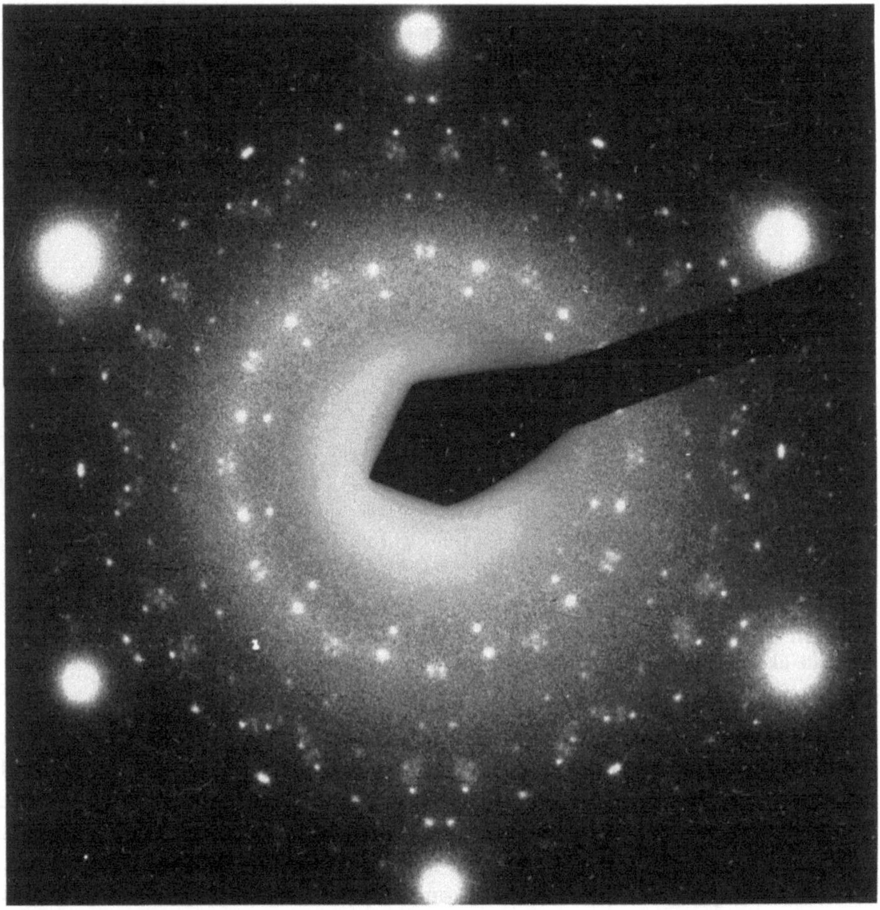

Fig. 1.14. Room-temperature selected area electron diffraction pattern of the graphite intercalation compound $C_{16}HNO_3 \cdot nH_2O$. (Clinard et al. 1983)

temperature selected area electron diffraction pattern of the 2nd-stage graphite intercalation compound of nitric acid $C_{16}HNO_3 \cdot nH_2O$, with $n \approx 0.14$ (Clinard et al. 1983). The high-intensity spots (shown by the arrows) originate from the graphite layers, while the weak spots are those of the hexagonal and modulated network of HNO_3, incommensurate with the layer of graphite. In this compound, the modulation is not due to the compromise between the nitric acid layer and the graphite substrate: it is induced by the array of water molecules relaxed to the nearest potential well of the graphite layer. The unit cell of the water molecule lattice is orthorhombic with $\vec{a} = 4\vec{a}_1 + \vec{a}_2$ and $\vec{b} = 2\vec{a}_1 + 7\vec{a}_2$ ($\vec{a}_1$ and $\vec{a}_2$ being the basic vectors of the graphite lattice; see Fig. 1.4).

The detailed treatment of incommensurate structures must be sought in more specialized literature (for instance: Pynn 1979; Bak 1982).

References

Amouric M, Baronnet A, Finck C (1978) Polytypisme et désordre dans les micas dioctaédriques synthétiques. Etude par imagerie de réseau. Mat Res Bull 13:627–634

Appelo CAY (1978) Layer deformation and crystal energy of micas and related minerals. I. Structural models for 1M and $2M_1$ polytypes. Am Mineral 63:782–792

Bailey SW (1966) Status of clay mineral structures. Clays Clay Mineral 26:3–21

Bailey SW (1980) Structure of layer silicates. In: Brindley GW, Brown G (eds) Crystal structures of clay minerals and their X-ray identification. Mineral Soc, London, pp 1–124

Bailey SW (1984) Crystal chemistry of the true micas. In: Bailey reviews in mineralogy, vol 13. Micas, 1–12. Mineral Soc Am, Chelsea, Mich, USA

Bak P (1982) Commensurate phases, incommensurate phases and the devil's staircase. Rep Progr Phys 45:587–629

Ben Brahim J, Armagan N, Besson G, Tchoubar C (1983) X-ray diffraction studies on the arrangement of water molecules in a smectite. I. Homogeneous two-water-layer Na-beidellite. J Appl Crystallogr 16:264–269

Ben Brahim J, Armagan N, Besson G, Tchoubar C (1986) Méthodes diffractométriques de caractérisation des états d'hydratation des smectites. Stabilité relative des couches d'eau insérées. Clay Mineral 21:111–124

Blatter CL, Robertson HE, Thomson GR (1973) Regularly interstratified chlorite-dioctahedral smectite in dike-intruded shales, Montana. Clays Clay Mineral 21:207–212

Bookin AS, Smoliar BB (1985) Simulation of bond lengths in coordination polyhedra of 2:1 layer silicates. In: Proc 5th Meet European clay groups, Prague 1983. Charles Univ Press, Prague, pp 51–56

Bouraoui A, Méring J (1964) Diamétre des couches atomiques élémentaires dans les carbones traités entre 1000 °C et 2000 °C. Carbon 1:465–469

Chukhrov FV, Gorshkov AI, Vitovskaya ES, Drits VA (1980) Crystallochemical nature of Co-Ni asbolane. Isv Akad Nauk SSSR Geol Ser 6:75–81 (in Russian)

Chukhrov FV, Gorshkov AI, Drits VA (1982) New structural variant of asbolane. Isv Akad Nauk SSSR Geol Ser 6:69–77 (in Russian)

Chukhrov FV, Gorshkov AI, Drits VA, Shterenberg BE, Sakharov BA (1983) Mixed-layer minerals asbolane-buserites. Isv Akad Nauk SSSR Geol Ser 5:91–100 (in Russian)

Clinard C, Tchoubar D, Tchoubar C, Rousseaux F, Fuselier H (1983) Structure, par diffraction des électrons et des rayons X, de la phase β de l'acide nitrique inséré dans le pyrographite. Synthetic Metals 7–1:333–336

Drits VA (1971) Crystallochemical features of trioctahedral micas. In: Epygenis and its mineral indicators. Ed.: AG Kossovskaya, Nauka, Moscow, pp 96–110 (in Russian)

Drits VA (1975) Structural and crystallochemical features of layer silicates. In: Crystal chemistry of minerals and geological problems. Ed. AG Kossovskaya, Nauka, Moscow, pp. 35–52 (in Russian)

Drits VA (1987) Electron diffraction and high resolution electron microscopy of mineral structures. Springer-Verlag, Berlin Heidelberg New York, 304 pp

Drits VA, Sakharov BA (1976) X-ray structure analysis of interstratified minerals. Nauka, Moscow, 255 pp (in Russian)

Drits VA, Alexandrova VA, Smolin PP (1975) Refinement of the crystal structure of talc. In: Crystal chemistry and geological problems. Ed: AG Kossovskaya, Nauka, Moscow, pp 99–105 (in Russian)

Drits VA, Sakharov BA, Plançon A, Ben Brahim J (1984) Distribution of layer types in mixed layer crystals of homogeneous composition. Krystallographia 29:350–355

Gard JA (1971) The electron-optical investigation of clays. Mineral Soc, London, 383 pp

Gay R, Gasparoux H (1965) Explication des diagrammes de diffraction des rayons X par les carbones graphités. In: Pacault A (ed) Les carbones, vol 1. Masson, Paris, pp 62–128

Grim RE (1968) Clay mineralogy. McGraw-Hill, New York, 596 pp

Guinier A (1964) Théorie et technique de la radiocristallographie. Chap. 13: Diffraction per les réseaux cristallins imparfaits. Dunod, Paris, pp 490–636

Hérold A (1979) Crystallochemistry of carbon intercalation components. In: Lévy F (ed) Physics and chemistry of materials with layered structures, vol 6. Reidel, Dordrecht, pp 323–421

Hirsch PB, Howie A, Nicholson RB, Pashley DW, Whelan MJ (1965) Electron microscopy of thin crystals. Butterworths, London, 549 pp

Jagodzinski H (1949) Eindimensionale Fehlordnung in Kristallen und ihr Einfluß auf die Röntgeninterferenzen. I. Berechnung des Fehlordnungsgrades aus den Röntgenintensitäten. Acta Crystallogr 2:201–207

Kodama H, Gatineau L, Méring J (1971) An analysis of X-ray diffraction line profiles of microcrystalline muscovites. Clays Clay Mineral 19:405–413

Lagrange P, Guérard D, Hérold A (1978) Sur la structure du composé KC_8. Ann Chim Fr 3:143–159

Maire J, Méring J (1970) Graphitization of soft carbons. In: Walker PP (ed) Chemistry and physics of carbon, vol 6. Dekker, New York, pp 125–189

Mélin J, Hérold A (1975) Thermal stability and structure of graphite-antimony pentachloride compounds. Carbon 13:357–362

Méring J, Longuet-Escart J (1954) Adaptation des méthodes de diffraction à l'étude de l'organisation des corps divisés. J Chim Phys 51:416–424

Novikov VM, Berkhin SI, Gorshkov AI, Drits VA, Organova NI, Rudnitskaya ES (1973) Interstratification chlorite-swelling chlorite. Isv Akad Nauk SSSR Geol Ser 8:38–47 (in Russian)

Organova NI, Drits VA, Dmitrik AL (1972) Structural study of tochilinite. I. Isometric varieties. Krystallographia 17–4:761–767 (in Russian)

Organova NI, Drits VA, Dmitrik AL (1973) One-layer valeriite. Dokl Akad Nauk 212:192–195

Plançon A, Tchoubar C (1975) Etude des fautes d'empilement dans les kaolinites partiellement désordonnées. I. Modèle ne comportant que des fautes par translation. J Appl Crystallogr 8:582–588

Pynn R (1979) Incommensurable structures. Nature (London) 281:433–437

Reynolds RC (1980) Interstratified clay minerals. In: Brindley GW, Brown G (eds) Crystal structures of clay minerals and their X-ray identification. Mineral Soc, London pp 229–304

Rothbauer R (1971) Untersuchung eines $2M_1$-Muskovits mit Neutronenstrahlen. Neues Jahrb Mineral Monatsh 44:143–154

Rousseaux F, Vangelisti R, Plançon A, Tchoubar D (1982) Etude quantitative, par diffraction des rayons X, des corrélations entre couches dans les composés graphite-chlorures ferriques de 1er et 2ème stades. Rev Chim Mineral 19:572–587

Sanz J, Serratosa JM (1983) ^{29}Si and ^{27}Al high-resolution MAS-NMR spectra of phyllosilicates. J Am Chem Soc 106:4790–4793

Smith JV, Yoder HS (1956) Experimental and theoretical studies of the mica-polymorphs. Mineral Mag 31:209–235

Solin SA (1982) The nature and structural properties of graphite intercalation compounds. Adv Chem Phys 49:455–532

Tchoubar C (1984) X-ray studies of defects in clays. Philos Trans R Soc London Ser A311: 259–269

Tchoubar C (1986) Quantitative determination of the fine structural features in clays, by modelling of the X-ray diffraction patterns. Mineral Petrogr Acta 29-A:35–54

Terner S, Bucek P (1979) Manganese oxide tunnel structures and their under growths. Science 203:456–458

Terner S, Bucek P (1981) Todorokite: a new family of naturally occurring manganese oxide. Science 212:1024–1027

Veblen DR, Bucek P, Burnham CW (1977) Asbestiform chain silicates: new minerals and structural groups. Science 198:359–365

Zvyagin BB (1967) Electron-diffraction analysis of clay mineral structures. Engl Transl. Plenum, New York (1967) 364 pp (Orig 1964 in Russian, Nauka, Moscow)

Zvyagin BB, Vrublevskaya ZV, Zhukhlistov AP, Sidorenko OV, Soboleva SV, Fedotov AF (1979) High voltage electron diffraction for the study of layer minerals. Nauka, Moscow, 224 pp (in Russian)

CHAPTER 2

Theory of the Diffraction Phenomenon Produced by Powders of Microcrystals with a Lamellar Structure

As seen in the preceding chapter, many lamellar crystals, be they natural or synthetic, are rich in structural defects of various types. Particles of those crystals are usually a few microns parallel to the layers and quite less than one micron in thickness (platelet particles). Because of this, X-ray diffraction techniques only yield powder spectra. Furthermore, a consequence of the presence of structural defects is a modification of the diffraction phenomenon itself and the formation of powder patterns which are, as a rule, notably different from those containing only Bragg reflections. The interpretation of these modified powder patterns requires *a new reading of the theory of diffraction of X-rays in the specific case of microcrystallized lamellar systems.*

Warren (1941) was among the first crystallographers to suggest an explanation for the diffraction produced by these systems: for the particular case of random layer lattices, the diffraction from each lamellar microcrystalline structure could be considered as resulting from interferences of waves diffracted by each of the parallel layers in the stack. For direct comparison with the classical description of the diffraction from an ideal triperiodic crystal, one can **assimilate the whole layer to a giant crystal motif** repeated more or less regularly along a single direction, which is the only one that need be studied.

In the present chapter, we shall examine the theory of the diffraction produced by crystal powders whose only particular feature is the small size of the grains. The problem associated with true structural defects (a "defect" being here *any difference with respect to the ideal triperiodic and perfect structure*) will only be considered in Chapters 3, 4, and 5.

2.1 Diffraction from an Isolated Layer of Finite Extent

Let us first consider a perfect triperiodic crystal. As was recalled in Chapter 1, the crystal is an arrangement of atoms in which it is possible to define an **atomic motif** within a **unit cell.** The triperiodic repetition in object-space (or real space) of this cell in the manner defined by the three vectors $\vec{a}_1$, $\vec{a}_2$, $\vec{a}_3$, will create the crystal. Referring the variables x, y, and z to the origin of the unit cell, we can use a function $\varrho(x,y,z)$ to describe the arrangement of atoms in the unit cell. The variables may be dimensionless numbers when they are expressed as fractions of the dimensions of the unit cell, or they may be in nanometers (or in Angström units). Depending on the crystal, the function $\varrho(x,y,z)$ will be more or less extensive, but in no case will it go beyond the boundaries of the unit cell.

A perfect triperiodic crystal of finite dimensions can then be represented by the function $C(\vec{r})$ defined as:

$$C(\vec{r}) = \varrho(x,y,z) * \left[\left\{ \sum_{P_1} \sum_{P_2} \sum_{P_3} \delta(\vec{r} - [P_1\vec{a}_1 + P_2\vec{a}_2 + P_3\vec{a}_3]) \right\} \cdot g(x,y,z) \right] , \quad (2.1)$$

where $\varrho(x,y,z)$, which represents the structural motif as previously defined, must be a function of the electron density in case of X-ray diffraction; $\vec{r}$ (a continuous variable) is the radius vector characterizing the position of a point in real space; P_1, P_2, P_3 are integers which may be negative, positive, or zero;

$\sum_{P_1} \sum_{P_2} \sum_{P_3} \delta(\vec{r} - P_1\vec{a}_1 - P_2\vec{a}_2 - P_3\vec{a}_3)$ is an infinite sum of Dirac distributions along the directions $[\vec{a}_1, \vec{a}_2, \vec{a}_3]$. This set of distributions defines the limitless triperiodic lattice of points in real space constructed by the vectors $\vec{a}_1, \vec{a}_2$ and $\vec{a}_3$; $g(x,y,z)$ is the *shape function* of the crystal, equal to 1 within and 0 outside the crystal, as defined by Ewald (1940). Multiplying the infinite sums of Dirac distributions by $g(x,y,z)$ implies that the triple sum only exists within the crystal (note in Eq. (2.1) that the symbol "$\cdot$" represents simple multiplication; "$*$" is the convolution operator) from its known properties, it can be deduced that writing

$$\varrho(x,y,z) * \left\{ \sum_{P_1} \sum_{P_2} \sum_{P_3} \delta(\vec{r} - P_1\vec{a}_1 - P_2\vec{a}_2 - P_3\vec{a}_3) \right\}$$

is equivalent to shifting the origin of the function ϱ to the different points in real space where the Dirac distributions are not nil, i.e., at the different nodes of the direct lattice.

Using the same formalism, it is easy to represent a perfect layer with a given array of atoms repeated periodically along two directions characterized by the translation vectors $\vec{a}_1$ and $\vec{a}_2$:

$$C'(\vec{r}) = \varrho(x,y,z) * \left[\left\{ \sum_{P_1} \sum_{P_2} \delta(\vec{r} - P_1\vec{a}_1 - P_2\vec{a}_2) \cdot g(x,y,0) \right\} \right] , \quad (2.2)$$

where $g(x,y,0)$ is now the shape function which bounds the layer in its own plane (polygon, circle, etc.).

Since, in general, atoms in the layer are at different z levels, $\varrho(x,y,z)$ is nonzero along z and the layer is therefore a **tridimensional array** of which only the **periodicity is two-dimensional.**

Let us determine the amplitude of the wave diffracted by one individual layer in the Fraunhofer diffraction conditions (usually valid in powder diffraction techniques). The amplitude is given by the Fourier transform (FT) of the preceding expression:

$$\Phi(\vec{s}) = \frac{F(\vec{s})}{\Omega} \cdot \left[\left\{ \sum_h \sum_k \delta(\vec{s}_0 - h\vec{A}_1 - k\vec{A}_2) \right\} * D(X, Y, \infty) \right] , \tag{2.3}$$

where $F(\vec{s})$, the **structure factor characterizing the scattering by the atomic motif,** is the *FT* of the electron density function $\varrho(x, y, z)$; $\vec{s}$ is the vector characterizing the position of a point in reciprocal space, such that $|\vec{s}| = 2 \sin\theta/\lambda$ in which λ is the wavelength of the electromagnetic radiation and θ is the Bragg angle of diffraction; $\sum_h \sum_k \frac{\delta(\vec{s}_0 - h\vec{A}_1 - k\vec{A}_2)}{\Omega}$ is a normalized set of Dirac functions which represents the *FT* of the double sum in Eq. (2.2); it is also the reciprocal lattice defined by the translation vectors $\vec{A}_1$ and $\vec{A}_2$; the four vectors $\vec{a}_1, \vec{a}_2, \vec{A}_1$ and $\vec{A}_2$ are such that $\vec{A}_i \times \vec{a}_j = \delta_{ij}$ where δ_{ij} is the Kronecker symbol; $\vec{s}_0$ is a vector in the $(\vec{A}_1, \vec{A}_2)$ plane; X, Y, Z are the components of the $\vec{s}$ vector along the reciprocal axes, with Z perpendicular to the $(\vec{A}_1, \vec{A}_2)$ plane; h and k are integers (Miller indices defining the nodes of the 2D reciprocal lattice); $\Omega = |\vec{a}_1 \times \vec{a}_2|$ is the area of the unit cell $(\vec{a}_1, \vec{a}_2)$; $D(X, Y, \infty)$ is the *FT* of $g(x, y, 0)$, known as the **shape transform** of the layer.

Function D is unlimited along Z, perpendicularly to the plane $(\vec{A}_1, \vec{A}_2)$. On the contrary, insofar as the extent of the layer in its own plane is large (the lengths along the directions $\vec{a}_1$ and $\vec{a}_2$ are greater than those of a few unit cells), function $D(X, Y, \infty)$ decreases along X and Y fairly rapidly and is vanishingly small for all integer values of h and k different from 0. Writing then $X = hA_1 + \varepsilon_1$ and $Y = kA_2 + \varepsilon_2$ yields:

$$D(X, Y, \infty) \approx D(\varepsilon_1, \varepsilon_2, \infty) .$$

Under these conditions, convoluting the double sum of Dirac distributions and the function D is equivalent to centering the function $D(\varepsilon_1, \varepsilon_2, \infty)$ on each of the hk nodes of the two-dimensional reciprocal lattice.

For a perfect isolated layer, the intensity of the diffracted beam normalized to a unit cell becomes:

$$I(\vec{s}) = \Phi(\vec{s})\Phi^*(\vec{s}) \cdot \Omega/\sigma ,$$

where $\Phi^*(\vec{s})$ is the complex amplitude conjugate of $\Phi(\vec{s})$ and σ is the area of the diffracting layer. After the proper substitutions:

$$I(\vec{s}) = \frac{F(\vec{s})F^*(\vec{s})}{\Omega\sigma} \left[\left\{ \sum_h \sum_k \delta(\vec{s}_0 - h\vec{A}_1 - k\vec{A}_2) \right\} * G(\varepsilon_1, \varepsilon_2, \infty) \right] , \tag{2.4}$$

where we have written $G(\varepsilon_1, \varepsilon_2, \infty) = D(\varepsilon_1, \varepsilon_2, \infty) D^*(\varepsilon_1, \varepsilon_2, \infty)$.

In the light of the preceding remarks concerning the size of the layer, it is clear that the diffracted beam intensity is localized along "cylinders" or "rods" of infinite length, of basal area $G(\varepsilon_1, \varepsilon_2)$, centered on the hk nodes of the reciprocal lattice and perpendicular to the plane of this lattice, as shown in Fig. 2.1.

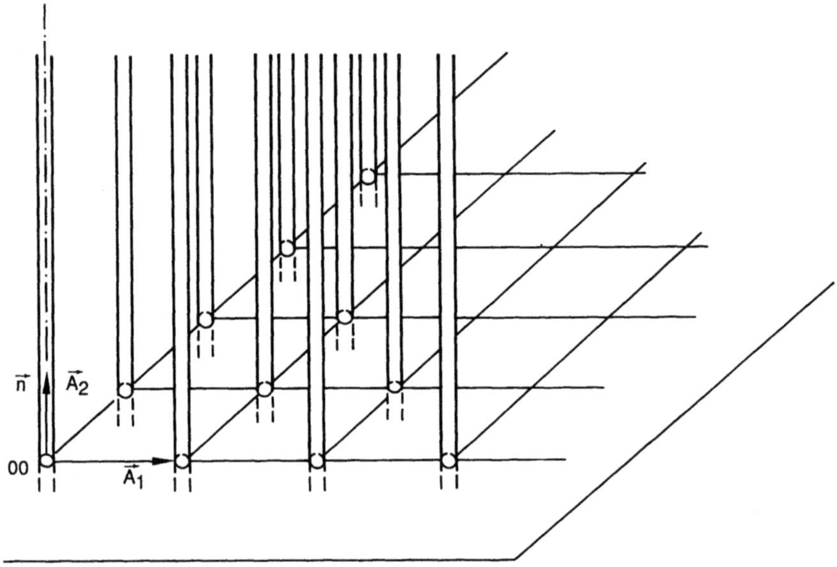

Fig. 2.1. Distribution of intensity in the case of a single layer of finite dimensions

Finally, we note that $F(\vec{s})$ is a slowly varying function of $\vec{s}$ and that it can therefore generally be replaced by a constant within the section of a rod (hk) at a given Z level. Hence, one can write $F(\vec{s}) = F(X, Y, Z) = F_{hk}(Z)$, so that the variation or the intensity within the rod is

$$I_{hk}(Z) = \frac{F_{hk}(Z)F^*_{hk}(Z)}{\Omega\sigma} G(\varepsilon_1, \varepsilon_2, \infty) = \frac{|F_{hk}(Z)|^2}{\Omega\sigma} G(\varepsilon_1, \varepsilon_2, \infty) .$$

This relation shows that, in the case of a single layer, the longitudinal variation of the intensity for each (hk) rod is given by the square of the modulus of the structure factor, with the transverse variation described by $G(\varepsilon_1, \varepsilon_2)$. $F_{hk}(Z)$ is calculated from the characteristics of the atomic motif by

$$F_{hk}(Z) = \sum_{n} f_n(s) \exp\left[2\pi i(hx_n + ky_n + Zz'_n)\right] , \qquad (2.5)$$

where f_n is the atomic scattering factor for the n-th atom at $s = 2\sin\theta/\lambda$, with coordinates x_n and y_n of the n-th atom expressed as fractions of the cell parameters while z'_n, the ordinate of the n-th atom in the direction perpendicular to the layer chosen as the origin, is expressed in nanometers.

2.2 Diffraction from a Defect-Free Stack of Identical Layers

2.2.1 General Description of the Diffraction

Consider an ordered stack of M identical layers with the same translation $\vec{r}_0$ between each pair of successive layers[5], and let us examine the general characteristics of the diffraction produced by this crystal. As developed above, we shall take each layer to be a giant atomic motif repeated in accordance with a given sequence along the stack.

Let us note $\vec{r}_m$ and $\vec{r}_{m'}$ the real space vectors from the origin to two homologous points respectively in layers m and m', themselves n-th nearest neighbors ($n = |m-m'|$) as in Fig. 2.2. The total amplitude of the beam diffracted by the stack can be expressed as

$$A(\vec{s}) = \sum_{m=1}^{M} \Phi_m(\vec{s}) \exp(2\pi i \vec{s}\cdot\vec{r}_m) \ ,$$

where Φ_m is the *structure factor* of the layer m, while $(2\pi\vec{s}\cdot\vec{r}_m)$ is the difference in phase between the diffracted and incident waves.

When normalized to a single layer, the corresponding intensity $i(\vec{s})$ is:

$$i(\vec{s}) = A(\vec{s})A^*(\vec{s})/M \ .$$

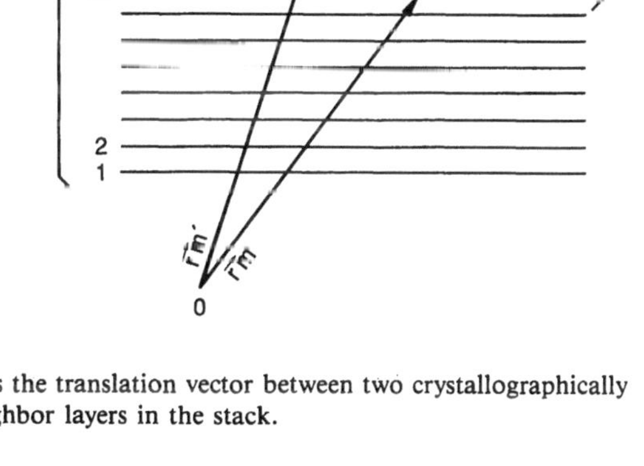

Fig. 2.2. Diagram of a crystal composed of M identical layers

[5] If the crystal is a n-layer polytype, $\vec{r}_0$ is the translation vector between two crystallographically homologous points connecting n-th neighbor layers in the stack.

Again, the same expression can be normalized with reference to a unit cell; this yields:

$$i(\vec{s}) = A(\vec{s})A^*(\vec{s})\Omega/M\sigma \ ,$$

or, after the appropriate substitutions and regrouping of terms,

$$i(\vec{s}) = \frac{\Omega}{M\sigma} \sum_{m=1}^{M} \sum_{m'=1}^{M} \Phi_m(\vec{s})\Phi^*_{m'}(\vec{s}) \exp[2\pi i \vec{s}\cdot(\vec{r}_m - \vec{r}_{m'})] \ , \tag{2.6}$$

in which $(\vec{r}_m - \vec{r}_{m'})$ is the translation between the n-th nearest neighbor layers m and m′. However, since the stack is free from defects, this same translation will be found between all the n-th nearest pairs in the single crystal, so that $|\vec{r}_m - \vec{r}_{m'}| = |\vec{r}_n|$. Furthermore, the structure factor Φ_m must be the same for all the layers, whatever the value of m, since they are identical. Hence $\Phi_m = \Phi_{m'} = \Phi$ and the intensity $i(\vec{s})$ becomes

$$i(\vec{s}) = \frac{\Omega}{M\sigma} \Phi(\vec{s})\Phi^*(\vec{s}) \left\{ M + \sum_{n=1}^{M-1} (M-n)[\exp(+2\pi i \vec{s}\cdot\vec{r}_n) + \exp(-2\pi i \vec{s}\cdot\vec{r}_n)] \right\}$$

or

$$i(\vec{s}) = \frac{\Omega}{\sigma} |\Phi(\vec{s})|^2 \left\{ 1 + 2\,\mathrm{Re}\left[\sum_{n=1}^{M-1} \frac{M-n}{M} \exp(2\pi i \vec{s}\cdot\vec{r}_n) \right] \right\} \ , \tag{2.7}$$

in which Re [. . .] represents the real part of the complex expression between the square brackets. The expression for the intensity $i(\vec{s})$ can be written as:

$$i(\vec{s}) = \Omega |\Phi(\vec{s})|^2 G/\sigma \ , \tag{2.8}$$

in which

$$G = 1 + 2\,\mathrm{Re}\left[\sum_{n=1}^{M-1} \frac{M-n}{M} \exp(2\pi i \vec{s}\cdot\vec{r}_n) \right] \ . \tag{2.9}$$

For perfect crystals, function G is known as the **interference function**, but it was called the *modulation function* by Méring (1949) when it referred to disordered structures. The notation is not the one usually adopted to represent the distribution of intensity in the diffracted beam, but this form is convenient to emphasize the effect of the internal structure of the layer (through the factor Φ) and of the stacking mode (through the factor G) on the variation of intensity of the diffracted wave. We shall see later how the presence of stacking faults or of different types of layers in the particle modify this relationship. However, it is important at this point of the development to verify that the expression does correspond to a distribution of the intensity over the hkl nodes in the case of an ordered stack of identical layers, and that the intensity of the diffracted beam at each node is directly correlated to the dimensions and shape of the lamellar crystal. We shall, for instance, find an expression for the variation of intensity (normalized to a single unit cell) along the Z axis of a rod centered on an hk node.

Let $\vec{s} = h\vec{A}_1 + k\vec{A}_2 + Z\vec{U}_z$ and $\vec{r}_n = n\vec{r}_0 = n(p_0\vec{a}_1 + q_0\vec{a}_2 + d_{001}\vec{u}_z)$ where $\vec{U}_z$ and $\vec{u}_z$ are the unit vectors along the Z and z axis and $(p_0\vec{a}_1 + q_0\vec{a}_1)$ is the translation between two adjacent layers in the direction parallel to their plane, while d_{001} is the distance between two homologous points in the direction perpendicular to the plane. The distribution of intensity along the rod (hk) is:

$$i_{hk}(Z) = \frac{\Omega}{M\sigma} |\Phi_{hk}(Z)|^2 \times \left\{ M + 2\,\mathrm{Re}\left[\sum_{n=1}^{M-1} (M-n) \exp\left[2\pi i n(hp_0 + kq_0 + Zd_{001})\right] \right] \right\} .$$

The explicit relation for the (00) rod is:

$$i_{00}(Z) = \frac{|F_{00}(Z)|^2}{\Omega\sigma} \left\{ 1 + 2\,\mathrm{Re}\left[\sum_{n=1}^{M-1} \frac{M-n}{M} \exp(2\pi i n Z d_{001}) \right] \right\} .$$

After putting $Q = \exp(2\pi i Z d_{001})$ and noting that the summation

$$S = \sum_{n=1}^{M-1} (M-n) Q^n$$

is equivalent to the sum of (M – 1) geometrical progressions, all with the same constant ratio Q and each containing respectively (M – 1), (M – 2), ... 1 terms, and after adding and substracting 1 to each of these series, we can write (Méring 1949):

$$S = \left\{ \sum_{n=1}^{M-1} S_{M-n} \right\} - (M-1) , \quad \text{where} \quad S_{M-n} = \frac{1-Q^{M-n+1}}{1-Q} .$$

By carring out the summation, we obtain $G_{00}(Z)$, the interference function along the Z axis, in the well-known form:

$$1 + 2\,\mathrm{Re}(S/M) = G_{00}(Z) = 1/M \frac{\sin^2 M\pi Z d_{001}}{\sin^2 \pi Z d_{001}} . \tag{2.10}$$

The principal maxima of this function are at $Z = l/d_{001}$, where l can be any integer; the first minima are at $\pm 1/Md_{001}$, on either side of these maxima (Fig. 2.3). The latter correspond to the $00l$ nodes of the reciprocal space and the diffracted beam is localized in the volumes centered on these nodes, while the value

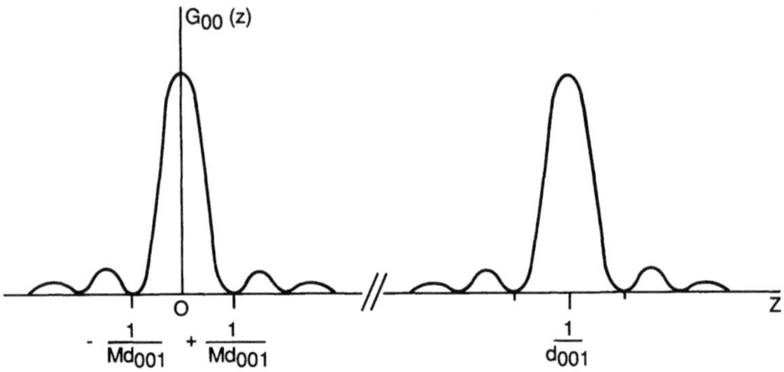

Fig. 2.3. The function $G_{00}(Z)$

of the intensity is given by the ordinate of the principal maxima of $G_{00}(Z)$. Note also that since the function is purely periodic, all the principal maxima have identical height and profile.

For any (hk) rod, $G_{hk}(Z)$ will have the same shape along Z as $G_{00}(Z)$ except for the fact that the principal maxima are now at

$$Z = \frac{1-(h\mathrm{p}_0+k\mathrm{q}_0)}{d_{001}} \,. \tag{2.11}$$

This illustrates the fact that the position of the $hk0$ nodes now depends on p_0 and q_0, i.e., on the mode of stacking of the layers, which will form an orthorhombic (when $\mathrm{p}_0 = \mathrm{q}_0 = 0$) or a monoclinic (when $\mathrm{p}_0 \neq 0$; $\mathrm{q}_0 = 0$) or a triclinic crystal (when $\mathrm{p}_0 \neq 0$; $\mathrm{q}_0 \neq 0$).

2.2.2 Effect of the Thinness of the Interferential Coherence Domains on the Intensity Distribution. Apparent Irrationality of the 00*l* Reflections

In the preceding section, we discussed the positions of the maxima defined by the interference function G. What is actually needed is a precise description of the **intensity** within the volume of each hkl node. This intensity is given by Eq. (2.7), in which $|\Phi(\vec{s})|^2$ is the square of the modulus of the amplitude of the diffracted beam within the volume of the hkl node.

As was explained in Section 2.1, the consequence of the spatial extension – which is usually fairly wide – of the interferential coherence domains is that, within the cross-section (X, Y) of the volume at each hkl node, the diffracted intensity varies as the product $|F_{hk}(Z)|^2 G(\varepsilon_1, \varepsilon_2)$, i.e., as the function $G(\varepsilon_1, \varepsilon_2)$. However, real crystals often give rise to a specific problem connected with the variation of intensity along the Z axis of the node owing to the fact that, in lamellar systems, the thickness of the interferential coherence domain is often reduced to that of a few layers. This specific problem will be illustrated with reference to the 00*l* reflections, for which the intensity along the Z axis, normalized to a single unit cell, is given by:

$$i_{00}(Z) = \frac{|F_{00}(Z)|^2}{\Omega\sigma} G_{00}(Z) = \frac{|F_{00}(Z)|^2}{\Omega M \sigma} \frac{\sin^2 M\pi Z d_{001}}{\sin^2 \pi Z d_{001}} . \tag{2.12}$$

Two cases are then possible:

1. The layers are composed of a single sheet of atoms, with identical z coordinates (as in graphite, for instance); the variation of the square of the structure factor modulus along Z is slow, as is the square of the atomic scattering factor. In this case, the variation of intensity $i_{00}(Z)$ closely follows $G_{00}(Z)$ and the intensity maxima occur at the same values of Z as do the maxima of $G_{00}(Z)$.
2. The case met most frequently is that of layers containing atoms with different z coordinates, as in the layer silicates. Figure 2.4 shows the variation of $|F_{00}(Z)|^2$ for a layer of montmorillonite with structural formula:

$$2[Na_x(Al_{2-x}Mg_x)Si_4O_{10}(OH)_2]$$

and with atomic coordinates as in Table 8.8. The position of the first maxima of $G_{00}(Z)$, corresponding to a constant interlayer distance $d_{001} = 0.955$ nm, is also shown in Fig. 2.4 by the dashed vertical lines. As long as M, the number of layers in the stack, is large (say M > 50), the profile of the principal maxima

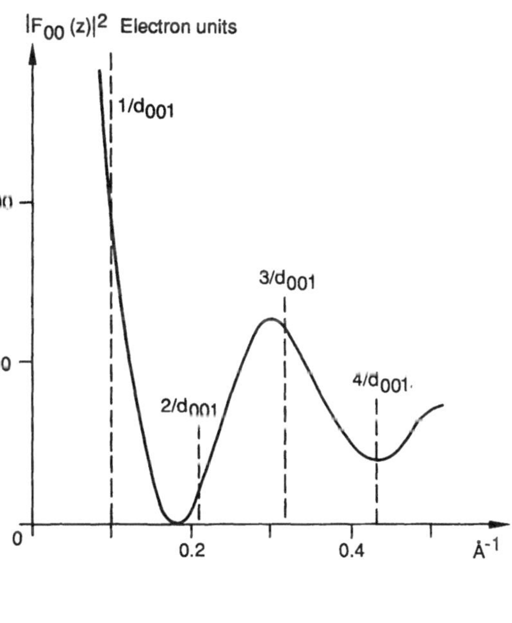

Fig. 2.4. The square of the structure factor for an anhydrous Na-smectite layer. (After Besson 1980)

of the function G is very narrow, with the width at half-height equal to $1/Md_{001}$; the variation along Z of the intensity differs only little from that of G, and the positions of the maxima of $i_{00}(Z)$ practically coincide with those of the principal maxima of $G_{00}(Z)$. On the contrary, when the number M of layers decreases, the principal maxima of G broaden and the variations of the intensity of $i_{00}(Z)$ show two peculiarities:

- the maxima are shifted towards increasing values of $|F_{00}(Z)|^2$
- the function $i_{00}(Z)$ is not symmetrical on either side of these maxima.

Figure 2.5 shows, on a larger scale, the behavior of $i_{00}(Z)$ around the 001 maximum for the montmorillonite when the stack contains 20 or 7 layers (curves a and b, respectively), calculated using expression (2.12). Obviously, small sizes of the domains exclude any direct exploitation of the position of the $00l$ maxima for the determination of d_{001}. Furthermore, for the different $00l$ values used, $|F_{00}(Z)|^2$ will have a positive or negative slope, which will shift the maxima towards high or low values of Z. In other words **the perfect periodicity of the 00*l* reflection positions will be destroyed.** This *irrationality* is only *apparent* and it should not be mistaken for the one shown by stacks containing a number of different interlayer distances (see Chap. 4). Hence, the only method which differentiates between the two sources of irrationality is the comparison between the experimental tracing of the $00l$ reflections and the positions and profile of the peaks calculated using the mathematical formalism described in the present section and in Chapter 5.

This apparent irrationality of the $00l$ reflections due to the small size of the interferential coherence domains in stacks of layer silicates was described by many authors, but was first studied by Ross (1968) and Reynolds (1968): the former gave a direct computation of the intensity of the diffracted wave along Z for crystals

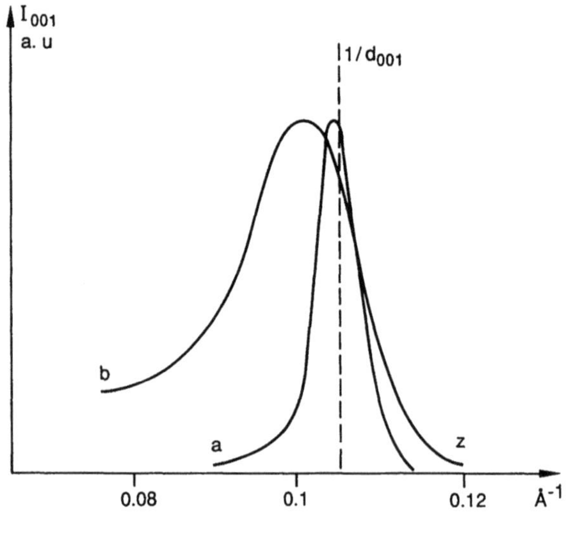

Fig. 2.5. The 001 reflection of a Na-smectite, with interlayer distance $d_{001} = 0.955$ nm, calculated for two values of M, the number of layers in the stack: M = 20 (*curve a*), and M = 7 (*curve b*). (After Besson 1980)

of montmorillonite, biotite, or muscovite stacks containing 2 to 30 layers, using the z coordinates of all the atoms in the crystal, while the latter, using the concept of layers, compared the experimental and calculated profiles of an illite containing a small number of layers. The precise interpretation of this phenomenon as presented above was given in 1976, both by Trunz and by Drits and Sakharov.

2.3 Diffraction by a Powder of Particles with Totally Random Orientation

The mathematical formalism presented in the preceding sections can be used to compute the distribution in reciprocal space of the intensity of the wave diffracted by a perfect but small-size single crystal. Since minerals with a lamellar structure mainly exist as conglomerates of microcrystals, XRD yields only powder patterns. We shall therefore give a mathematical description of the intensity distribution of the diffraction originating from a powder of microcrystals of fault-free stacks of identical layers, first when the particles are randomly oriented (forming an "isotropic powder"), then when there is partial privileged orientation of the particles.

2.3.1 General Expression for the Intensity of the Wave Diffracted by an Isotropic Powder

The diffraction of X-rays by randomly oriented stacks was discussed theoretically by Laue (1932), then by Warren (1941), and by Wilson (1949). These first approaches to the mathematical formalism were continued by Brindley and Méring (1951) who used numerical integration to solve some of the problems associated with powders containing randomly oriented particles with layers of particular polygonal shapes. Recently, more widely applicable solutions to problems associated with the mode of stacking of layers and the size of the interferential coherence domains have been proposed (Maire and Méring 1970; Rousseaux and Tchoubar 1975; Plançon et al. 1980). The theoretical description of powder patterns which will be elaborated in the present section will be based on these results and, more specifically, on the numerical integration method of Brindley and Méring.

The problem here is to find an expression $I(\vec{s})$ for the intensity of the beam diffracted at an angle 2θ (with $s = 2\sin\theta/\lambda$) by a set of similar crystals with random orientation starting from the distribution of the intensity $i(\vec{s})$ of the wave diffracted in reciprocal space by a single microcrystal. Note that we have chosen to express I as a function of the modulus of the diffusion vector $\vec{s}$ rather than as a function of the angle 2θ as is more often the case. There are two advantages to this choice: the resulting expressions are independent of the wavelength of the incident radiation, and the position of the hkl reflections is given directly by the reciprocal of the parametral d_{hkl} values.

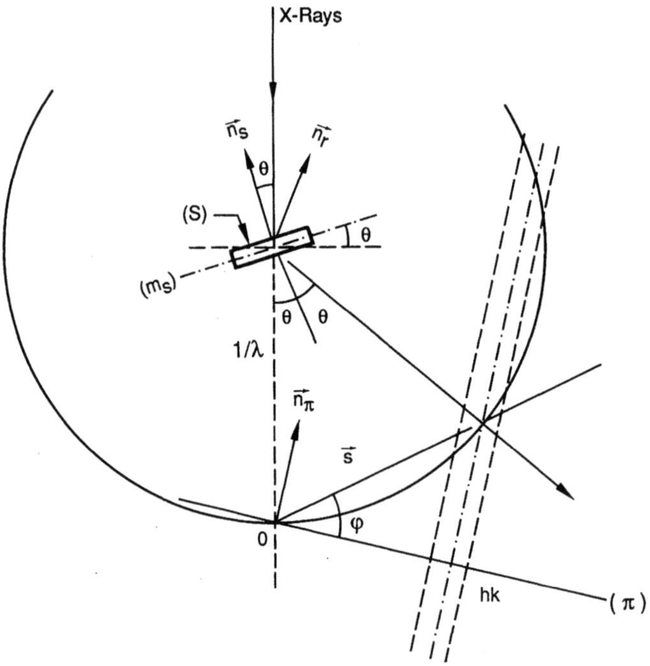

Fig. 2.6. Ewald construction for an (hk) rod for recordings in $\theta-2\theta$ transmission conditions; (m_s) is the median plane of the sample (S); for the other symbols, see text

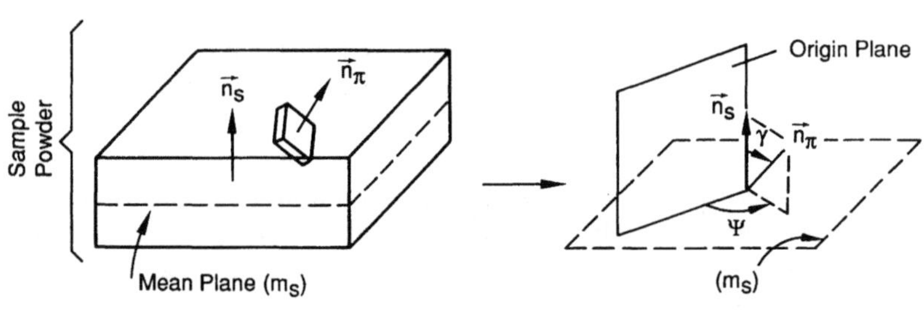

Fig. 2.7. Definition of the angular variables γ and ψ characterizing the orientation of a particle within the powder

The powder pattern can be built up using the Ewald construction. In Fig. 2.6, we show the intersection of an (hk) rod (from a particle in the powder) with the Ewald sphere of radius $1/\lambda$, where λ is the wavelength of the incident radiation. The rod is perpendicular to the planes in the stack, i.e., to the $(\vec{A}_1, \vec{A}_2)$ plane (π) in Fig. 2.6. Let $\vec{n}_\pi$ be the unit vector normal to this plane and $\vec{n}_s$ be the unit vector perpendicular to the median plane of the powder sample, or to the surface of the sample holder. Any particle in the set can then be characterized by two angles: γ, between $\vec{n}_\pi$ and $\vec{n}_s$ in the $(\vec{n}_\pi, \vec{n}_s)$ plane, and the azimuthal angle ψ (see Fig. 2.7).

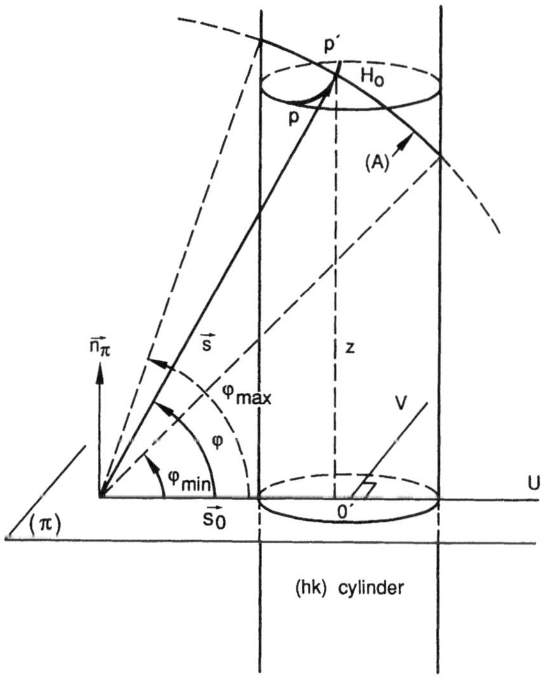

Fig. 2.8. Diagram of an (hk) rod in reciprocal space and its intersection with the integration sphere of radius s. (A) is the trace of the sphere of the radius s intersecting the (hk) cylinder; PP' is the intersection of (A) with the (hk) cylinder at the ordinate Z

Insofar as the powder is perfectly isotropic, the number of stacks with given values of γ and ψ within a certain volume of sample is constant, whatever γ and ψ be. Hence, computing the powder pattern can be done using only a single particle and varying its orientation with respect to the incident beam. Thus, the global intensity I(s), at the tip of the vector $\vec{s}$ corresponding to the angle 2θ, is the sum of the individual intensities $[i(\vec{s})]_{hk}$ at the different points of intersection of the hk rods with the Ewald sphere at the end of vector $\vec{s}$ when the reciprocal lattice is rotated in all directions around the origin O. However, as shown by Brindley and Méring, the same result is obtained by keeping the reciprocal lattice fixed and considering the intersection of each (hk) rod with spheres centered on O but with increased values of the radius s (see Fig. 2.8). The problem is then transformed into that of computing, for each value of s, the intensity borne by the portion of the sphere intersecting the (hk) rod or the volume around the hkl node. For the (hk) rod considered, one can write, in the case of an isotropic powder (cf. Sect. 2.4.2):

$$I_{hk}(s) = \int [i(\vec{s})]_{hk} \frac{dA}{4\pi s^2}, \tag{2.13}$$

in which dA is an element of the surface for the sphere of radius s, and $[i(\vec{s})]_{hk}$ is the intensity, within the (hk) rod, at the point corresponding to the tip of the vector $\vec{s}$. For a small-sized and defect-free stack, the value of $[i(\vec{s})]_{hk}$ as given by Eqs. (2.3) and (2.7), is:

$$[i(\vec{s})]_{hk} = \frac{1}{\Omega\sigma} |F_{hk}(Z)|^2 G(\varepsilon_1, \varepsilon_2) G_{hk}(Z) \tag{2.14}$$

and $I_{hk}(s)$ then becomes:

$$I_{hk}(s) = \frac{1}{\Omega\sigma} \int |F_{hk}(Z)|^2 G_{hk}(Z) G(\varepsilon_1, \varepsilon_2) \frac{dA}{4\pi s^2} . \tag{2.15}$$

The factor $4\pi s^2$, in the denominator of the expression above, is known as the Lorentz factor for powder patterns. It is valid for isotropic powders and its presence results in a decrease of the intensity as s increases.

Since $G(\varepsilon_1, \varepsilon_2)$ is only significant when ε_1 and ε_2 are small as compared to A_1 and A_2 (cf. Sect. 2.1), Warren (1941), then Brindley and Méring (1951) showed that the integration required to obtain $I_{hk}(s)$ [in Eq. (2.13)] could be simplified by using the "tangent cylinder" approximation.

2.3.2 The Tangent Cylinder Approximation

Let O′U and O′V be two axes in the plane (π), respectively parallel and perpendicular to the vector $\vec{s}_0 = h\vec{A}_1 + k\vec{A}_2$, as shown in Fig. 2.8. The intersection of the spherical segment (A), of radius s, with the perpendicular cross-section of the (hk) cylinder, is represented by the arc PP′ and the point H on this arc corresponding to the tip of the vector $\vec{s} = \overrightarrow{OH}$. Since the cross-section $G(\varepsilon_1, \varepsilon_2)$ of the (hk) rod is small, the tangent cylinder approximation consists in replacing the sphere s within the (hk) rod by a cylinder, also of radius s, with its axis passing through O and parallel to O′V. This is equivalent to assuming that H and H′, respectively on the arc PP′ and on the tangent segment CC′, coincide (see Fig. 2.9). If φ now be the angle in the $(\vec{n}_\pi, \vec{s}_0)$ plane between $\vec{s}$ and (Π), the coordinate (U, V, W) of point H are, within the approximation used: $U = s\cos\varphi - s_0; V; W = Z = s\sin\varphi$; within the same approximation, the surface element dA for the spherical segment now coincides with a corresponding element of the tangent cylinder so that $dA = s\, d\varphi\, dV$. Equation (2.15) now becomes:

$$I_{hk}(s) = \frac{1}{4\pi\Omega\sigma s} \iint |F_{hk}(Z)|^2 G_{hk}(Z) G(U, V)\, d\varphi\, dV .$$

The computation of $I_{hk}(s)$ involves evaluating the function:

$$T(U) = \int_A G(U, V)\, dV ,$$

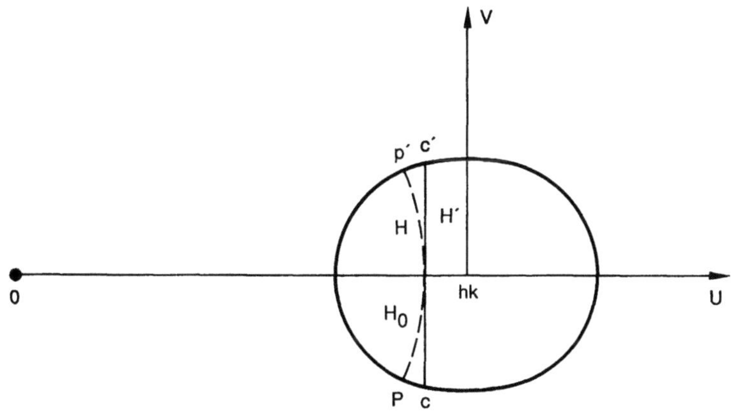

Fig. 2.9. The tangent cylinder approximation. *PP′* is the arc of circle with radius $s = OH_0$; *CC′* is a portion of the generatrix of the cylinder parallel to the axis O′V and of radius s

which depends only on the size and shape of the layers in their own plane (or, equivalently, on the size and shape of the interferential coherence domain in the same plane). The intensity is then obtained by calculating

$$I_{hk}(s) = \frac{1}{4\pi\Omega\sigma s} \int |F_{hk}(Z)|^2 G_{hk}(Z) T(U)\, d\varphi \ , \tag{2.16}$$

in which the angle φ is connected to the hk indices of the rod and to the ordinate Z by the relation $\sin\varphi = Z/s$, and the integration limits correspond to the intersections in the plane $(\vec{n}_\Pi, \vec{s}_0)$ of the sphere of radius s with the (hk) rod. As shown by Brindley and Méring (1951), coherence domains shaped like regular polygons contribute significantly to the intensity only within an (hk) cylinder of radius U_{max}, with $U_{max} = 1.2/r$ where r is the radius of the disc whose surface is equal to that of the regular polygon, as shown in the next section.

2.3.3 Physical Significance of and General Expression for T(U)

By definition, the function T(U) is the normalized projection of the cross-section of the (hk) rod in the direction of the vector $\vec{s}_0$. Knowing that $G(U,V) = D(U,V)D^*(U,V)$, let $P(u_0, v_0)$ be the *FT* of the product DD^*:

$$P(u_0, v_0) = \iint D(U,V) D^*(U,V) \exp[-2\pi i(U u_0 + V v_0)]\, dU\, dV \ .$$

In this expression, the axes O′U and O′V in the plane of the layer are respectively parallel to the directions Ou and Ov, previously defined. For $v_0 = 0$, the relation above becomes:

$$P(u_0,0)=\iint D(U,V)D^*(U,V)\exp(-2\pi i U u_0)\,dU\,dV \tag{2.17}$$

or

$$P(u_0,0)=\int T(U)\exp(-2\pi i U u_0)\,dU\ .$$

Inversely, we can see that T(U) is the *FT* of the Patterson function $P(u_0,0)$:

$$T(U)=\int P(u_0,0)\exp(2\pi i U u_0)\,du_0\ . \tag{2.18}$$

Taking now into account the fact that D(U, V) is the *FT* of the shape function g(u, v), and the fundamental properties of the *FT* in general, i.e., that:

- $D^*(U,V)$ is the *FT* of $g(-u,-v)$;
- the *FT* of the product DD^* is the semi-convolution $g(u,v)*g(-u,-v)$,

we have

$$P(u_0,v_0)=\iint g(u,v)\,g(u_0+u,v_0+v)\,du\,dv$$

and

$$P(u_0,0)=\iint g(u,v)\,g(u_0+u,v)\,du\,dv\ . \tag{2.19}$$

This last expression shows that the Patterson function $P(u_0,0)$ is the area common to the diffracting layer and to its "ghost" obtained by translating the layer by $|u_0|$ in the direction of vector $\vec{s}_0$, as in Fig. 2.10.

From Fig. 2.10, it is easy to see that one can write:

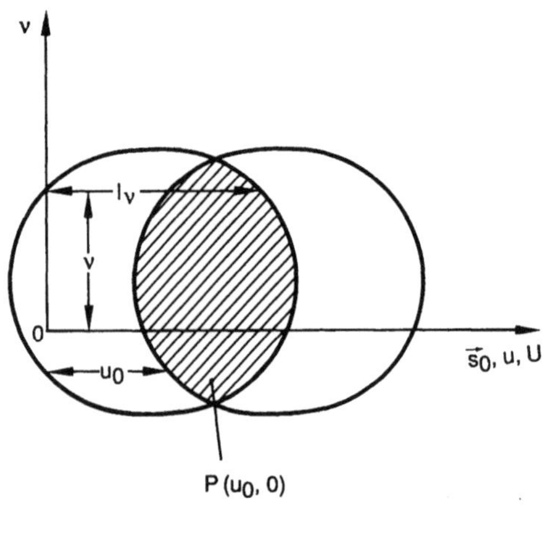

Fig. 2.10. Geometrical representation of $P(u_0,0)$

$$P(u_0, 0) = \int [l_v - |u_0|]\,dv \quad ,$$

in which l_v is the length of the chord parallel to $\vec{s}_0$ at the ordinate v. Function T(U) now becomes:

$$T(U) = \int\int (l_v - |u_0|) \exp(2\pi i U u_0)\,du_0\,dv \quad ,$$

where the limits of integration are the boundaries of the crystal or of the interferential coherence domain. The first integration, between the limit $u_0 = \pm l_v$, yields:

$$T(U) = \frac{1}{2(\pi U)^2} \int (1 - \cos 2\pi U l_v)\,dv = \frac{1}{(\pi U)^2} \int \sin^2(\pi U l_v)\,dv \ . \qquad (2.20)$$

This last expression permits the calculation of T(U) if the size, the shape, and the orientation with respect to the vector $\vec{s}_0$ of the diffracting layer s are known. This was done by Brindley and Méring for coherence domains in the shape of simple regular polygons (square, hexagon) or of a circle. The functions T(U) are then:

Square, side a, with $\vec{s}_0$ parallel to a side

$$T(U) = a^3 \left(\frac{\sin \pi a U}{\pi a U}\right)^2 \ .$$

Square, side a, with $\vec{s}_0$ parallel to a diagonal

$$T(U) = \frac{a^3}{(\pi a U)^2 \sqrt{2}} \left[1 - \frac{\sin(2\pi a U \sqrt{2})}{2\pi a U \sqrt{2}}\right] \ .$$

Hexagon, side a, with $\vec{s}_0$ parallel to a side

$$T(U) = \frac{2\sqrt{3}\,a^3}{(2\pi a U)^2} \left[1 - \frac{\sin(2\pi a U)}{2\pi a U} \{2\cos(2\pi a U) - 1\}\right] \ .$$

Hexagon, side a, with $\vec{s}_0$ perpendicular to a side

$$T(U) = \frac{2a^3}{(2\pi a U)^2} \left[2 - \cos(2\pi a U \sqrt{3}) - \frac{\sin(2\pi a U \sqrt{3})}{2\pi a U \sqrt{3}}\right] \ .$$

Circle, radius r

$$T(U) = \frac{r^3}{(\pi r U)^2} \left\{1 - J_0(4\pi r U) + 2 \sum_n \left[\frac{1}{(n-1)(n+1)} J_n(4\pi r U)\right]\right\} \ .$$

In this last expression, J_0 and J_n are Bessel functions of order 0 and n respectively, with n even and different from 0.

Table 2.1 gives the values of $T(U)/r^3$ calculated by Brindley and Méring as a function of the dimensionless parameter r U, where **r is the radius of the circle with area equal to that of the regular polygon,** since calculations show that the regular polygon and its equivalent circle have the same influence on the intensity distribution along the (hk) rod.

It should be noted that using any of the expressions given above for T(U) supposes that all the particles in the powder – or all the interferential coherence domains – have the same a or r dimension. Obviously, particles in an actual powder will show a distribution of these parameters so that it will be necessary to use a weighting factor in the computation of the intensity of the diffracted beam. If P(a) is the weighting function (the proportion in weight) of particles having a given size "a", Eq. (2.16) for the diffracted intensity should be modified to:

$$I_{hk}(\bar{s}) = \frac{1}{4\pi\Omega\sigma s} \sum_a P(a) \int_A |F_{hk}(Z)|^2 G_{hk}(Z) T_a(U) \, d\varphi \ ,$$

in which $T_a(U)$ is the value of T(U) for a given value of "a". However, a fairly good approximation for the intensity $I_{hk}(s)$ can generally be obtained more simply through the weighted mean $\overline{T_M(U)} = T(U)/\sigma$. The intensity of the diffracted beam then becomes:

$$I_{hk}(s) = \frac{1}{4\pi\Omega s} \int_A |F_{hk}(Z)|^2 G_{hk}(Z) \overline{T_M(U)} \, d\varphi \ . \qquad (2.21)$$

Furthermore, if the distribution of the dimensions which characterizes the polygons is sufficiently narrow, $\overline{T_M(U)}$ can be considered to refer to a mean polygon with characteristic dimension $\bar{a}$ (or $\bar{r}$ for the equivalent circle) where $\bar{a}$ (or $\bar{r}$) is a weighted mean. This simplified approach is usually quite adequate to deal with the powder patterns from most microcrystalline layer systems.

Assimilating the diffracting domain to a circle of equal area, as suggested by Brindley and Méring, is a legitimate approximation only if the domain is shaped like a regular polygon. In other cases, computing T(U) from the general function (2.20) is more complicated. The influence of shape anisometry on the powder patterns from randomly stacked layers was studied by Wilson (1949), then by Rousseaux and Tchoubar (1975) who studied rectangular domains. The results obtained were extrapolated by the authors to ribbon-shaped particles, as reported in the next section.

Table 2.1. Values of $T(U)/r^3$ for three shapes of the interferential coherence domains. (After Brindley and Méring (1951))

	$T(U)/r^3$				
rU	○	⬡ →	⬡ →	□ →	◇ →
0.0	5.33	5.37	5.32	5.56	5.24
0.02	5.31	5.36	5.30	5.52	5.17
0.04	5.25	5.31	5.24	5.46	5.08
0.06	5.14	5.19	5.13	5.35	4.95
0.08	4.98	5.03	4.97	5.18	4.80
0.10	4.79	4.83	4.78	4.98	4.61
0.12	4.57	4.60	4.56	4.75	4.37
0.14	4.34	4.34	4.31	4.49	4.09
0.16	4.03	4.06	4.04	4.21	3.79
0.18	3.79	3.78	3.75	3.91	3.49
0.20	3.48	3.48	3.44	3.60	3.18
0.22	3.16	3.16	3.12	3.28	2.87
0.24	2.85	2.85	2.80	2.95	2.56
0.26	2.54	2.54	2.48	2.60	2.26
0.28	2.24	2.24	2.16	2.26	1.96
0.30	1.94	1.96	1.86	1.94	1.68
0.32	1.66	1.69	1.58	1.64	1.44
0.34	1.41	1.44	1.33	1.37	1.24
0.36	1.18	1.21	1.11	1.12	1.06
0.38	0.98	1.01	0.92	0.90	0.91
0.40	0.80	0.84	0.75	0.70	0.78
0.42	0.64	0.69	0.60	0.53	0.68
0.44	0.51	0.56	0.47	0.38	0.60
0.46	0.41	0.45	0.37	0.26	0.53
0.48	0.33	0.37	0.29	0.18	0.47
0.50	0.28	0.31	0.23	0.13	0.44
0.55	0.19	0.18	0.18	0.04	0.39
0.60	0.165	0.15	0.20	0.02	0.36
0.65	0.20	0.15	0.23	0.08	0.32
0.70	0.225	0.16	0.27	0.17	0.28
0.75	0.24	0.17	0.28	0.245	0.24
0.80	0.22	0.15	0.26	0.26	0.20
0.85	0.195	0.14	0.205	0.24	0.16
0.90	0.145	0.12	0.16	0.20	0.14
0.95	0.10	0.105	0.10	0.135	0.135
1.00	0.08	0.09	0.06	0.075	0.13
1.05	0.07	0.08	0.055	0.02	0.12
1.10	0.065	0.07	0.06	0.00	0.11
1.15	0.05	0.06	0.065	0.00	0.10
1.20	0.04	0.055	0.08	0.01	0.09

Note: The relative orientation of the interferential coherence domains and the vector $\bar{s}_0$ are indicated by the figures and the arrow over each column.

2.3.4 Computation of T(U) for Rectangular Interferential Coherence Domains

Consider a coherence domain in the shape of a rectangle, with length L and width l, and let β be the angle defined by $\tan\beta = l/L$ as in Fig. 2.11. The orientation of the rectangle with respect to the direction of the vector $\vec{s}_0$ (parallel to the axes Ou or O'U) can be described by the angle α between L and s_0. By symmetry, α can be limited to $0 \leqslant \alpha \leqslant \pi/2$.

Depending on whether $\alpha > \beta$, $\alpha < \beta$ or $\alpha = \beta$, the expression for T(U) will take different forms.

a) $\alpha > \beta$ (Fig. 2.11). Equation (2.16) must be broken down into the sum

$$T(U) = T_1(U) + T_2(U) + T_3(U) \; .$$

The first and last term originate in the integration performed respectively over triangles AA'B and CC'D. Since the triangles are congruent and their orientations are equivalent as far as the integrations are concerned, $T_1(U) = T_3(U)$. The computations of $T_1(U)$ gives:

$$T_1(U) = \frac{1}{2(\pi U)^2} \int_0^{v_1} (1 - \cos 2\pi U l_v) \, dv \; ,$$

where $v_1 = l \cos\alpha$ and $l_v = \dfrac{v l}{\cos\alpha \sin\alpha}$,

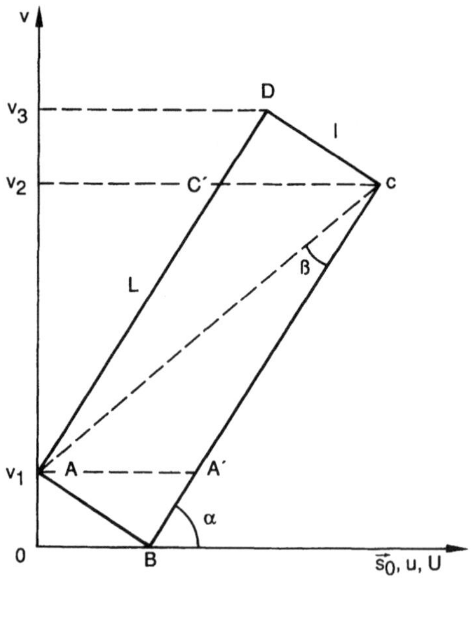

Fig. 2.11. The rectangular domain, with $\alpha > \beta$. (After Rousseaux and Tchoubar 1975)

yielding finally:

$$T_1(U) = \frac{1}{2(\pi U)^2}\left[l\cos\alpha - \frac{\cos\alpha\sin\alpha}{2\pi U}\sin\left(\frac{2\pi U l}{\sin\alpha}\right)\right].$$

The term $T_2(U)$ represents the integration over the parallelogram AA′CC′ in which the chord l_v remains constant when v varies between v_1 and v_2. Starting from:

$$T_2(U) = \frac{1}{2(\pi U)^2}\int_{v_1}^{v_2}(1-\cos 2\pi U l_v)\,dv\ ,$$

with $v_2 = L\sin\alpha$ and $l_v = \frac{l}{\sin\alpha} = \text{const},$

we find

$$T_2(U) = \frac{1}{(\pi U)^2}(L\sin\alpha - l\cos\alpha)\sin^2\left(\frac{\pi U l}{\sin\alpha}\right). \qquad (2.22)$$

The general expression for T(U), namely:

$$T(U) = 2T_1(U) + T_2(U)$$

becomes:

$$T(U) = \frac{1}{(\pi U)^2}\left[l\cos\alpha - \frac{\cos\alpha\sin\alpha}{2\pi U}\sin\left(\frac{2\pi U l}{\sin\alpha}\right) + (L\sin\alpha - l\cos\alpha)\sin^2\left(\frac{\pi U l}{\sin\alpha}\right)\right]. \qquad (2.23)$$

When L becomes much greater than l (slat-shaped particles, or ribbons), both $T_1(U)$ et $T_3(U)$ become negligible compared to $T_2(U)$ and (2.23) is reduced to Eq. (2.22)

b) $\alpha<\beta$ (Fig. 2.12)
The development, quite analogous to the one detailed above, leads to:

$$T(U) = \frac{1}{(\pi U)^2}\left[L\sin\alpha - \frac{\cos\alpha\sin\alpha}{2\pi U}\sin\left(\frac{2\pi U L}{\cos\alpha}\right) + (l\cos\alpha - L\sin\alpha)\sin^2\left(\frac{\pi U L}{\cos\alpha}\right)\right]. \qquad (2.24)$$

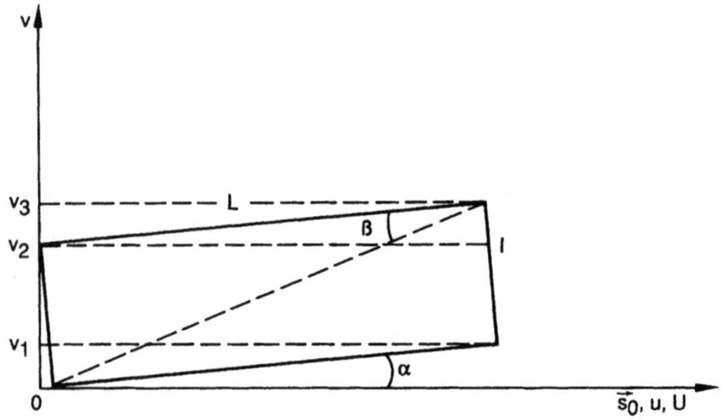

Fig. 2.12. The rectangular domain, with $\alpha < \beta$. (After Rousseaux and Tchoubar 1975)

c) $\alpha = \beta$

The diagonal of the rectangle is now parallel to $\vec{s}_0$ and $v_1 = v_2 = l \cos \alpha = L \sin \alpha$. The expression for T(U) simplifies to the following relations:

$$T(U) = \frac{L \sin \alpha}{(\pi U)^2} \left[1 - \frac{\sin \alpha}{2\pi U l} \sin \left(\frac{2\pi U l}{\sin \alpha} \right) \right] \tag{2.25}$$

$$T(U) = \frac{l \cos \alpha}{(\pi U)^2} \left[1 - \frac{\sin \alpha}{2\pi U l} \sin \left(\frac{2\pi U L}{\cos \alpha} \right) \right] \tag{2.26}$$

2.4 Diffraction from a Powder of Partially Oriented Particles

Underlying Section 2.3 is the assumption that the particles in the powder have fully random orientation. Yet, the layer or ribbon morphology of the actual particles often leads to partial privileged orientation in the powder, and this anisotropy in the spatial orientation of the particles results in modifications both of the relative intensities and of the profiles of the powder pattern reflections. It is therefore necessary to take this partial orientation into account when computing the theoretical shape and intensities of the various types of reflections.

The effect of this privileged orientation in pyrocarbons was studied by Guntert and Cvikevich (1964). Later, Taylor and Norrish (1966) and Reynolds (1976) also studied this effect in relation with the intensity of the $00l$ reflections from oriented deposits of clay minerals. They were followed by Ruland and Tompa (1968) who developed a more general theory usable with profiles of diffraction bands from particles having no cylindrical symmetry in their orientation distribution and applied the theory to carbon fibers. However, the approximations used by these authors are hardly applicable if, simultaneously, the spatial extension of the layers in their own plane is small and their inner structure is complex, as in lamellar silicates. This problem was examined by de Courville et al. (1979) who specifically looked into the influence of orientation on patterns obtained in the symmetrical $(\theta - 2\theta)$ transmission conditions and scaled to absolute values. They considered the frequently encountered particular case of moderately oriented powders. A last generalization was elaborated by Plançon (1980) who studied the effect of orientation on patterns obtained in the non symmetrical transmission mode.

We shall present the methods of de Courville et al. and of Plançon, which use a classical approach and have the advantage of maintaining a close link, at all stages, with the physical parameters which characterize the structure of the system. This method has been successfully used in the study of various powders, as will be shown by a few examples in Chapter 8.

2.4.1 Definition of the Spatial Distribution of the Particles in a Powder

Let us refer once more to Fig. 2.7 in which the orientation of a given particle was characterized by the angle γ between the normal $\vec{n}_\Pi$ to plane (Π) and the normal $\vec{n}_s$ to the mean plane of the specimen, and the azimuthal angle of rotation ψ between the $(\vec{n}_\Pi, \vec{n}_s)$ plane and the reference plane containing $\vec{n}_s$. To each constant direction of the vector $\vec{n}_s$ correspond, in the powder, a large number of particles, each of them more or less rotated in its own (Π) plane. The orientation of each of these particles can be completely defined using the angles γ and ψ, and a new variable α. For each particle, α is the angle between the planes $(\vec{n}_\Pi, \vec{\Delta}_0)$ and $(\vec{n}_\Pi, \vec{\Delta})$ where $\vec{\Delta}_0$ and $\vec{\Delta}$, both in the plane (Π), are respectively parallel to a fixed arbitrary reference direction and to a given crystallographic axis (Fig. 2.13). These three angles now constitute the direction angles for the particle.

The probability of finding a particle with a normal $\vec{n}_\Pi$ in the domain $d\gamma$, $d\psi$, $d\alpha$ about a set of given values of γ, ψ and α is now:

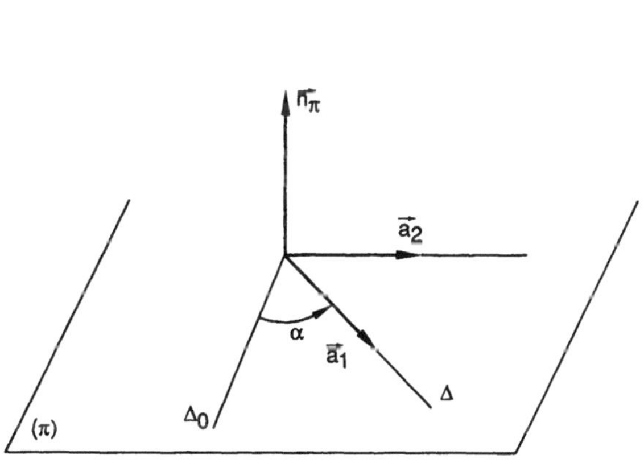

Fig. 2.13. The angle α used to define the rotation of the particles in their own plane. Δ_0 is the fixed reference direction parallel to a chosen crystallographic axis, here $\vec{a}_1$

$$d^3p = p(\gamma, \psi, \alpha) \sin\gamma \, d\gamma \, d\psi \, d\alpha \ , \tag{2.27}$$

where $p(\gamma, \psi, \alpha)$ is the probability that the direction $\vec{n}_\Pi$ for the particle shall be in the solid angle $d^2\Omega = \sin\gamma \, d\psi \, d\gamma$, as defined in Fig. 2.14, and that this particle be such that its $(\vec{n}_\Pi, \vec{\Delta})$ plane is at an angle between α and $\alpha + d\alpha$ with the plane $(\vec{n}_\Pi, \vec{\Delta}_0)$. If the rotation of the particles in their own plane is independent on the position of vector $\vec{n}_\Pi$ with respect to $\vec{n}_s$, it is possible to define two independent probabilities:

- on one hand, the probability $p(\alpha)$ that the plane $(\vec{n}_\Pi, \vec{\Delta})$, characteristic of a given particle, shall make an angle within α and $\alpha + d\alpha$ with the plane $(\vec{n}_\Pi, \vec{\Delta}_0)$;
- on the other hand, the probability $p(\gamma, \psi)$ that the vector $\vec{n}_\Pi$ shall be within the solid angle $d^2\Omega$.

The normalization condition is that the sum of the probabilities of finding the particle oriented in any given direction must be equal to 1, or

$$\int_{\alpha=0}^{2\pi} p(\alpha) d\alpha \int_{\psi=0}^{2\pi} \int_{\gamma=0}^{\pi} p(\gamma, \psi) \sin\gamma \, d\gamma \, d\psi = 1 \ . \tag{2.28}$$

If the particles have a regular polygonal shape, then $p(\alpha) = 1/2\pi$, whatever the value of α, owing to the isotropy of rotation in their own plane. Furthermore, the variables γ and ψ are usually independent in a powder, so that $p(\gamma, \psi) = p(\gamma)p(\psi)$ and the normalization condition becomes:

$$\int_{\psi=0}^{2\pi} p(\psi) d\psi \int_{\gamma=0}^{\pi} p(\gamma) \sin\gamma \, d\gamma = 1 \ . \tag{2.29}$$

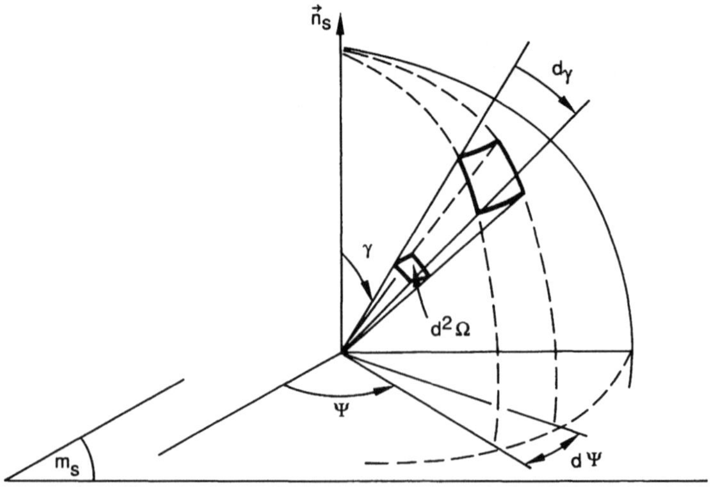

Fig. 2.14. Definition of the solid angle $d^2\Omega$

The further consequence of the shape isotropy of the particles in their own plane is the occurrence of azimuthal isotropy expressed in the fact that $p(\psi)$ is independent of the angle ψ. The sole remaining anisotropy in the powder is the presence of a privileged orientation of the particles with respect to the mean plane (m_s) of the specimen, so that $p(\gamma)$ is now a function of γ. Since the condition for normalization (2.27) is merely:

$$\int_{\gamma=0}^{\pi} p(\gamma)\sin\gamma\,d\gamma = 1 \ , \qquad (2.30)$$

the problem is reduced to determining the relation [between the probability $p(\gamma)$ and the angle γ] which will eventually be used in the computation of the intensity of the diffracted beam. Remembering that $\gamma = 0$ and $\gamma = \pi/2$ when the particles are respectively parallel and perpendicular to the mean plane of the specimen, this can be done experimentally by studying the variation of the integrated intensity from a given $00l$ reflection as the orientation of the sample is changed with respect to the incident X ray beam (the method will be more fully described in Chap. 7). In order to give a correct description of the influence of privileged orientation among the particles on the intensity of the reflections, account must be taken of the experimental conditions under which the X-ray powder pattern was obtained since, for any given angle 2θ, the number of particles participating in the diffraction differs according to the mounting used. The ones most frequently favored for powder diffractometry thus give:

- Transmission powder patterns obtained with angle θ (characteristic of the specimen mean plane or of the sample holder, within respect with the incident beam as in Fig. 2.6) equal to half the diffraction angle 2θ; this is the symmetrical $\theta-2\theta$ mounting generally used with proportional counter and stepping motors techniques. The influence of orientation in this case was described by de Courville et al. (1979).
- Transmission powder patterns with position of the specimen fixed and independent of the diffraction angle 2θ; this is the nonsymmetrical mounting preferably used with localized linear or curved detectors. This case was described by Plançon (1980).
- Reflection powder patterns with a $\theta-2\theta$ symmetrical or nonsymmetrical mounting.

2.4.2 Diffraction from a Partially Oriented Powder in a Symmetrical $\theta-2\theta$ Transmission Mounting

This is the case depicted in the Ewald diagram of Fig. 2.6 in Sect. 2.3.1. In this symmetrical $\theta-2\theta$ arrangement, the angle between the normal $\vec{n}_s$ to the mean plane of the sample and the direction of the incident beam is always half the angle

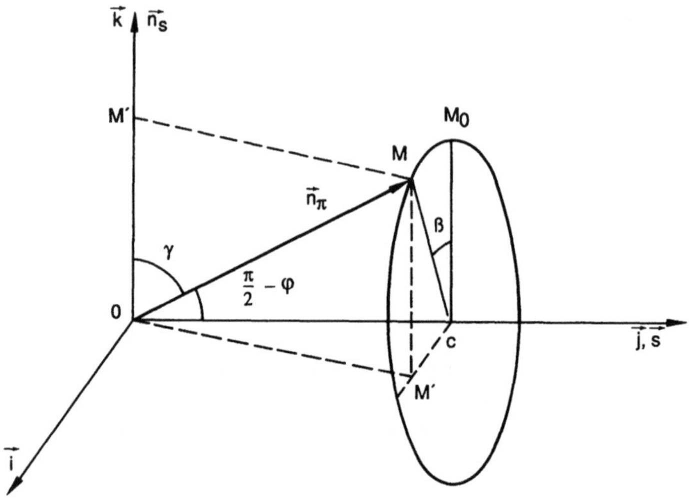

Fig. 2.15. The relation between the various angles which define the orientation of a particle characterized by its $\vec{n}_{\Pi}$

of diffraction 2θ. This is tantamount to saying that $\vec{n}_s$ is always perpendicular to the vector $\vec{s}$ at the end of which the intensity $I_{hk}(s)$ of the diffracted beam is calculated. Let φ be the angle in the $(\vec{n}_s, \vec{s}_0)$ plane between the vector $\vec{s}$ and the (Π) plane of a diffracting particle (Fig. 2.6); all the particles with (Π) tangent to the cone with axis $\vec{s}$ and half vertex-angle φ give the same contribution to the intensity of the wave diffracted at angle 2θ. The normals $\vec{n}_{\Pi}$ to these particles form a cone with axis $\vec{s}$ and half vertex-angle $(\pi/2-\varphi)$ as shown in Fig. 2.15.

Let M be the tip of one of the vectors $\vec{n}_{\Pi}$ of this cone. The angle γ between $\vec{n}_s$ and $\vec{n}_{\Pi}$ varies as M rotates around $\vec{s}$ on a circle of radius $\cos\varphi$ perpendicular to $\vec{s}$. Let C be the center of the circle and β the angle between CM and CM_0, where M_0 is the point at which the circle cuts the $(\vec{n}_s, \vec{s})$ plane. The position of a normal $\vec{n}_{\Pi}$ is completely defined if φ and β are given and the probability of finding a normal $\vec{n}_{\Pi}$ corresponding to the direction angles φ and β is written $p(\varphi, \beta)$. However, since the experimental results yield the probability $p(\gamma, \varphi)$, the problem is now correlating $p(\varphi, \beta)$ with $p(\gamma, \psi)$.

We therefore define a set of orthogonal axes $\vec{i}, \vec{j}, \vec{k}$ such that $\vec{k} = \vec{n}_s$, $\vec{j} = \vec{s}/s$ and $\vec{i}$ is perpendicular to the $(\vec{j}, \vec{k})$ plane. When expressed as a function of this set of axes, $\vec{n}_{\Pi}$ becomes:

$$\vec{n}_{\Pi} = \vec{i}\,CM'' + \vec{j}\,OC + \vec{k}\,M''M$$

where

$$CM'' = CM \sin\beta = \cos\varphi \sin\beta \ , \quad OC = \sin\varphi \quad \text{and}$$

$$M''M = CM \cos\beta = \cos\varphi \cos\beta$$

so that

$$\vec{n}_\Pi = \vec{i}\cos\varphi\sin\beta + \vec{j}\sin\varphi + \vec{k}\cos\varphi\cos\beta \ . \quad (2.31)$$

In the symmetrical $\theta - 2\theta$ mounting, and since $\cos\gamma$ is given by the scalar product $\vec{n}_s \cdot \vec{n}_\Pi$, we have

$$\cos\gamma = \cos\varphi\cos\beta \ , \quad (2.32)$$

so that the angle γ between $\vec{n}_s$ and $\vec{n}_\Pi$ corresponding to given values of φ and β is

$$\gamma = \arccos(\cos\varphi\cos\beta) \ .$$

We shall now determine the probability d^3p of finding a particle with a normal $\vec{n}_\Pi$ within the solid angle $d^2\Omega$ attached to the variations $d\varphi$ and $d\beta$ of angles φ and β, when its crystallographic direction Δ (defined in Section 2.4.1) is within the angular sector $\alpha, \alpha + d\alpha$. Recalling that the radius of the circle is $\cos\varphi$, we have immediately $d^2\Omega = \cos\varphi\, d\varphi\, d\beta$, as can be seen in Fig. 2.16. The probability of finding $\vec{n}_\Pi$ within the angle $d^2\Omega$ is $p(\varphi, \beta, \alpha)$ and if the particles are randomly oriented in their own plane (isotropic distribution), $p(\alpha) = 1/2\pi$ and

$$d^3p = \frac{d\alpha}{2\pi}\, p(\varphi,\beta)\, d^2\Omega = \frac{1}{2\pi}\, p(\varphi,\beta)\cos\varphi\, d\alpha\, d\beta\, d\varphi \ . \quad (2.33)$$

Summing over β now yields the probability d^2p of finding particles with normals within the solid angle $d\Omega$ included between the two cones with half vertex-angles $\pi/2 - \varphi$ and $\pi/2 - (\varphi + d\varphi)$ shown in Fig. 2.16. Hence,

$$d^2p = \frac{1}{2\pi}\cos\varphi\, d\alpha\, d\varphi \int_{\beta=0}^{2\pi} p(\varphi,\beta)\, d\beta \ .$$

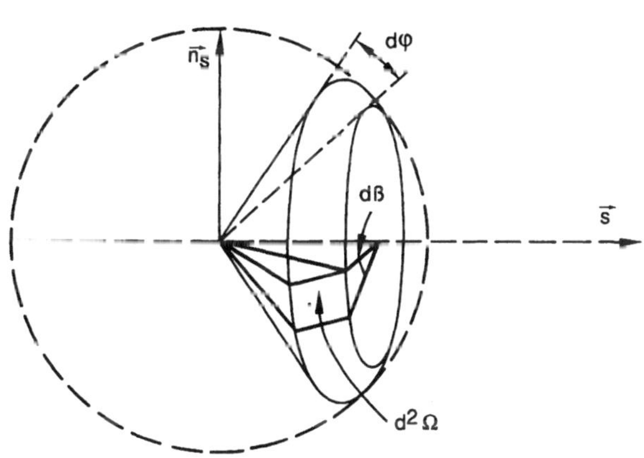

Fig. 2.16. The solid $d^2\Omega$, corresponding to given values of β and α

Let us define

$$\bar{p}(\varphi) = \frac{1}{2\pi} \int_0^{2\pi} p(\varphi, \beta) \, d\beta \; . \tag{2.34}$$

In the case of powders with azimuthal isotropy around n_s, and for a given φ_0, integrating the function $p(\varphi, \beta)$ between the limits $\beta = 0$ and $\beta = 2\pi$ is equivalent to integrating the experimental function $p(\gamma)$ between the limits φ_0 and $\pi - \varphi_0$, with γ, φ, and β related to each other as in Eq. (2.32). In powders, and more particularly when the specimen is composed of platelets, the particles are usually distributed symmetrically with respect to the mean plane of the specimen. It is therefore sufficient to determine the experimental value of $p(\gamma)$ in the interval $(0, \pi/2)$ and to obtain the rest of the function by symmetry from $p(\gamma) = p(\pi - \gamma)$. The function – which must be normalized – a maximum for $\gamma = 0$ and a minimum for $\gamma = \pi/2$, and its slope is zero at both these values, as shown in Fig. 2.17. Replacing the integration over β in (2.34) consists in determining, for each value of φ_0, the area shaded in Fig. 2.17 between the limits symmetrical with respect to $\gamma = \pi/2$; after division by 2π, d^2p becomes

$$d^2p = \bar{p}(\varphi) \cos\varphi \, d\varphi \, d\alpha \; .$$

All the particles with the same probability d^2p of having their normal within the solid angle $d\Omega$ contribute equally to the diffraction from the powder, and it is therefore possible to write:

$$d^2 I_{hk}(s) = [i(\vec{s})]_{hk} d^2p \; , \tag{2.35}$$

in which $[i(\vec{s})]_{hk}$, the diffracted beam intensity (normalized to a single unit cell) from one of these particles, is given by. Eqs. (2.7) and (2.14). Integrating Eq. (2.35) then gives the total intensity:

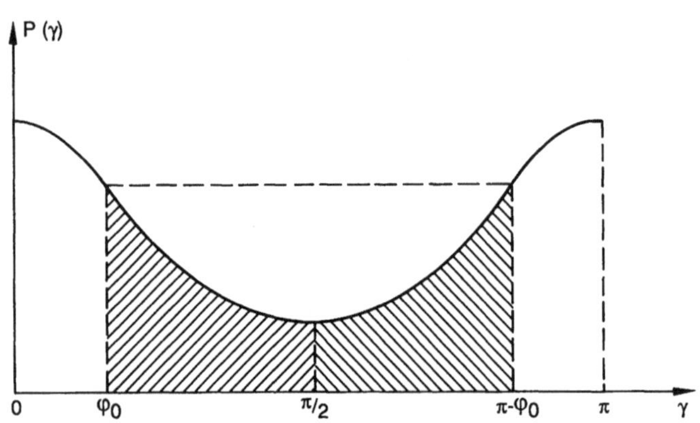

Fig. 2.17. Physical significance of $\bar{p}(\varphi)$

$$I_{hk}(s) = \iint [i(\vec{s})]_{hk} \bar{p}(\varphi) \cos\varphi \, d\varphi \, d\alpha \ . \tag{2.36}$$

The first integration – which is over the variable φ – is performed using the method of Brindley and Méring (1951), keeping the direction of the plane (Π) of the particles in a constant direction and varying the direction of $\vec{s}$. The element of area dA swept by $\vec{s}$ within the cylinder (hk) when φ and α vary by $d\varphi$ and $d\alpha$ is shown in Fig. 2.18 and is

$$dA = s^2 \cos\varphi \, d\varphi \, d\alpha \ .$$

The expression (2.36) for $I_{hk}(s)$ now becomes:

$$I_{hk}(s) = \iint [i(\vec{s})]_{hk} \bar{p}(\varphi) \frac{dA}{s^2} \ , \tag{2.37}$$

in which the limits of integration φ_{min} and φ_{max}, which vary with the radius s of the sphere, correspond to the intersection in the (n_π, s_0) plane of the (hk) rod with the sphere.

The second integration, over α, corresponds to the computation of T(U) using the tangent cylinder approximation, as detailed in Section 2.3.3. The intensity of the wave diffracted at the angle 2θ by all rods with given values of the indices h and k becomes, after replacing $[i(\vec{s})]_{hk}$ by (2.14):

$$I_{hk}(s) = \frac{1}{\Omega \sigma s} \int |F_{hk}(Z)|^2 G_{hk}(Z) T(U) \bar{p}(\varphi) \, d\varphi \ . \tag{2.38}$$

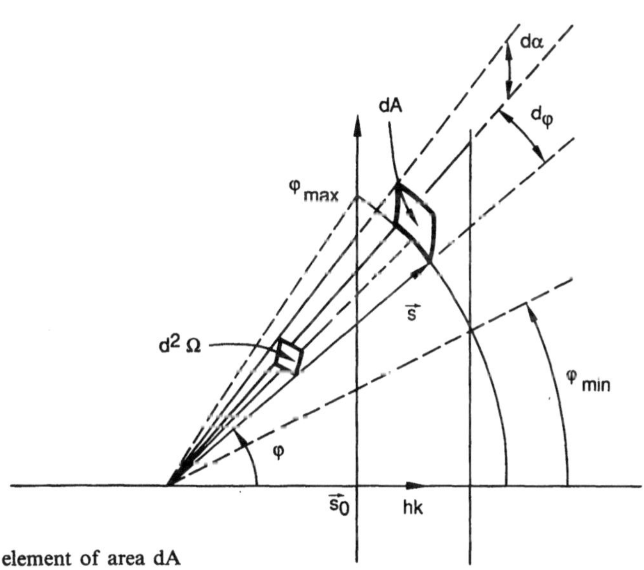

Fig. 2.18. Scheme of the element of area dA

This result is equivalent to that of Brindley and Méring given in (2.16); it also includes the effect of mutual orientation among the particles with respect to the mean plane of the specimen. Note also that $\bar{p}(\varphi)$ is constant and equal to $1/4\pi$ in isotropic powders, as given by the normalization conditions of Section 2.4.1, so that the two expressions are strictly equivalent in that case.

2.4.3 Diffraction from a Partially Oriented Powder in an Asymmetrical Transmission Mounting or in a Reflection Arrangement

When the diffraction pattern is obtained by transmission of the X-ray beam through a fixed specimen, in an asymmetrical $\theta-2\theta$ mouting, or by reflection, whether the arrangement be symmetrical or not, the principle of computation for the intensity of the wave diffracted by the partially oriented powder is identical with the one presented in the preceding section: $I_{hk}(s)$ is still given by (2.38), but the method used to obtain $\bar{p}(\varphi)$ is different.

As before, each value of $\bar{p}(\varphi)$ is obtained from Eq. (2.34), with integration over β of the function $p(\varphi,\beta)$ replaced by integration of $p(\gamma)$. However, each type of mounting entails specific relationships between γ and the direction angles β and φ, since the normal $\vec{n}_s$ to the mean plane of the sample is perpendicular to vector $\vec{s}$ only in the symmetrical $\theta-2\theta$ transmission mounting. In all other cases, the angle between $\vec{n}_s$ and $\vec{s}$ can take any value, as shown in Fig. 2.19, which illustrates an asymmetrical $\theta-2\theta$ arrangement. In the figure, $\vec{n}_{sym}$ is the normal perpendicular to $\vec{s}$, and the angle between $\vec{n}_{sym}$ and $\vec{n}_s$ is ω. The new relation between γ, φ and β is shown in Fig. 2.20 in which the same system of coordinate axes is used as in Fig. 2.15.

The vector $\vec{n}_\Pi$ is still described by Eq. (2.32), namely

$$\vec{n}_\Pi = \vec{i}\cos\varphi\sin\beta + \vec{j}\sin\varphi + \vec{k}\cos\varphi\cos\beta$$

but the vector $\vec{n}_s$ in the $(\vec{j},\vec{k})$ plane is now

$$\vec{n}_s = -\vec{j}\sin\omega + \vec{k}\cos\omega$$

and

$$\cos\gamma = \vec{n}_s\cdot\vec{n}_\Pi = \cos\beta\cos\varphi\cos\omega - \sin\varphi\sin\omega \ , \tag{2.39}$$

so that

$$\gamma = \arccos\left[\cos\beta\cos\varphi\cos\omega - \sin\varphi\sin\omega\right] \ .$$

Although β is always bounded by $0 \leqslant \beta < 2\pi$ in a powder with azimuthal isotropy around $\vec{n}_s$, the limiting values of γ, which correspond to the limits of β, depend on the value of the angle ω between the vectors $\vec{n}_{sym}$ and $\vec{n}_s$:

1. For the reflection or transmission arrangements asymmetrical in $\theta-2\theta$, ω can have any value, say ω_0, and the integration limits for γ, as given by Eq. (2.39), are

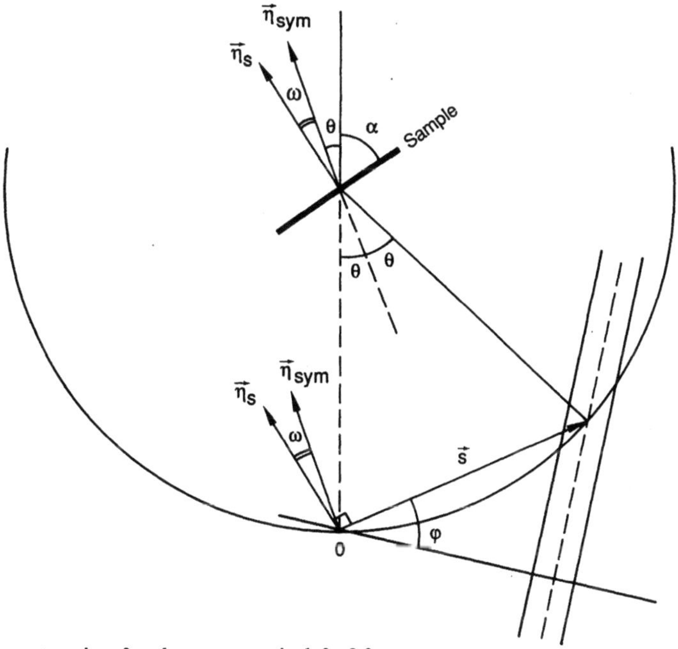

Fig. 2.19. The Ewald construction for the asymmetrical $\theta-2\theta$ arrangement

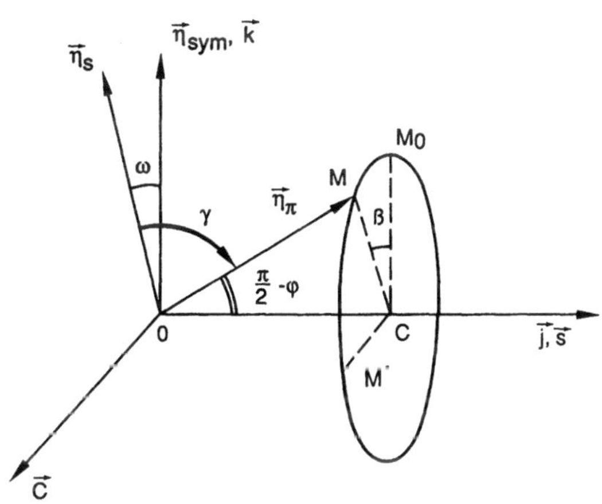

Fig. 2.20. The angular relationships involving the vectors n_s' and n_Π in the asymmetrical transmission mounting

$$\gamma_{min} = \varphi_0 + \omega_0 \quad \text{and} \quad \gamma_{max} = \pi - \varphi_0 + \omega_0 \ ,$$

where φ_0 is a given value of φ. Note that, in opposition to the results obtained with the symmetrical mounting, the limits of γ are not symmetrical with respect to $\gamma = \pi/2$;

2. In the particular case of the symmetrical $\theta-2\theta$ reflection mounting, $\omega = -\pi/2$ since $\vec{n}_s$ is parallel to $\vec{s}$ and, for each particular value φ_0 of the angle φ, all particles with their normals $\vec{n}_\Pi$ at an angle $\pi/2-\varphi_0$ with $\vec{s}$ show the same angle γ_0 between the vector $\vec{n}_\Pi$ and the normal $\vec{n}_s$ to the specimen, so that $\gamma = \pi/2-\varphi_0$; under these conditions, the sum over γ of $p(\gamma)$ reduces to a single value $p(\gamma_0)$ and the integral in Eq. (2.34) becomes

$$\bar{p}(\varphi) = \frac{1}{2\pi}\int_{\beta=0}^{2\pi} p(\varphi,\beta)\,d\beta = \frac{1}{2\pi}\,p(\gamma_0) = \frac{1}{2\pi}\,p(\pi/2-\varphi_0)\ . \tag{2.40}$$

2.4.4 Diffraction from a Partially Oriented Powder in the Particular Case of the (00) Rod

The expression in Eq. (2.38) becomes particularly simple when it is applied to the (00) rod, i.e., to the $00l$ reflections. Thus, rewriting it without using the function T(U) gives:

$$I_{00}(s) = \frac{1}{\Omega\sigma}\iint |F_{00}(Z)|^2 G_{00}(Z)\,G(U,V)\,\bar{p}(\varphi)\,\frac{dA}{s^2}\ . \tag{2.41}$$

For the (00) rod, the scattering vector $\vec{s}$ has only a single component along Z and the integral in (2.41) is calculated on the surface common to the sphere of radius Z and to the (00) rod, which has its origin at the center O of the sphere (Fig. 2.21). Insofar as the cross-section of the rod is small (this condition was discussed in Sects. 2.1 and 2.3.1), the tangent cylinder approximation can be used again.

In general, and for any $00l$ reflection, the node in reciprocal space at which the intensity of the diffracted wave is sought is sufficiently far from the origin for the radius s of the integration sphere to be much larger than the diameter of the rod[6]. It follows that angle φ between the plane (Π) of any particle participating in the diffraction and any vector $\vec{s}$ with its tip within the rod will differ but little from $\pi/2$. For all values of s, $\bar{p}(\varphi)$ will be quite close to $\bar{p}(\pi/2)$ which can itself be related to the experimentally determined values of $p(\gamma)$ through Eq. (2.39) and the normalization conditions:

$$\bar{p}(\pi/2) = \frac{p(\gamma_0)}{2\pi}\ ,$$

where $\gamma_{\sigma} = \pi/2+\omega_0$, in which ω_0 is the angle between the normals $\vec{n}_{sym}$ and $\vec{n}_s$ in the symmetrical $\theta-2\theta$ arrangement.

In the light of the arguments above, and remembering that $dA = dU\,dV$ (cf. Fig. 2.8), the intensity of the diffracted wave given in (2.41) becomes:

[6] This is no longer the case in **small-angle X-ray diffraction** and in an ideally oriented deposit.

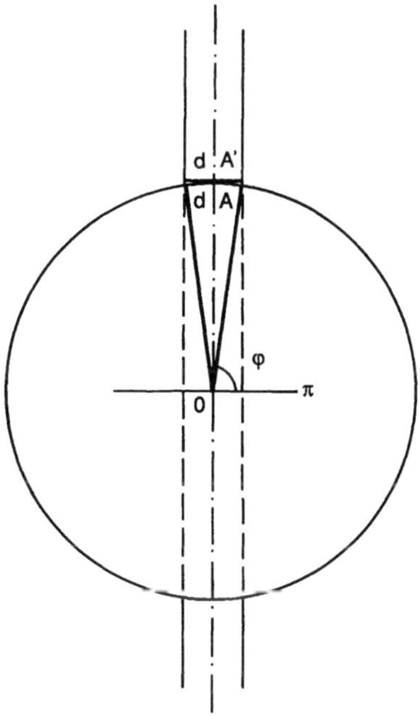

Fig. 2.21. The tangent cylinder approximation applied to the (00) rod

$$I_{00}(Z) = \frac{1}{2\pi\Omega\sigma Z^2} p(\pi/2+\omega_0)\,|F_{00}(Z)|^2 G_{00}(Z) \iint G(U,V)\,dU\,dV \ .$$

Furthermore, using the expressions given in Eqs. (2.17), (2.18) and (2.19), we obtain:

$$\iint G(U,V)\,dU\,dV = P(0,0) = \iint g(u,v)\,g(u,v)\,du\,dv = \sigma \ ,$$

in which σ is the area of the diffracting domain, so that, finally, the intensity $I_{00}(Z)$ is:

$$I_{00}(Z) = \frac{1}{2\pi\Omega Z^2} p(\pi/2+\omega_0)\,|F_{00}(Z)|^2 G_{00}(Z) \ . \tag{2.42}$$

When the particles in the powder have fully random orientations, $p(\gamma_0)$ – which can be expressed as $p(\gamma_0) = p(\pi/2+\omega_0)$ – must be constant and equal to 1/2 for any value of ω_0. The intensity, normalized to a single unit cell, is given by the well-known expression:

$$I_{00}(Z) = \frac{1}{4\pi\Omega Z^2} |F_{00}(Z)|^2 G_{00}(Z) \ , \tag{2.43}$$

in which $G_{00}(Z)$, as indicated in (2.10), is

$$G_{00}(Z) = \frac{1}{M} \frac{\sin^2 \pi M Z d_{001}}{\sin^2 \pi Z d_{001}} .$$

It is worth pointing out that Eqs. (2.42) and (2.43) differ only by the form of the expression $p(\pi/2 + \omega_0)$ which is **the value at a specific point** of the function $p(\gamma)$, and **not its integral** over the interval $(\gamma_{min}, \gamma_{max})$. Again, $p(\pi/2 + \omega_0)$ is the probability of finding, in the partially oriented powder, particles with plane (Π) at an angle $(\pi/2 + \omega_0)$ with the plane of the holder. By modifying the orientation of the latter with respect to the incident X-ray beam, the angle ω_0 and $p(\pi/2 + \omega_0)$ are both changed, with consequent modification of the intensity of the $00l$ reflections considered. We shall see in Section 6.5.2 how this is used in the experimental determination of $p(\gamma)$.

References

Besson G (1980) Structure des smectites dioctaédriques. Paramètres conditionnant des fautes d'empilement des feuillets. Thèse de Doctorat, Orléans University

Brindley, GW, Méring J (1951) Diffraction des rayons X par les structures en couches désordonnées. Acta Crystallogr 4:441–447

de Courville J, Tchoubar D, Tchoubar C (1979) Détermination expérimentale de la fonction d'orientation. Son application au calcul des bandes de diffraction. J Appl Crystallogr 12:332–338

Drits VA, Sakharov BA (1976) X-ray structure analysis of interstratified minerals. Nauka, Moscow, 255 pp (in Russian)

Ewald PP (1940) X-ray diffraction by finite and imperfect crystal lattices. Proc Phys Soc 52:167–178

Guntert DJ, Cvikevich S (1964) Preferred orientation and its effects on the (hk) reflections in X-ray patterns of pyrolytic graphites. Carbon 1:309–313

Laue M (1932) Cross lattice spectra. Z Kristallogr 82:127–141

Maire J, Méring J (1970) Graphitization of soft carbons. In: Walter PP (ed) Chemistry and physics of carbon, vol 6. Dekker, New York, pp 125–189

Méring J (1949) L'interférence des rayons X dans les systèmes à stratification désordonnée. Acta Crystallogr 2:371–377

Plançon A (1980) The calculation of intensities diffracted by a partially oriented powder with a layer structure. J Appl Crystallogr 13:524–528

Plançon A, Besson G, Pons CH, Tchoubar C (1980) Calcul du profil des bandes de diffraction produites par des empilements de feuillets de nature différente. J Appl Crystallogr 13:148–153

Reynolds RC (1968) The effect of particle size on apparent lattice spacings. Acta Crystallogr A24:319–320

Reynolds RC (1976) The Lorentz factor for basal reflections from micaceous minerals in oriented powder aggregates. Am Mineral 61:484–491

Ross M (1968) X-ray diffraction effects by non ideal crystals of biotite, muscovite, montmorillonite. Z Kristallogr 126:80–97

Rousseaux F, Tchoubar D (1975) Méthode d'analyse du profil de bande produite par des feuillets diffractants ayant une forme anisométrique. J Appl Crystallogr 8:365–371

Ruland W, Tompa H (1968) The effect of preferred orientation on the intensity distribution of (*hk*) interferences. Acta Crystallogr A24:93–99
Taylor RM, Norrish K (1966) The measurement of orientation distribution and its application to quantitative X-ray diffraction analysis. Clay Mineral 6:127–141
Trunz V (1976) Influence of crystalline size on the apparent basal spacings of kaolinite. Clays Clay Mineral 24:84–87
Warren BE (1941) X-ray diffraction in random layer lattices. Phys Rev 9:693–698
Wilson AJC (1949) X-ray diffraction by random layers: ideal line profiles and determination of structure amplitudes from observed line profiles. Acta Crystallogr 2:245–251

CHAPTER 3

Diffraction from Lamellar Crystals with Stacking Faults

The principal structural defects which can occur or coexist in lamellar systems were described in the first chapter. More specifically, stacking faults were classified in three categories:

- random translation or rotation defects;
- well-defined defects due to translations, rotations, or enantiomorphism of the layers;
- fluctuations in the positions of the layers.

In the present chapter, we shall examine the effects of these stacking faults on powder diffraction diagrams, limiting ourselves to the particular case of crystals composed of identical layers, or of layers sufficiently alike so that they can all be characterized by the same mean unit cell, leaving for Chapters 4 and 5 the case of different layers.

When the layers in a crystal are alike, most stacking faults occurring in planes parallel to the layers generally do not greatly affect the interlayer distance which will, at most, fluctuate around a mean value. In other words, insofar as the layers are identical, the presence of stacking faults due to rotations or translations of layers in their own plane will have little affect on the $00l$ reflections, and the visualization of these defects is best obtained through the hkl series.

We know that the mode of succession of the stacking defects affects the spatial distribution of the intensity in the diffracted beams. We also stated in Chapter 1 that there may sometimes be among these defects rules of correlation described by the parameter R characteristic of the range of interactions, and different from or equal to zero depending on whether there are interactions or not. The present chapter will be limited to the study of lamellar systems with uncorrelated defects (R = 0), while Chapter 5 will deal with correlated defects.

3.1 General Expression for the Diffraction Produced by Stacks of Layers with Position Defects

3.1.1 Mathematical Description of the Diffraction

The intensity of the beam diffracted at the angle 2θ by a partially oriented powder and characterized by $|\vec{s}| = s = 2\sin\theta/\lambda$ was given, in the preceding chapter, by Eq. (2.12):

$$I_{hk}(s) = \frac{1}{4\pi} \int [i(\vec{s})]_{hk} \frac{dA}{s^2} ,$$

in which $[i(\vec{s})]_{hk}$ is the intensity (normalized to the unit cell) at the tip of vector $\vec{s}$ within the (hk) rod of a defect-free stack of M layers. This intensity is itself obtained from Eq. (2.6):

$$i(\vec{s}) = \frac{\Omega}{M\sigma} \sum_{m=1}^{M} \sum_{m'=1}^{M} \Phi_m(\vec{s})\Phi^*_{m'}(\vec{s}) \exp[2\pi i \vec{s}\cdot(\vec{r}_m - \vec{r}_{m'})] .$$

The only difference in the computation of the intensity of the diffracted beam issuing from a powder with position defects in the stacks, as compared to the computation described in Chapter 2, lies in the expression giving $[i(\vec{s})]_{hk}$. Going back to Eq. (2.6), let us combine the terms in which $m = m'$, thus obtaining:

$$i(\vec{s}) = \frac{\Omega}{M\sigma} \left\{ \sum_{m=1}^{M} \Phi_m(\vec{s})\Phi^*_m(\vec{s}) + \sum_{\substack{m=1 \\ m\neq m'}}^{M} \sum_{m'=1}^{M} \Phi_m(\vec{s})\Phi^*_{m'}(\vec{s}) \exp[2\pi i \vec{s}\cdot(\vec{r}_m - \vec{r}_{m'})] \right\} .$$

By introducing now the characteristics associated with the nth-nearest neighbor and the vector $\vec{r}$ such that $|\vec{r}_n| = |\vec{r}_m - \vec{r}_{m'}|$ and $n = |m-m'|$:

$$i(\vec{s}) = \frac{\Omega}{M\sigma} \left\{ \sum_{m=1}^{M} |\Phi_m(\vec{s})|^2 + 2\,\mathrm{Re}\left[\sum_{n=1}^{M-1} \sum_{m=1}^{M-n} \Phi_m(\vec{s})\Phi^*_{m+n}(\vec{s}) \exp(2\pi i \vec{s}\cdot\vec{r}_n) \right] \right\} . \quad (3.1)$$

If the stack contains position faults, then:

- the presence of a rotation fault between the layer m and its nth-nearest neighbor can be described by the structure factor of the (m+n)th layer rotated in the reference system of the mth layer;

- as a result of the presence of a translation defect between two nth-nearest neighbors, there may be many possible translations $\vec{r}_{n1}, \vec{r}_{n2}, \ldots, \vec{r}_{nj}$ for a given value of n, with respective probability of occurrence, $p_{n1}, p_{n2}, \ldots p_{nj}$.

For each pair of nth-nearest neighbors, the diffraction is therefore the mean of the interferences corresponding to all possible configurations. This notion of mean value can be expressed by writing:

$$\overline{|\Phi(\vec{s})|^2} = \frac{1}{M} \sum_{m=1}^{M} |\Phi_m(\vec{s})|^2$$

and

$$\overline{\Phi(\vec{s})\Phi_n^*(\vec{s}) \exp(2\pi i \vec{s}\cdot\vec{r}_n)} = \frac{1}{M-n} \sum_{m=1}^{M-n} \Phi_m(\vec{s})\Phi_{m+n}^*(\vec{s}) \exp(2\pi \vec{s}\cdot\vec{r}_n) ,$$

so that Eq. (3.1) now becomes:

$$i(\vec{s}) = \frac{\Omega}{\sigma} \left\{ \overline{|\Phi(\vec{s})|^2} + 2\,\mathrm{Re}\left[\sum_{n=1}^{M-n} \frac{M-n}{M} \overline{\Phi(\vec{s})\Phi_n^*(\vec{s}) \exp(2\pi i \vec{s}\cdot\vec{r}_n)} \right] \right\} . \quad (3.2)$$

Two points should now be emphasized:

1. As given above, Eq. (3.2) is quite general; it will describe the diffraction produced by a stack of layers, whether they be identical or not, in the absence or in the presence of position defects, the latter being correlated or not. More particularly, this same relationship will be used in Chapter 5 as a starting point for the description of the diffraction due to a stack of different layers whose sequence may be random or which may show correlation between the faults.
2. When dealing with a fault-free stack containing identical layers, all the structure factors Φ are identical; the concept of mean value then becomes meaningless and Eq. (3.2) can be simplified to Eq. (2.7) of the preceding chapter.

3.1.2 The Matrix Formalism

The mean values needed in Eq. (3.2) are best handled with the help of a matrix formalism. This was first developed by Hendricks and Teller (1942), who only considered the case of lamellar crystals with an infinite number of layers, and later extended by Kakinoki and Komura (1952) to crystals with a limited number of layers. However, in both papers, different structure factors were given to two identical layers which only differed by the value of the translation vector parallel to their own plane with respect to a given set of axes. This resulted in the appearance of matrices of extremely high rank which rapidly became unmanageable. Simpler formalisms were later described (Allegra 1961a, b, c, 1964; Kakinoki and Komura

1962; Plançon and Tchoubar 1976), which considerably decreased the size of the matrices. The present development will be based on these different contributions.

Let us suppose that, with respect to a single frame of reference, there be g possible rotations for the layers in a stack. Using now Eq. (3.2), we can express, for each rotation, the amplitude of the diffused wave issuing from a rotated layer, i.e., the different structure factors $\Phi_1, \Phi_2, \ldots \Phi_g$, each associated with a relative proportion $W_1, W_2, \ldots W_g$. We now define the following factors:

[W] is a diagonal (g×g) matrix in which the element (ii) represents the relative proportion W_i of the layers with the structure factors Φ_i.

[Φ] is a square matrix of rank g, in which each (ij)th element (ith row, jth column) is the product $\Phi_j \Phi_i^*$.

Thus, when there are only three different structure factors (g = 3), these matrices are:

$$[W] = \begin{bmatrix} W_1 & 0 & 0 \\ 0 & W_2 & 0 \\ 0 & 0 & W_3 \end{bmatrix} \quad \text{and} \quad [\Phi] = \begin{bmatrix} \Phi_1\Phi_1^* & \Phi_2\Phi_1^* & \Phi_3\Phi_1^* \\ \Phi_1\Phi_2^* & \Phi_2\Phi_2^* & \Phi_3\Phi_2^* \\ \Phi_1\Phi_3^* & \Phi_2\Phi_3^* & \Phi_3\Phi_3^* \end{bmatrix} .$$

In accordance with the rules of matrix multiplication, the product [Φ] [W] will be:

$$[\Phi][W] = \begin{bmatrix} W_1\Phi_1\Phi_1^* & W_2\Phi_2\Phi_1^* & W_3\Phi_3\Phi_1^* \\ W_1\Phi_1\Phi_2^* & W_2\Phi_2\Phi_2^* & W_3\Phi_3\Phi_2^* \\ W_1\Phi_1\Phi_3^* & W_2\Phi_2\Phi_3^* & W_3\Phi_3\Phi_3^* \end{bmatrix}$$

The *trace* or *spur* of this matrix is the sum of the elements on the principal diagonal, i.e., with i = j. When applied to the above matrix, this yields the term $\overline{|\Phi(\vec{s})|^2}$ needed in Eq. (3.2) since

$$\mathrm{Tr}\{[\Phi][W]\} = \sum_{i=1}^{g=3} W_i \Phi_i \Phi_i^* = \overline{|\Phi(\vec{s})|^2} . \tag{3.3}$$

Let us now express the second mean value in Eq. (3.2) in terms of this matrix formalism. In the general case, the passage from a layer, characterized by the structure factor Φ_i, to its rotated nth-neighbor layer, characterized by a structure factor Φ_j, can be obtained by means of N_n translations (N_n may vary with n). Let $(\vec{r}_{ij}^{\,k})_n$ be one of these N_n different translations ($1 \leqslant k \leqslant N_n$), and let $(p_{ij}^k)_n$ be its probability of occurrence. The second mean term in Eq. (3.2) will now be:

$$\overline{\Phi(\vec{s})\Phi_n^*(\vec{s})\exp(2\pi i\vec{s}\cdot\vec{r}_n)} = \sum_{i=1}^{g} W_i\Phi_i \sum_{j=1}^{g}\sum_{k=1}^{N_n} (p_{ij}^k)_n \Phi_j^* \exp[2\pi i\vec{s}\cdot(\vec{r}_{ij}^{\,k})_n] . \tag{3.4}$$

or, after putting

$$(Q_{ij})^n = \sum_{k=1}^{N_n} (p_{ij}^k)_n \exp[2\pi i \vec{s} \cdot (\vec{r}_{ij}^k)_n] \ , \tag{3.5}$$

then:

$$\overline{\Phi(\vec{s})\Phi^*(\vec{s})\exp(2\pi i \vec{s}\cdot\vec{r}_n)} = \sum_{i=1}^{g} W_i \Phi_i(\vec{s}) \sum_{j=1}^{g} \Phi_j^*(\vec{s})(Q_{ij})^n \ . \tag{3.6}$$

Note that $(\vec{r}_{ij}^k)_n$, found in Eqs. (3.4) and (3.5), is one of the translations among the N_n translations considered. Thanks to the index k attached to it, it can be unambiguously identified. Although this is not explicitly shown in the notation used, the "global" translation $(\vec{r}_{ij}^k)_n$ must depend on the specific sequence and the "elementary translations" of the layers between the first layer Φ_i and its nth-neighbor Φ_j. Furthermore, the determination of $(Q_{ij})^n$, as given by (3.5), requires the knowledge of the different probabilities $(p_{ij}^k)_n$; this implies assumptions on possible correlations linking the elementary translations between successive layers. The general case will be examined in Chapter 5.

When there is no correlation between the translations (in which case the parameter R which describes the range of interaction between two translations will be equal to 0), the determination of $(Q_{ij})^n$ is much simplified, as shown by Plançon and Tchoubar (1976), since it is then possible to describe the passage from layer Φ_i to the nth-neighbor layer Φ_j as equivalent to the passage from layer Φ_i to the layer Φ_l, the (n−1)th-neighbor, followed by the passage from the layer Φ_l to its first neighbor Φ_j, for all possible values of l, as shown hereafter.

Consider one of the possible displacements, namely $(\vec{r}_{il}^{k'})_{n-1}$, between a layer Φ_i and its (n−1)th-nearest neighbor Φ_l, followed by $(\vec{r}_{lj}^{k''})_1$, one of the possible displacements between layer Φ_l and Φ_j, the next layer (Fig. 3.1). We can write

$$(\vec{r}_{ij}^k)_n = (\vec{r}_{il}^{k'})_{n-1} + (\vec{r}_{lj}^{k''})_1 \ .$$

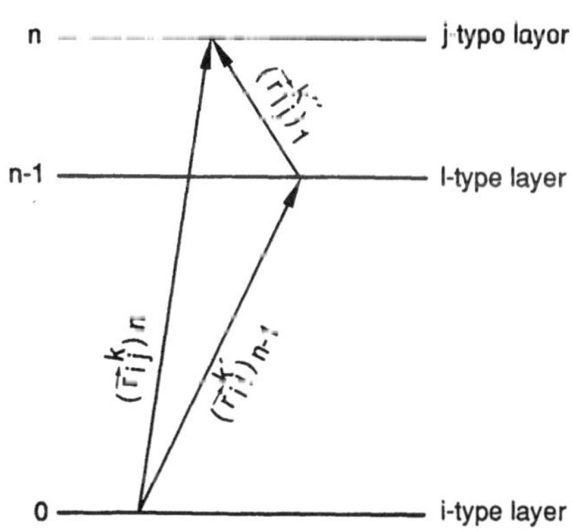

Fig. 3.1. Components of the displacement $(r_{ij}^k)_n$ between an i-type and nth-nearest, j-type layer

The probability attached to this displacement is given by

$$(p_{ij}^{k})_n = (p_{il}^{k'})_{n-1}(p_{lj}^{k''})_1$$

so that

$$\begin{aligned}(p_{ij}^{k})_n \exp[2\pi i\vec{s}\cdot(\vec{r}_{ij}^{k})_n] &= (p_{il}^{k'})_{n-1}(p_{lj}^{k''})_1 \exp\{2\pi i\vec{s}\cdot[(\vec{r}_{il}^{k'})_{n-1}+(\vec{r}_{lj}^{k''})_1]\} \\ &= (p_{il}^{k'})_{n-1}\exp[2\pi i\vec{s}(\vec{r}_{il}^{k'})_{n-1}](p_{lj}^{k''})_1 \exp[2\pi i\vec{s}(\vec{r}_{lj}^{k''})_1] \ .\end{aligned}$$

Putting now this relation in Eq. (3.5) and remembering that all possible values of l must be taken into account, we have:

$$(Q_{ij})^n = \sum_{l=1}^{g}\sum_{k'=1}^{N_{n-1}}\sum_{k''=1}^{N_1}(p_{il}^{k'})_{n-1}\exp[2\pi i\vec{s}\cdot(\vec{r}_{il}^{k'})_{n-1}](p_{lj}^{k''})_1\exp[2\pi i\vec{s}\cdot(\vec{r}_{lj}^{k''})_1] \ .$$

Since the sums over k′ and k″ are independent, the last expression can also be written as

$$\begin{aligned}(Q_{ij})^n = \sum_{l=1}^{g}&\left\{\sum_{k'=1}^{N_{n-1}}(p_{il}^{k'})_{n-1}\exp[2\pi i\vec{s}\cdot(\vec{r}_{il}^{k'})_{n-1}]\right\} \\ &\times\left\{\sum_{k''=1}^{N_1}(p_{lj}^{k''})_1\exp[2\pi i\vec{s}\cdot(\vec{r}_{lj}^{k''})_1]\right\} ,\end{aligned}$$

or:

$$(Q_{ij})^n = \sum_{l=1}^{g}(Q_{il})^{n-1}(Q_{lj})^1 \ . \tag{3.7}$$

Let us now introduce matrix [Q], of rank g, in which the (ij)th element is equal to:

$$(Q_{lj})^1 = \sum_{k''=1}^{N_1}(p_{lj}^{k''})_1\exp[2\pi i\vec{s}\cdot(\vec{r}_{lj}^{k''})_1] \ . \tag{3.8}$$

The matrix is

$$[Q] = \begin{bmatrix}\sum\limits_{k''}(p_{11}^{k''})_1\exp[2\pi i\vec{s}\cdot(\vec{r}_{11}^{k''})_1]\ldots & \sum\limits_{k''}(p_{1g}^{k''})_1\exp[2\pi i\vec{s}\cdot(\vec{r}_{1g}^{k''})_1] \\ \vdots & \vdots \\ \sum\limits_{k''}(p_{g1}^{k''})_1\exp[2\pi i\vec{s}\cdot(\vec{r}_{g1}^{k''})_1]\ldots & \sum\limits_{k''}(p_{gg}^{k''})_1\exp[2\pi i\vec{s}\cdot(\vec{r}_{gg}^{k''})_1]\end{bmatrix} \ . \tag{3.9}$$

By virtue of Eq. (3.7), each $(Q_{ij})^n$ term becomes the (ij)th element of matrix $[Q]^n$.

Considering now the matrices [W] and [Φ] already defined and the product $[\Phi][W][Q]^n$, we find that the trace of the product matrix is equal to the r.h.s. of (3.6), hence that

$$\overline{\Phi(\vec{s})\Phi_n^*(\vec{s})\exp[2\pi i\vec{s}\cdot\vec{r}_n]} = \mathrm{Tr}\{[\Phi][W][Q]^n\} \ . \tag{3.10}$$

Replacing now the mean values by the equivalent expressions given by (3.3) and (3.10) involving the matrices [Φ], [W] and [Q], Eq. (3.2) becomes:

$$i(\vec{s}) = \frac{\Omega}{\sigma}\left[\mathrm{Tr}\{[\Phi][W]\} + 2\,\mathrm{Re}\sum_{n=1}^{M-1}\frac{M-n}{M}\,\mathrm{Tr}\{[\Phi][W][Q]^n\}\right] \ . \tag{3.11}$$

It is possible, at this point, to justify the use of the matrix formalism. The determination of $i(\vec{s})$, the intensity of the diffracted beam, requires the computation of the mean value $\overline{\Phi\Phi_n^*\exp[2\pi i\vec{s}\cdot\vec{r}_n]}$ for each of the sets of pairs of nth-neighbor layers.

- The direct computation is performed by applying Eq. (3.4) to nth-nearest neighbors and by determing all the $(\vec{r}_{ij}^{\,k})_n$ possible translations, **each of which is a sum of (n−1) elementary translations**, as well as the cognate probabilities $(p_{ij}^k)_n$, each of which is also a **sum of coefficients.** This having been done, the process must be repeated for the (n+1)th-nearest neighbors, aggravated by the fact that there are now n rather than (n−1) elementary translations, with an identical increase in the number of probabilistic coefficients. This process occurs every time n changes by one unit.
- In the matrix formalism, once the matrix $[Q]^n$ has been written for the pairs of nth-neighbors, the passage to the pairs of (n+1)th-neighbors is done quite simply by multiplying $[Q]^n$ by the matrix [Q] characteristic of first-neighbors.

Hence, using the matrix formalism is not merely a convenient way of writing the relations which describe the diffraction produced by disordered sequences of layers; it also results in a considerable reduction of the amount of computation required in the practical applications of the theoretical results.

Returning to Eq. (3.11), let us rewrite it as

$$i(\vec{s}) = \frac{\Omega}{\sigma}\,\mathrm{Tr}\left\{[\Phi][W] + 2\,\mathrm{Re}\,[\Phi][W]\sum_{n=1}^{M-1}\frac{M-n}{M}[Q]^n\right\} \ . \tag{3.12}$$

Since the elements of the matrix [W] and the different elements $\Phi_i\Phi_i^*$ involved in $\mathrm{Tr}\{[\Phi][W]\}$ are all real, Eq. (3.12) can be simplified by noting that

$$\mathrm{Tr}\{[\Phi][W]\} = \mathrm{Tr}\{\mathrm{Re}\,[\Phi][W]\} \ ,$$

so that

$$i(\vec{s}) = \frac{\Omega}{\sigma} \operatorname{Tr} \operatorname{Re} \left[[\Phi][W] \left\{ [I] + 2 \sum_{n=1}^{M-1} \frac{M-n}{M} [Q]^n \right\} \right] , \tag{3.13}$$

in which [I] is the unit matrix of order g.

The summation $s = \sum_{n=1}^{M-1} (M-n)[Q]^n$ is carried out along the same lines as described in Section 2.2.1, noting that s is equivalent to the sum of (M−1) geometrical progressions of common ratio [Q] which respectively include (M−1), (M−2), ... 1 terms. This leads to

$$s = \frac{[I]}{[I]-[Q]} \{M[I] - [Q]([I] + [Q] + \ldots + [Q]^{M-1})\} - M[I]$$

or

$$s = \frac{M[Q]}{[I]-[Q]} + \frac{[Q]^{M+1} - [Q]}{([I]-[Q])^2} .$$

The expression giving the intensity becomes

$$i(\vec{s}) = \frac{\Omega}{\sigma} \operatorname{Tr} \operatorname{Re} \left\{ [\Phi][W] \left[[I] + \frac{2[Q]}{[I]-[Q]} + \frac{2}{M} \frac{[Q]^{M+1} - [Q]}{([I]-[Q])^2} \right] \right\} \tag{3.14}$$

and after introducing

$$[R] = [I] + \frac{2[Q]}{[I]-[Q]} + \frac{2}{M} \frac{[Q]^{M+1} - [Q]}{([I]-[Q])^2} ,$$

we obtain

$$i(\vec{s}) = \frac{\Omega}{\sigma} \operatorname{Tr} \operatorname{Re} \{[\Phi][W][R]\} . \tag{3.15}$$

Equations (3.12) and (3.15) were given by Kakinoki and Komura (1952).

Finally, by bringing in Eqs. (2.12), (2.14), (2.15) and (2.38), the intensity of the wave diffracted along each (*hk*) rod by a partially oriented powder of lamellar crystals with stacking defects is given by

$$I_{hk}(s) = \frac{1}{\Omega \sigma s} \int \bar{p}(\varphi) \operatorname{Tr} \operatorname{Re} \{[F][W][R]\} T(U) \, d\varphi , \tag{3.16}$$

in which [F] is a square matrix, analogous to [Φ] but in which each element $(\Phi_i^* \Phi_j)$ is replaced by $(F_{hk})_i^* (F_{hk})_j$, where F_{hk} is the structure factor corresponding to the lattice $(\vec{a}, \vec{b})$, calculated using Eq. (2.5) for the indices h, k.

We can note that Eq. (3.16) was established on the supposition that the powder is composed of substacks which each contain the same number of layers. Evidently, actual powders contain substacks with different numbers of layers. In the general case, one must then use a *weighting fraction* $\alpha(M)$ for the stacks containing M layers and Eq. (3.16) is then replaced by

$$I_{hk}(s) = \frac{1}{\Omega \sigma s} \sum_{M} \alpha(M) \int \bar{p}(\varphi) \, \mathrm{Tr} \, \mathrm{Re}\{[F][W][R]\} T(U) \, d\varphi \, . \tag{3.17}$$

Exploiting this relationship requires expliciting the matrices [F], [W] and [R] – or [Q], which is included in [R]. The expressions giving these matrices depend on the model chosen for the stacks, as shown in the following section.

3.2 Diffraction Produced by Stacks Containing Rotation or Translation Faults Without Mutual Interaction

3.2.1 Effects of Random Rotation or Translation Stacking Defects on the Diffraction

Let us now consider a stack in which all the identical layers have the same *A O* but are subject to random translation defects, and let us try to see how this affects Eq. (3.14), which gives the intensity of the diffracted beam.

- Since all the (ij) elements of $[\Phi]$ are identical, the matrix is reduced to $\Phi(\vec{s})\Phi^*(\vec{s})$.
- Similarly, each element $(Q_{ij})^1$ of the matrix [Q], as defined in Eq. (3.8), is a measure of the possible phase difference between two adjacent identical layers which may be affected by N possible translations. All elements of the matrix [Q] are equal to

$$q = \sum_{k''=1}^{N} (p^{k''}) \exp(2\pi i \vec{s} \cdot \vec{r}^{k''}) \tag{3.18}$$

in which $\vec{r}^{k''}$ is one of the N possible translations between first-neighbor layers.

Taking these facts into account, Eq. (3.14) becomes:

$$i(\vec{s}) = \frac{\Omega}{\sigma} \Phi(\vec{s})\Phi^*(\vec{s}) \, \mathrm{Re} \left[1 + \frac{2q}{1-q} + \frac{2}{M} \frac{q^{M+1} - q}{(1-q)^2} \right] \, . \tag{3.19}$$

Let P_A now be the probability of finding a random translation defect between two adjacent layers; P_A is, in fact, the sum of a large number of probabilities p_A of meeting any particular translation $\vec{r}_A$. For isotropic distribution of P_A

$$\sum p_A \exp(2\pi i \vec{s}\cdot\vec{r}_A) = 0 \ .$$

Furthermore, let $\vec{r}_0$ be the "correct" translation corresponding to the defect-free stack of two first-neighbor layers and let $(1-P_A)$ be the probability of its occurrence. Writing now $\vec{t}_0$ and d_{001} for the ordered translation and the interplanar distance respectively, we can express $\vec{r}_0$ as $\vec{r}_0 = \vec{t}_0 + \vec{u}_z d_{001}$ and the relationship (3.18) becomes:

$$q = (1-P_A)\exp(2\pi i \vec{s}\cdot\vec{r}_0) = (1-P_A)\exp[2\pi i \vec{s}\cdot(\vec{t}_0 + \vec{u}_z d_{001})] \ . \quad (3.20)$$

This shows that the result of the presence, in a stack, of random translation defects between adjacent layers is that, *statistically*, these defects break up the interferential coherence of the substacks between which they occur. The same result will be brought about by rotation faults between adjacent layers. In both cases, *it is as if the substacks separated by the defects diffracted independently from each other.* More particularly, if these defects exist between *all* the layers of a stack (in which case $P_A = 1$), **the diffraction produced is the same as the one which would be observed to issue from a single isolated layer,** with an intensity weighted by the number of layers in the stack.

Furthermore, if we let $q = U\exp(i\omega)$ and if Eq. (3.19) be put in the same form as (2.8), namely

$$i(\vec{s}) = \Omega\,|\Phi|^2 G/\sigma \ ,$$

we obtain an expression for the modulation function G which was given by Méring (1949):

$$G = \frac{1-U^2}{1+U^2-2U\cos\omega} + \frac{2U}{M}\,\frac{[2U-(1+U^2)\cos\omega](1-U^M\cos M\omega)-(1-U^2)U^M\sin M\omega\sin\omega}{(1+U^2-2U\cos\omega)^2} \quad (3.21)$$

which, for $P_A = 1$, gives $G = 1$ since $U = 0$. Hence Eqs. (3.19) and (3.20) lead correctly to the conclusion that the diffraction is reduced to that of a single layer.

Let us now evaluate the consequence of the presence of random stacking faults on the reflection profiles in a powder diagram:

- In the absence of stacking faults, the diagram from powdered lamellar crystals is a succession of hkl Bragg peaks more or less broadened as a function of the size of the crystallites or of the interferential coherence domains. This is illustrated in Fig. 3.2 showing the profiles of the $02l$ and $11l$ lines calculated for a microcrystalline mica $2[K_{0.8}(Si_{3.5}Al_{0.5})(Al_{1.7}Mg_{0.3})O_{10}(OH)_2]$ having perfect stacking of dioctahedral cis-vacant layers by Drits et al. (1984) (cf. Chap. 1 for the description of such a layer).

Fig. 3.2. Calculated XRD powder pattern for a dioctahedral cis-vacant mica, in the total absence of stacking defects. **a** 021 reflections; **b** 13*l* reflections. The intensity is scaled both in electron units (e.u.) and in arbitrary units (a.u.). (After Drits et al. 1984)

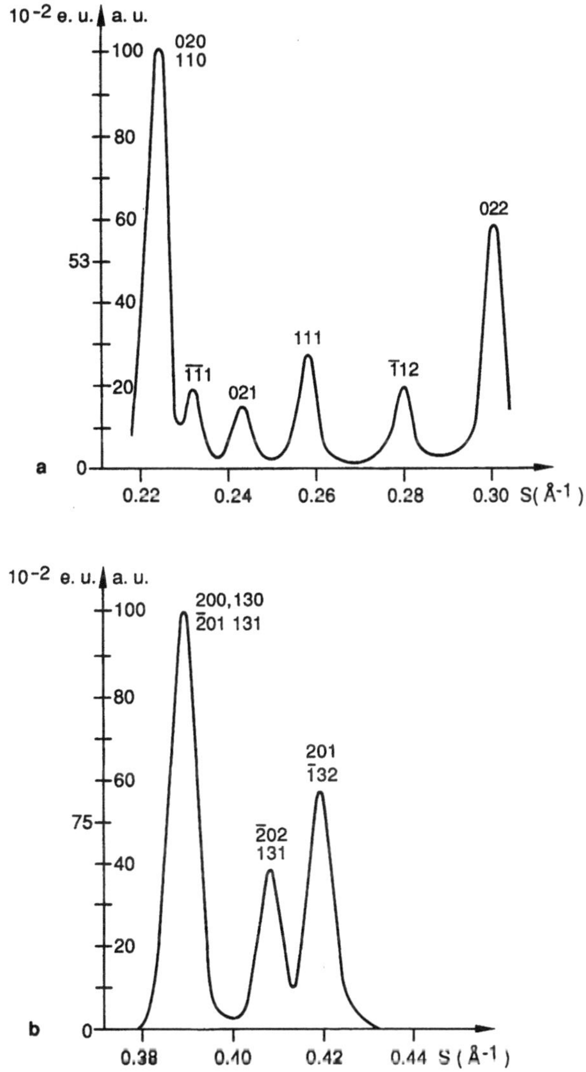

– Let now random, noncorrelated defects occur in such a stack. As the probability P_A of having these defects increases, there is a gradual evolution of the set of *hkl* reflections corresponding to given values of *h* and *k* towards a global asymmetrical profile called an *hk diffraction band*. This profile is characterized by a rapid increase of the diffracted intensity near the value $s_0 = |h\vec{a}^* + k\vec{b}^*|$, followed by a slow decrease exhibiting modulations which get weaker as P_A is closer to 1 till, for $P_A = 1$, the profile is no longer modulated and becomes a *pure band* proper to a crystal with bidimensional

(2D) lattice. Figure 3.3 shows this evolution for the same angular domain of the reflections $02l$, $11l$ as in Fig. 3.2, corresponding to the profile of the composite (02,11) band when $P_A = 0.5$ and $P_A = 1$ (Figs. 3.3a and b respectively).

We may finally note that the maximum of an *hk* diffraction band is always shifted towards the large angles with respect to the value s_0. This is due to the fact that, statistically, all the layers in the stack diffract independently from each other when $P_A = 1$, which is equivalent to saying that the intensity of the diffracted

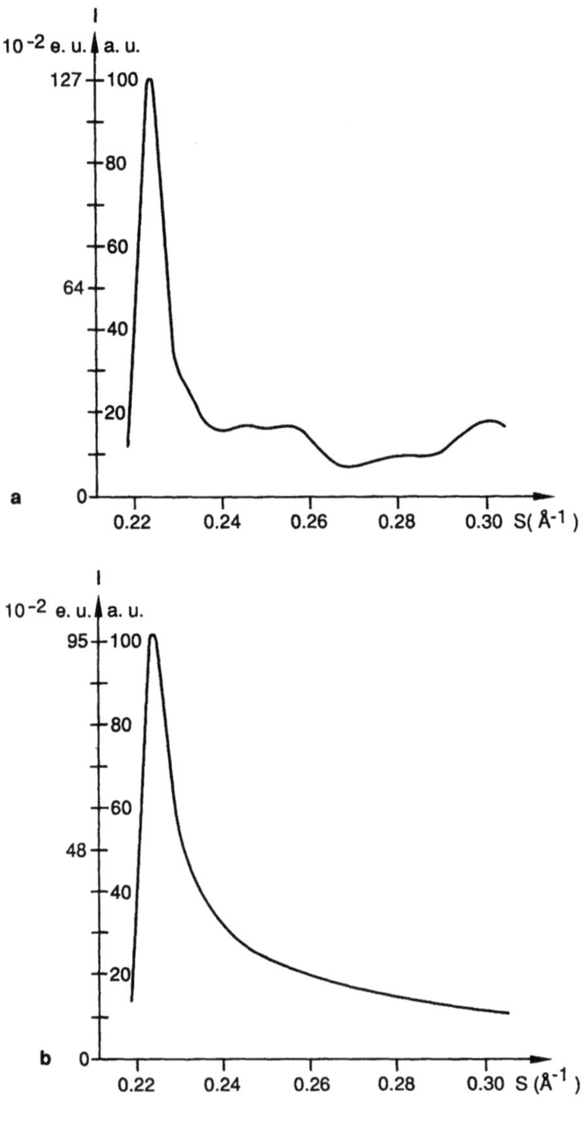

Fig. 3.3. Calculated (02,11) powder diffraction band for a cis-vacant dioctahedral mica containing randoms stacking faults with the probability p_A of occurrence between adjacent layers. Curve **a** $P_A = 0.5$; curve **b** $P_A = 1$. (After Drits et al. 1984)

beams is localized in the rods centered on the hk nodes of a plane lattice reciprocal to the $(\vec{a}, \vec{b})$ lattice. This intensity is proportional to $|F_{hk}(Z)|^2 G(\varepsilon_1, \varepsilon_2)$ (cf. Chap. 2.2). It is continuous in Z and varies along the cross-section of each rod as $G(\varepsilon_1, \varepsilon_2)$, characteristic of the extension, in the plane of the layers, of the interferential coherence domains. Hence, if the powder diagram is interpreted in terms of the intersection of the rods with spheres of increasing radii s (Chap. 2.3.1 and Fig. 2.8), it is found that the intensity intercepted by each rod reaches a maximum value for $s > s_0$. The difference between the maximum of an hk band and s_0 depends on the spatial extension L of the interferential coherence domains in the plane of the layers. As shown by Warren (1941), the angular shift is given by $\Delta(\sin\theta) = 0.16\,\lambda/L$, where λ is the wavelength of the incident X-radiation.

It is important to note that the presence of this shift in the position of the diffraction maxima in the powder diagrams of 2D-crystallites prevents any determination of the value of the parameters a and b from direct measurements of the position of the maxima.

3.2.2 Effect of Well-Defined Translation Defects on the Diffraction

When the position defects in a stack are well-defined translation faults, the relative positions of two identical, nonrotated and first-neighbor layers are linked:

- either by the correct translation vector $\vec{r}_0 = \vec{t}_0 + \vec{u}_z d_{001}$, which has a probability of occurrence p_0.
- or by one of the N different well-defined translations described by the vector $\vec{r}_j = \vec{t}_0 + \vec{t}_j + \vec{u}_z d_{001}$ where $\vec{t}_j$ is the translation defect in the plane of the layer, with p_j as its probability of occurrence.

The intensity of the beam diffracted by such a stack is once more described by Eq. (3.19) in which q, which is still given by Eq. (3.18), should be written as

$$q = p_0 \exp[2\pi i \vec{s}\cdot(\vec{t}_0 + \vec{u}_z d_{001})] + \sum_{j=1}^{N} p_j \exp[2\pi i \vec{s}\cdot(\vec{t}_0 + \vec{t}_j + \vec{u}_z d_{001})] \ .$$

The translations $\vec{t}_j$ can be defined with respect to the lattice $(\vec{a}, \vec{b})$ and can be classified by categories, each one of the latter being itself defined by $(\pm m\vec{a}/v \pm n\vec{b}/w)$, where m and n are positive integers which can vary from one pair of adjacent layers to the other, whereas v and w, also positive integers, are specific to each type of stacking model. For each category of defect corresponding to a pair of values v and w, the N translations $\vec{t}_j$ can be reclassified and expressed as

- $\vec{t}_1 = +\vec{a}/v$ and $\vec{t}_2 = -\vec{a}/v$, corresponding to $m = 1$ and $n = 0$;
- $\vec{t}_3 = +\vec{b}/w$ and $\vec{t}_4 = -\vec{b}/w$, corresponding to $m = 0$ and $n = 1$;

- $\vec{t}_5 = +\vec{a}/v+\vec{b}/w$; $\vec{t}_6 = -\vec{a}/v-\vec{b}/w$; $\vec{t}_7 = -\vec{a}/v+\vec{b}/w$ and $\vec{t}_8 = +\vec{a}/v-\vec{b}/w$, corresponding to various cases in which $m = 1$ and $n = 1$;
- etc.

If P_T now be the probability of finding the set of these N translation defects between adjacent layers and if, furthermore, it is assumed that the probabilities of occurrence p_j are all equal (which implies that $p_j = P_T/N$ and $p_0 = 1-\sum p_j = (1-P_T)$, the expression giving q becomes:

$$q = \exp\,[2\pi i\vec{s}(\vec{t}_0+\vec{u}_z d_{001})]\left[(1-P_T)+\frac{P_T}{N}\sum_{j=1}^{N}\exp\,(2\pi i\vec{s}\cdot\vec{t}_j)\right].$$

Remembering now that $\vec{s} = h\vec{a}^*+k\vec{b}^*+Z\vec{u}_z$ and regrouping the exponential terms corresponding to $\vec{t}_1$ and $\vec{t}_2$, $\vec{t}_3$ and $\vec{t}_4$, ..., by pairs, the expression for q can be rearranged so as to involve the sum of N/2 cosine terms, namely:

$$q = \exp\,[2\pi i\vec{s}\cdot\vec{r}_0]\left\{(1-P_T)\right.$$
$$\left.+\frac{2P_T}{N}\left[\cos\frac{2\pi h}{v}+\cos\frac{2\pi k}{w}+\ \dots\ +\cos 2\pi\left(\frac{mh}{v}+\frac{nk}{w}\right)+\dots\right]\right\}. \quad (3.22)$$

This expression is analogous to the one in Eq. (3.20) dealing with random defects; hence, using once more $q = U\exp\,(i\omega)$ yields:

$$U = (1-P_T)+\frac{2P_T}{N}\left[\cos\frac{2\pi h}{v}+\cos\frac{2\pi k}{w}+\ \dots\ +\cos 2\pi\left(\frac{mh}{v}+\frac{nk}{w}\right)+\dots\right].$$

As noted by Méring (1949), we see that translation defects will have different effects, depending on the relationship between the indices h and k of the rods and the parameters v and w:

- if h and k are respectively multiples of v and w, all the cosine terms are equal to 1 and the portioning out of the intensity along the (hk) rods is **totally insensitive to the presence of the defects** ($\pm m\vec{a}/v\pm n\vec{b}/w$);
- if neither h nor k are multiples of v and w, the defects ($\pm m\vec{a}/v\pm n\vec{b}/w$) are **equivalent to random defects.**

Using as example the microcrystalline mica already met in Section 3.2.1 and assuming the presence of stacking faults defined by $\pm nb/3$, the reflections $20l$, $13l$, $06l$, $33l$, ... which all have $k = 3n$, are totally insensitive to the presence of these defects, while the reflections $02l$, $11l$, $04l$, $22l$, gradually change towards (02,11), (04,22), ... diffraction bands in which the modulation decreases as P_T

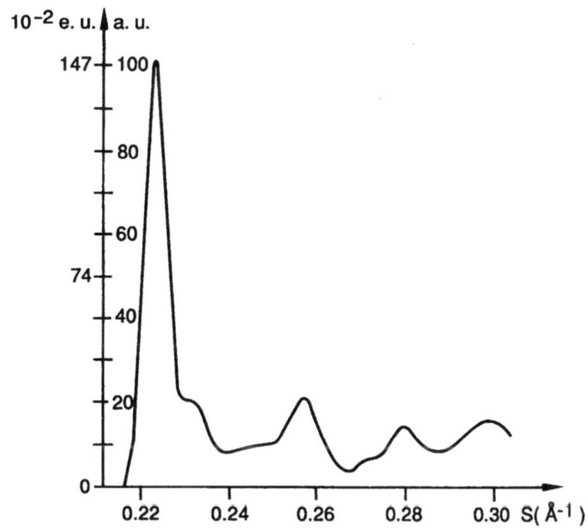

Fig. 3.4. The (02,11) powder diffraction band computed for a dioctahedreal cis-vacant mica with $\pm\vec{b}/3$ translation stacking defects, with the probability of occurrence $P_T = 0.5$. (After Drits et al. 1984)

approaches 2/3. Figure 3.4 shows the profile of the (02, 11) band for $P_T = 0.5$. When $P_T = 2/3$, the probabilities of finding the translations $\vec{t}_0, \vec{t}_0 + \vec{b}/3$ and $\vec{t}_0 - \vec{b}/3$ between adjacent layers are equal; this corresponds to maximum disorder and the (02, 11) band then has exactly the same profile as the one presented in Fig. 3.3 b.

Thus, the diffraction diagram from powdered lamellar crystals with only well-defined translation faults is characterized by the simultaneous presence, on one hand, of Bragg *hkl* reflections corresponding to certain values of *h* and/or *k* and, on the other hand, by the presence of *hk* bands analogous to those produced by a structure with a 2D-tendency.

3.2.3 Effect of Well-Defined Rotation Defects on the Diffraction

In order to simplify the presentation, we shall envisage the case of a stack with a single type of rotation defect between adjacent layers, namely a rotation by $2n\pi/v$, where v is a specific and well-defined integer and n an integer which can vary from one pair of layers to the other. With reference to a fixed set of reference coordinates, we may meet (n+1) types of layers:

- those which are not rotated with respect to a reference orientation, which have a unit cell structure factor F_0 and a probability W_0 of occurrence in the stack;

- those which are rotated by $+2\pi/v$, $4\pi/v$, ... $2j\pi/v$, ... $2n\pi/v$, with structure factors $F_1, F_2, \ldots, F_j, \ldots, F_n$, and respective abundances W_1, $W_2, \ldots$, $W_j, \ldots, W_n$.

For two successive layers labeled "m" and "m+1" in the stack, we may have the following possibilities:

- both layers may have identical *A O* (although both may be rotated) with respect to the reference direction; the probability attached to this occurrence is p_{ii};
- layer m and layer m+1 are both rotated with respect to the reference direction but not by the same amount; the two layers are said to be of type i and j and to have structure factors F_i and F_j respectively, while the probability of meeting this situation is p_{ij}.

Because of the presence of these rotations, a more specific definition of the *ordered translation* is needed. Up till now, the translation $\vec{t}_0$, which is wholy parallel to the plane $(\vec{a}, \vec{b})$, was said to be ordered or "correct" if it led from layer m to layer (m+1), both layers being identical, not rotated, with mutual positions corresponding to the ones they would have in a perfect triperiodic structure. *The translation $\vec{t}_0$ is therefore normalized with respect to the axes $\vec{a}$ and $\vec{b}$ of layer m.* If we wish to retain an analogous definition for ordered translations between adjacent layers in the case of a stack with rotation defects, we are led to consider (n+1) "ordered" translations which could occur between layer m and layer (m+1), defined with respect to the orientation of layer m:

- the ordered translation will be $\vec{t}_0$ when layer m is not rotated with respect to the reference direction;
- the ordered translations will be $\vec{t}_1, \vec{t}_2, \ldots \vec{t}_j, \ldots \vec{t}_n$ when layer m is rotated by $+2\pi/v, 4\pi/v, \ldots$ etc. ... with respect to the reference direction. These $\vec{t}_j$ translations are combinations of $\vec{t}_0$ with rotations $+2\pi/v, 4\pi/v$, etc. ....

We shall now establish the consequences of this model on the expressions for matrices [F], [W], [φ] and [P], knowing that [Q] = [φ] [P] where [φ] is a diagonal (g×g) matrix whose elements represent the phase differences, while [P] is a square matrix of rank g in which the elements are the probabilities of passing from a layer with the structure factor Φ_i to an adjacent layer with the structure factor Φ_j.

Matrix [F]

In this matrix, each element (ij) corresponds to the passage from an i-type layer to a neighboring j-type layer. This is expressed by $(F_i^* F_j)$ and the matrix is

$$[F] = \begin{bmatrix} F_0^* F_0 & F_0^* F_1 & \ldots & F_0^* F_n \\ F_1^* F_0 & F_1^* F_1 & \ldots & F_1^* F_n \\ \vdots & \vdots & & \vdots \\ F_n^* F_0 & F_n^* F_1 & \ldots & F_n^* F_n \end{bmatrix} \tag{3.23}$$

For a given couple of indices hk, the structure factor F_0 is computed with the help of Eq. (2.5),

$$F_{hk}(Z) = \sum_j f_j \exp\left[2\pi i(hx_j + ky_j + Zz'_j)\right] ,$$

whereas the individual structure factors $F_1, F_2 \ldots$ are obtained from the same relation by using the same 2D lattice rotated by $2\pi/v, 4\pi/v, \ldots$ etc. with respect to the fixed referential.

Matrix [φ]

The diagonal matrix [φ] which expresses the phase difference due to the passage from an i-type to a j-type layer is given by

$$[\varphi] = \begin{bmatrix} \exp(2\pi i\vec{s}\cdot\vec{r}_0) & 0 & \ldots & 0 \\ 0 & \exp(2\pi i\vec{s}\cdot\vec{r}_1) & \ldots & \vdots \\ \vdots & \vdots & & \vdots \\ 0 & 0 & \ldots & \exp(2\pi i\vec{s}\cdot\vec{r}_n) \end{bmatrix} \qquad (3.24)$$

in which $\vec{r}_0 = \vec{t}_0 + \vec{u}_z d_{001}$ and $\vec{r}_j = \vec{t}_0 + \vec{t}_j + \vec{u}_z d_{001}$.

Matrices [W] and [P]

The fact that, in the present chapter, we are considering lamellar systems without interactions between the defects (R = 0) must be kept in mind when writing out the explicit expressions for matrices [W] and [P]. The consequence is that the probability of having a j-type (m+1)th layer following an i-type layer is independent of the nature (or, in the present case, of the rotation) of layer m, and that the probability therefore only depends on the relative proportions of j-type layers, so that $p_{ij} = W_j$.

If P_R be the probability of finding the set of translations **other than** $\vec{t}_0$ between a pair of adjacent layers in the stack and if, furthermore, we assume that all the $\vec{t}_j \neq \vec{t}_0$ translations are equally probable, matrices [W] and [P] are

$$[W] = \begin{bmatrix} 1-P_R & 0 & \ldots & 0 \\ 0 & P_R/n & \ldots & 0 \\ \vdots & \ldots & \vdots & \ldots & \vdots \\ \vdots & \ldots & \vdots & \ldots & \vdots \\ 0 & 0 & \ldots & P_R/n \end{bmatrix} \qquad (3.25)$$

and

$$[P] = \begin{bmatrix} (1-P_R) & P_R/n & \dots & P_R/n \\ (1-P_R) & P_R/n & \dots & P_R/n \\ \vdots & \vdots & \dots & \vdots \\ \vdots & \vdots & \dots & \vdots \\ (1-P_R) & P_R/n & \dots & P_R/n \end{bmatrix} \tag{3.26}$$

In contrast to the cases already presented, it is not easy to predict, by mere inspection, the influence of well-defined rotation defects on powder diagrams. By computation, the effects of rotations of the type $2\pi n/v$ (with fixed v) on the intensity along an (hk) rod are found to depend on the relationship between h, k, and v, as described in the case of well-defined translation stacking faults. However, although the effects are similar, it is possible to distinguish between the two categories of defects (rotation or translation) since each of them leads to specific modifications of the intensity.

Referring once more to the microcrystalline mica described at the end of Section 3.2.2, let us suppose that the stacks contain well-defined faults due to $2\pi n/3$ rotations. Figure 3.5 shows the profile of the (02,11) band for a probability of occurrence of such a defect between adjacent layers $P_R = 2/3$, corresponding, for this model, to maximum possible disorder since, for any pair of nearest-neighbor layers, the probabilities that layer (m+1) shall be rotated by +120°, or −120°, or not rotated with respect to layer m are all equal to 1/3. Comparing Figs. 3.5 and 3.3b brings out a difference in the profiles of the two (02,11) bands: in the model with

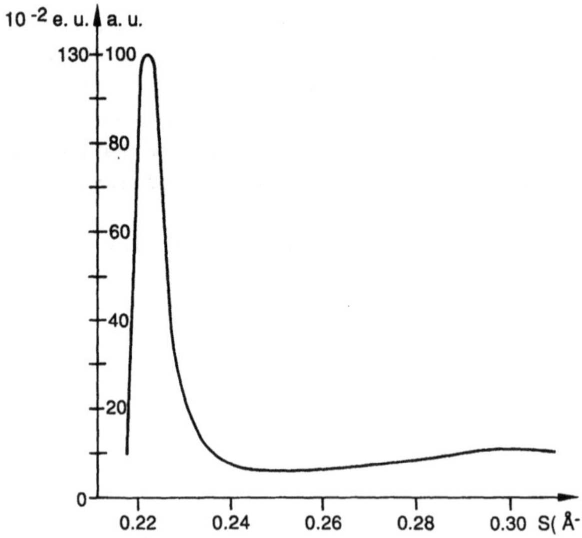

Fig. 3.5. Computed profile of the (02,11) diffraction band for a cis-vacant dioctahedral mica with $\pm 2\pi/3$ rotational stacking defects with a probability of occurrence $P_n = 2/3$

translation faults, the band decreases monotonically towards the large angles from its maximum value, whereas a minimum is found on the curve obtained from the model with $2\pi n/3$ rotations.

3.3 Diffraction Produced by Stacks with Defects Due to Fluctuations in the Positions of the Layer

As indicated at the beginning of this chapter, layers in a stack can also be subject to defects due to slight fluctuations of the translation vectors around a mean value $\langle\vec{r}\rangle$. It was also shown (Sect. 1.6.6) that defects due to fluctuations in the positions of the layers can be subdivided into two categories, leading to systems with disorders of the first or of the second type. We shall now successively examine the mathematical models used to describe these two categories of systems and the specific modifications of the powder diagrams due to the presence of fluctuations in the position of layers.

As a general case, we shall consider a crystal composed of identical layers subject only to position fluctuation defects. Let us first suppose that there may be many possible well-defined translations between adjacent layers, that $\langle\vec{r}_j\rangle$ is the mean value and p_j the probability of occurrence of the vector characteristic of a j-type translation, and that this vector may be affected by a slight fluctuation so that the actual translation and the mean value are linked by the relationship

$$\vec{r}_j = \langle\vec{r}_j\rangle + \vec{\varepsilon}_j \; , \tag{3.27}$$

in which $\vec{\varepsilon}_j$ is the deviation with respect to $\langle\vec{r}_j\rangle$. The vectors $\langle\vec{r}_j\rangle$ and $\vec{\varepsilon}_j$ are such that

$$\langle\vec{r}_j\rangle = \langle\vec{t}_j\rangle + \bar{d}_{001}\,\vec{u}_z \; , \tag{3.28}$$

where $\langle\vec{t}_j\rangle$ is the mean translation of the layers in their own plane and $\bar{d}_{001}$ is their mean interlayer distance. The deviation $\vec{\varepsilon}_j$ can be expressed as a function of the parameters $(\vec{a},\vec{b})$ of the unit cell by:

$$\vec{\varepsilon}_j = \varepsilon_{jx}\vec{a} + \varepsilon_{jy}\vec{b} + \varepsilon_{jz}\vec{u}_z \; , \tag{3.29}$$

where ε_{jx} and ε_{jy} are the deviations between the actual translation vector $\vec{t}_j$ and the mean translation $\langle\vec{t}_j\rangle$ in the directions of the crystallographic axes $\vec{a}$ and $\vec{b}$. Similarly, ε_{jz} is the difference between the actual interlayer separation and the mean interlayer distance $\bar{d}_{001}$, expressed in nanometers.

The mean phase difference of the wave diffracted by adjacent layer pairs in the whole crystal can be expressed as

$$q^{(1)} = \sum p_j \overline{\exp(2\pi i \vec{s}\cdot\vec{r}_j)} \; , \tag{3.30}$$

where the bar over the exponential term implies that we must take the mean value of the terms corresponding to the actual individual translations occurring between adjacent layers. Replacing now $\vec{r}_j$ by its equivalent given in Eq. (3.27), we find

$$q^{(1)} = \sum_j p_j \exp[2\pi i \vec{s}\cdot\langle\vec{r}_j\rangle]\,\overline{\exp(2\pi i \vec{s}\cdot\vec{\varepsilon}_j)} \ . \tag{3.31}$$

Knowing that $\vec{s} = h\vec{a}^* + k\vec{b}^* + Z\vec{U}_z$, and assuming that, along the three crystallographic directions, the fluctuations are independent of each other, we find

$$\overline{\exp(2\pi i \vec{s}\cdot\vec{\varepsilon}_j)} = \overline{\exp[2\pi i(h\varepsilon_{jx} + k\varepsilon_{iy} + Z\varepsilon_{jz}]}$$

$$= \overline{\exp(2\pi i h\varepsilon_{jx})}\cdot\overline{\exp(2\pi i k\varepsilon_{jy})}\cdot\overline{\exp(2\pi i Z\varepsilon_{jz})} \ .$$

Thus, for this type of three-dimensional disorder, the problem can be broken down to three independent expressions, each involving a single dimension.

In order to obtain the mean values of the exponential terms, we must know the distribution law for the deviation ε_j as well as the nature of the interactions between the fluctuations (disorders of the first or of the second type). Regarding the law of distribution of the variations ε_j, we can assume that it is a normal Gaussian along each axis:

$$\gamma(\varepsilon_j) = \frac{1}{\sqrt{2\pi}\,\sigma_j} \exp\left(-\frac{\varepsilon_j^2}{2\sigma_j^2}\right) , \tag{3.32}$$

in which $\sigma_j = (\overline{\varepsilon_j^2})^{1/2}$ is the standard (or root mean square) deviation characteristic of the width at half-height of the distribution $\gamma(\varepsilon_j)$. It follows that:

$$\overline{\exp[2\pi i \vec{s}\cdot\vec{\varepsilon}_j]} = \int_{-\infty}^{+\infty} \exp(2\pi i s\cdot\varepsilon_j)\gamma(\varepsilon_j)\,d\varepsilon_j$$

$$= \frac{1}{\sqrt{2\pi}\,\sigma_j} \int_{-\infty}^{+\infty} \exp(2\pi i s\cdot\varepsilon_j) \exp\left(-\frac{\varepsilon_j^2}{2\sigma_j^2}\right) d\varepsilon_j$$

$$= \frac{1}{\sqrt{2\pi}\,\sigma_j} \int_{-\infty}^{+\infty} \cos(2\pi s\cdot\varepsilon_j) \exp\left(-\frac{\varepsilon_j^2}{2\sigma_j^2}\right) d\varepsilon_j = \exp(-2\pi^2 s^2 \sigma_j^2) \tag{3.33}$$

and Eq. (3.31) takes the general form

$$q^{(1)} = \sum_j p_j \exp(2\pi i \vec{s}\cdot\langle\vec{r}_j\rangle) \exp(-2\pi^2(\vec{s}\vec{\sigma}_j)^2) \tag{3.34}$$

in which

$$(\vec{s}\vec{\sigma}_j)^2 = h^2\sigma_{jx}^2 + k^2\sigma_{jy}^2 + Z^2\sigma_{jz}^2 \qquad (3.35)$$

with σ_{jx} and σ_{jy} expressed as fractions of the *a*, *b* unit cell parameters and σ_{jz} expressed in nanometers.

The computation of the phase differences between pairs of nth-nearest neighbors [i.e., of the different $q^{(n)}$ which are needed to calculate the intensity of the diffracted beam] requires the knowledge of the interaction existing among the fluctuations of the mutual positions of the layers in the stack.

3.3.1 Position Fluctuations Leading to a Disorder of the First Type

The problem will be simplified if we assume that only a single type of translation, with mean value $\langle\vec{r}\rangle$, is possible between adjacent layers. In accordance with the definition of a disorder of the first type (cf. Sect. 1.6.6), the total translation between two nth-nearest neighbor layers is $n\langle\vec{r}\rangle$, but, from one pair to another, the actual translation differs from this mean value and the law of distribution of these variations is independent of n. In other words, Eq. (3.27) is valid for all values of n and we must have

$$\vec{r}_n = n\langle\vec{r}\rangle + \vec{\varepsilon} \ ,$$

and the mean phase difference between the waves diffracted by two nth-nearest neighbor layers is:

$$q^{(n)} = \overline{\exp(2\pi i\vec{s}\cdot\vec{r}_n)} = \exp(2\pi i\vec{s}\cdot n\langle\vec{r}\rangle) \int_{-\infty}^{+\infty} \exp(2\pi i\vec{s}\cdot\vec{\varepsilon})\gamma(\varepsilon)\,d\varepsilon$$

or

$$q^{(n)} = \exp(2\pi i\vec{s}\cdot n\langle\vec{r}\rangle)\exp(-2\pi^2(\vec{s}\vec{\sigma})^2) \ . \qquad (3.36)$$

This shows that in contrast with the result in Section 3.1.2, the phase differences $q^{(n)}$ cannot be obtained by merely raising $q^{(1)}$ to the nth power since the r.h.s. of Eq. (3.36) is not equal to the nth power of the r.h.s. of Eq. (3.34)[7].

The relationships in (3.37) and (3.36) will give us the modulation functions $G_{hk}(Z)$ along a rod indexed (*hk*), as defined in Section 2.2.1 and normalized to a single layer, for a powder composed of stacks of different thicknesses with position defects due to fluctuations constituting a disorder of the first type (Guinier 1964)

[7] When many types of translations t_j exist in the stack between adjacent layers, with corresponding probabilities p_j, the computation is only possible if the fluctuations of the positions of the layers follow a single Gaussian law for all values of n and irrespective of the type of translation. Eq. (3.36) then becomes

$$q^{(n)} = \exp(-2\pi^2(\vec{s}\vec{\sigma})^2)\left[\sum_j p_j \exp(2\pi i\vec{s}\cdot\langle\vec{r}_j\rangle)\right]^n \ . \qquad (3.37)$$

$$G_{hk}(Z) = 1+2\exp(-2\pi^2(\vec{s}\vec{\sigma})^2)\ \mathrm{Re} \sum_M \alpha(M) \sum_{n=1}^{M-1} \frac{M-n}{M} \exp(2\pi i \vec{s}\cdot n\langle\vec{r}\rangle_n)\ . \quad (3.38)$$

We see that the maxima of intensity are found at the same Z ordinate – given by Eq. (2.11) – as the ones found in the absence of fluctuations. However, the presence of this type of defect increases the diffuse background and, simultaneously, decreases the intensity of the individual reflections, as would thermal motion. From the form of Eq. (3.38), we can also see that the loss of intensity increases exponentially along the rod with the square of the z value. Lastly, by taking Eq. (3.35) into account, we find from Eq. (3.36) that the decrease in intensity will itself be augmented when the values of h and k increase.

As an example, let us find the decrease in intensity of the $00l$ reflections for two different values of the fluctuation in the mean interlayer distance. Writing the modulation function as

$$G_{00}(l) = 1+2\exp\left(-2\pi^2 l^2 \frac{\sigma^2}{\bar{d}_{001}^2}\right) \mathrm{Re} \sum_M \alpha(M) \sum_{n=1}^{M-1} \frac{M-n}{M} \exp(2\pi i Z n \bar{d}_{001}) \quad (3.39)$$

and choosing first $\sigma = 0.03\,\bar{d}_{001}$, we find that for interferential coherence domains sufficiently thick ($M > 10$ or 15 layers), the unity in the r.h.s. is negligible compared to the exponential term. Computation now yields reductions of 2%, 7%, and 15% for the maxima of the reflections 001, 002, and 003 respectively, as compared with the values obtained in the absence of fluctuation defects. If $\sigma = 0.07\,\bar{d}_{001}$ the decreases in intensity are found to be 9%, 32%, and 60%.

3.3.2 Position Fluctuations Leading to a Disorder of the Second Type

We once more consider a stack of identical layers in which the translation vector is subject to fluctuations which follow a Gaussian distribution, without long range correlation among the defects which therefore yield a disorder of the second type. The mean phase difference between the beams diffracted by two nth-nearest neighbors is now

$$q^{(n)} = \overline{\exp(2\pi i \vec{s}\cdot\vec{r}_n)} = \exp(2\pi i \vec{s}\cdot n\langle\vec{r}\rangle)\ \overline{\exp(2\pi i \vec{s}\cdot\vec{\varepsilon}_n)}\ . \quad (3.40)$$

As shown in Section 1.6.6, the deviation $\vec{\varepsilon}_n$ depends on n, and the dispersion of the actual translations with respect to $n\langle\vec{r}\rangle$ is $\sqrt{n}\,\sigma$. Hence

$$q^{(n)} = \exp(2\pi i \vec{s}\cdot n\langle\vec{r}\rangle)\exp(-2\pi^2 s^2 n \sigma^2) = [q^{(1)}]^n \quad (3.41)$$

and the intensity of the diffracted waves can be obtained directly from (3.19) provided each q term corresponding to the difference in phase of the beams diffracted by adjacent layers, as computed for a defect-free stack, be multiplied by $\exp(-2\pi^2 s^2 \sigma^2)$. From Eq. (3.41), we also see that $q^{(n)}$ is then simply obtained by raising $q^{(1)}$ to the nth power, as was done for the stack described in Section 3.1.2.

This conclusion is still valid when the substacks contain many types of translations between nearest-neighbor layers. As stated at the beginning of Section 3.3, when the three spatial components of the position fluctuations are independent, the complete study of these fluctuations is reduced to the study of each of these components. Let us therefore suppose that there are two possible translations, with mean values $\langle\vec{t}_1\rangle$ and $\langle\vec{t}_2\rangle$ and corresponding probabilities p_1 and p_2 respectively, along a given direction in space. Furthermore, let the sequence of translations be random (R = 0), with fluctuations characterized by Gaussian distributions with standard deviations σ_1 and σ_2. The probability of having an actual translation vector $\vec{t}^{(n)}$ between two nth-nearest neighbor layers is $P[t^{(n)}]$. For two nearest neighbor layers, we have

$$P[t^{(1)}] = \frac{p_1}{\sqrt{2\pi}\,\sigma_1} \exp\left(-\frac{|t^{(1)} - \langle t_1\rangle|^2}{2\sigma_1^2}\right) + \frac{p_2}{\sqrt{2\pi}\,\sigma_2} \exp\left(-\frac{|t^{(1)} - \langle t_2\rangle|^2}{2\sigma_2^2}\right), \qquad (3.42)$$

while, for second-nearest layers, the probability of finding them related by the vector $\vec{t}^{(2)}$ is given by the convolution product

$$\begin{aligned} P[t^{(2)}] &= \int P[t^{(1)}]\, P(t^{(2)} - t^{(1)})\, dt^{(1)} \\ &= \int \left[\frac{p_1}{\sqrt{2\pi}\,\sigma_1} \exp\left(-\frac{|t^{(1)} - \langle t_1\rangle|^2}{2\sigma_1^2}\right) + \frac{p_2}{\sqrt{2\pi}\,\sigma_2} \exp\left(-\frac{|t^{(1)} - \langle t_2\rangle|^2}{2\sigma_2^2}\right)\right] \\ &\quad \times \left[\frac{p_1}{\sqrt{2\pi}\,\sigma_1} \exp\left(-\frac{|t^{(2)} - t^{(1)} - \langle t_1\rangle|^2}{2\sigma_1^2}\right)\right. \\ &\quad \left. + \frac{p_2}{\sqrt{2\pi}\,\sigma_2} \exp\left(-\frac{|t^{(2)} - t^{(1)} - \langle t_2\rangle|^2}{2\sigma_2^2}\right)\right] dt^{(1)} \end{aligned} \qquad (3.43)$$

in which the upper limit of the integral is infinite. The phase difference of the waves diffracted by the second-nearest layers is

$$q^{(2)} = \int P(t^{(2)}) \exp(2\pi i \vec{s} \cdot \vec{t}^{(2)})\, dt^{(2)} \; . \qquad (3.44)$$

Integrations of the term including only p_1 in (3.43) and (3.44) yields:

$$\frac{p_1^2}{2\pi\sigma_1^2}\iint \exp(2\pi i s\cdot t^{(2)})\exp\left(-\frac{|t^{(1)}-\langle t_1\rangle|^2}{2\sigma_1^2}\right)\exp\left(-\frac{|t^{(2)}-t^{(1)}-\langle t_1\rangle|^2}{2\sigma_1^2}\right)dt^{(2)}dt^{(1)}$$

$$=\frac{p_1^2}{2\pi\sigma_1^2}\int \exp\left(-\frac{|t^{(1)}-\langle t_1\rangle|^2}{2\sigma_1^2}\right)$$

$$\times\left[\int \exp(+2\pi i s\cdot t^{(2)})\exp\left(-\frac{|t^{(2)}-t^{(1)}-\langle t_1\rangle|^2}{2\sigma_1^2}\right)dt^{(2)}\right]dt^{(1)}$$

$$=\frac{p_1^2}{\sqrt{2\pi}\sigma_1}\exp(-2\pi^2s^2\sigma_1^2)\int \exp\left(-\frac{|t^{(1)}-\langle t_1\rangle|^2}{2\sigma_1^2}\right)\exp[2\pi i s\cdot(t^{(1)}-\langle t_1\rangle)]\,dt^{(1)}$$

$$=p_1^2\exp(-4\pi^2s^2\sigma_1^2)\exp(4\pi i s\cdot\langle t_1\rangle)=\{p_1\exp(-2\pi^2s^2\sigma_1^2)\exp(2\pi i s\cdot\langle t_1\rangle)\}^2\ .$$

The other terms in (3.44) are integrated in the same way, so that we finally obtain

$$q^{(2)}=[p_1\exp(-2\pi^2s^2\cdot\sigma_1^2)\exp(2\pi i s\cdot\langle t_1\rangle)+p_2\exp(-2\pi^2s^2\sigma_2^2)\exp(2\pi i s\cdot\langle t_2\rangle)]^2\ .$$

It can be shown that the phase difference corresponding to any number of well-defined translations with fluctuations between nth-nearest neighbors in the 3D-case is

$$q^{(n)}=\left[\sum_j p_j\exp(-2\pi^2(\vec{s}\vec{\sigma}_j)^2)\exp(2\pi i\vec{s}\cdot\langle\vec{r}_j\rangle)\right]^n\ . \tag{3.45}$$

For fluctuations of position affecting a j-type translation vector between pairs of layers and constituting a disorder of the second type, the contribution of these pairs to the intensity of the diffracted wave decreases as s increases. In other words, the **apparent** probability of finding a j-type translation between two given layers decreases inversely with s. The same conclusion could have been reached qualitatively for disorders of the first type by writing (3.37) as

$$q^{(n)}=\left\{\sum_j p_j[\exp(-2\pi^2(\vec{s}\vec{\sigma})^2)]^{1/n}\exp(2\pi i\vec{s}\cdot\langle\vec{r}_j\rangle)\right\}^n\ . \tag{3.46}$$

We shall now examine the effect of the presence of this kind of defect on the powder diagram, once again with reference to $00l$ reflections; Fig. 3.6 corresponds to the microcrystalline mica already described in Sections 3.2.1 and 3.2.3. We suppose that there is a Gaussian distribution of fluctuations characterized by squared standard deviation σ^2 (such as $\sigma^2=0.1$ Å^2) affecting the values of the basal distance $\bar{d}_{001}=10$ Å. The profiles of the bands were calculated with the help of the mathematical formalism described in the following paragraphs. Furthermore, all the interferential coherence domains were assumed to be composed of ten layers and to have the same thickness.

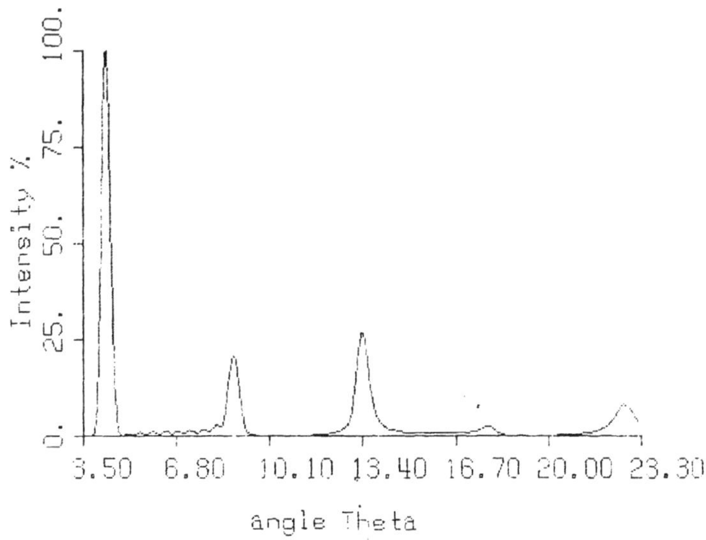

Fig. 3.6. Calculated profiles of the $00l$ reflections for a microcrystalline mica with a thickness M of the coherent domains equal to ten layers and mean interlayer distance $\langle d_{001}\rangle = 10$ Å affected by fluctuations characterized by a squared mean deviation $\sigma^2 = 0.1$ Å^2

Figure 3.6 shows that the width of the reflexions increases with l. This broadening is different from the one observed with defect-free but thin crystals when the profiles are examined as a function of the angle of diffraction 2θ, since the width at half height is then proportional to $1/\cos\theta$ (cf. the Scherrer formula in Sect. 1.6.7). On the other hand, if these same $00l$ reflections are plotted vs. the modulus of the scattering vector ($s = 2\sin\theta/\lambda$), they will have identical widths, since the crystal is perfect and the width is given by Eq. (2.10). This is shown in Fig. 3.7, which gives the $00l$ reflections for the same mica as in Fig. 3.6 in the absence of fluctuations in the basal distances.

It should also be pointed out that if the standard deviation σ_j be taken as independent of the type of translation defect ($\sigma_l = \sigma$), and if σ is sufficiently small, Eq. (3.45) can be expressed as

$$q^{(n)} = \left\{[1-2\pi^2(h^2\sigma_x^2+k^2\sigma_y^2+Z^2\sigma_z^2)] \sum_j p_j \exp(2\pi i \vec{s}\cdot\langle\vec{r}_j\rangle)\right\}^n . \qquad (3.47)$$

The practical consequences of the existence of such small fluctuations on the diffraction as described by Eq. (3.47) will be more fully discussed in Section 3.3.4. For the time being, we shall merely examine the case of the $00l$ reflections discussed by Houska and Warren (1954) using a simplified form of (3.47). The method given by these authors and later modified by Maire and Méring (1970) gives direct access to the mean standard deviation σ; it will therefore now be detailed.

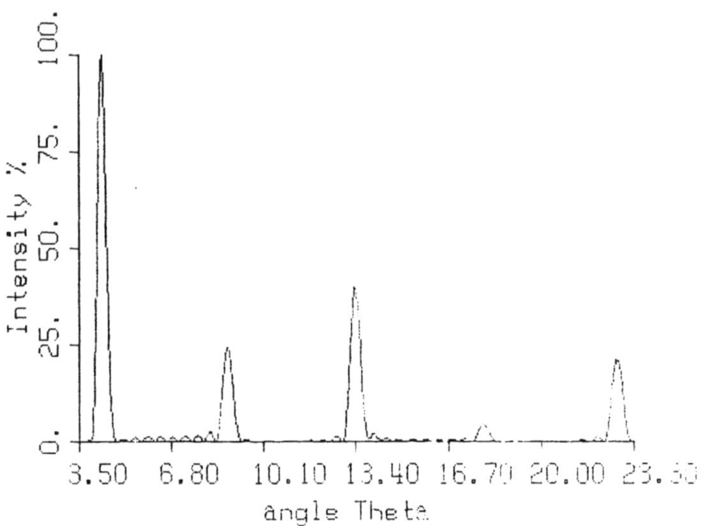

Fig. 3.7. Calculated profiles of the 00*l* reflections for a microcrystalline mica with a fluctuation of the interlayer distance $d_{001} = 1$ nm and with M = 10

3.3.3 Determination of the Mean Standard Deviation of the Fluctuations Affecting the Interlayer Distances by Direct Profile Analysis of the 00*l* Reflections

Consider a powder composed of stacks in which the interlayer distances are affected by fluctuations which constitute a disorder of the second type. Starting from Eqs. (2.42), (3.38), (3.39) and (3.45), it is possible to write the expression giving the distribution of intensity along the (00) rod as

$$I_{00}(Z) = \frac{\bar{p}(\pi/2)}{2\pi\Omega Z^2} \,|F_{00}(Z)|^2 G'_{00}(Z) \tag{3.48}$$

in which $G'_{00}(Z)$ is the modulation function normalized to a single layer, incorporating both the distribution of the thicknesses and the fluctuations in the basal distances, namely

$$G'_{00}(Z) = 1 + 2\,\mathrm{Re}\left\{\sum_{M} \alpha(M) \sum_{n-1}^{M-1} \frac{M-n}{M} \overline{[\exp(2\pi i Z z)]}^{\,n}\right\} \tag{3.49}$$

with $z = \bar{d}_{001} + \varepsilon_z$.

Let l be the integral indices corresponding to the maxima of the function $G'_{00}(Z)$. For any particular value of Z, we can write $Z = l'/\bar{d}_{001}$ in which l' is a continuous index which can be made to correspond to the maxima of $G'_{00}(Z)$ by letting $l' = l+2v$, with $0 \leqslant v \leqslant 1/2$.

The mean of the exponential term then becomes:

$$\overline{[\exp(2\pi i Z z)]^n} = \exp(2\pi i n v)\ \overline{\left[\exp\left(2\pi i \frac{l'\varepsilon_z}{\bar{d}_{001}}\right)\right]^n} .$$

If the intensity of the wave diffracted along the Z axis of the (00) rod is mainly concentrated close to the maxima of $G'_{00}(Z)$, then:

- on one hand, the intensity must be nonzero for narrow ranges of values of v close to 0 (v is much less than 1/2);
- on the other hand, the value of $\exp(2\pi i l'\varepsilon_z/\bar{d}_{001})$ must be close to unity and $l'\varepsilon_z \ll \bar{d}_{001}$ (i.e., $\varepsilon_z \ll \bar{d}_{001}$).

It is then possible to replace l' by the integral value l in the exponential term and to retain the first two terms in the series expansion of the exponential, thus obtaining

$$\overline{\left[\exp\left(2\pi i \frac{l'\varepsilon_z}{\bar{d}_{001}}\right)\right]^n} \approx \left(1-2\pi^2 l^2 \frac{\overline{\varepsilon_z^2}}{\bar{d}_{001}^2}\right)^n = \left(1-2\pi^2 l^2 \frac{\sigma^2}{\bar{d}_{001}^2}\right)^n .$$

The $G'_{00}(Z)$ function now becomes:

$$G'_{00}(Z) = 1+2\,\mathrm{Re}\left[\sum_M \alpha(M) \sum_{n=1}^{M-1} \frac{M-n}{M}\left(1-2\pi^2 l^2 \frac{\sigma^2}{\bar{d}_{001}^2}\right)^n \exp(2\pi i n v)\right] . \tag{3.50}$$

In this expression[8], the order of summation can be changed provided the limits are properly defined.

For a given value of n, let us write H_n, the probability of having an nth-layer neighboring an arbitrarily chosen layer, as

$$H_n = \sum_{M=n}^{\infty} \alpha(M) \frac{M-n}{M} . \tag{3.52}$$

The function $G'_{00}(Z)$ then becomes

$$G'_{00}(Z) = 1+2\,\mathrm{Re}\left[\sum_{n=1}^{\infty} H_n \left(1-2\pi^2 l^2 \frac{\sigma^2}{\bar{d}_{001}^2}\right)^n \exp(2\pi i n v)\right] . \tag{3.53}$$

[8] Note that writing $G'_{00}(Z)$ as in (3.50) is equivalent to having for the general expression (3.47)

$$q^{(n)} = \left[\left(1-2\pi^2 l^2 \frac{\sigma^2}{\bar{d}_{001}^2}\right) \exp(2\pi i Z \bar{d}_{001})\right]^n \tag{3.51}$$

Hence, the profile of a basal reflexion indexed l_0 can be expressed as a Fourier series with coefficients $A_n(00l)$ given by

$$A_n(001) = H_n\left(1-2\pi^2 l^2 \frac{\sigma^2}{d_{001}^2}\right)^n \tag{3.54}$$

as shown by Houska and Warren (1954). Note that the smaller the values of l, the better the approximation adopted in this computation. It should also be pointed out that the parameter σ^2, characteristic of the fluctuations, can be obtained directly from the Fourier coefficients corresponding to the first two $00l$ reflections by using

$$\ln\left[A_n(001)\right] - \ln\left[A_n(002)\right] \approx 6\pi^2 n \frac{\sigma^2}{d_{001}^2}\ . \tag{3.55}$$

Since the difference of the two logarithms is a linear function of n, the slope is directly proportional to σ^2. Once this is known, the probabilities H_n are obtained by using Eq. (3.54).

The method presented here and used in this order was given by Maire and Méring (1960) and, independently, by Digonskii and Krylov (1960). It is, in fact, a modified form of the method of Houska and Warren (1954) in which the determination of the coefficients H_n precedes that of σ^2. These last authors use:

$$\ln\left[A_n(00l)\right] = \ln H_n + 2n\pi^2 l^2 \frac{\sigma^2}{d_{001}^2}$$

in which $A_n(00l)$ are normalized values. An extrapolation to $l = 0$ of the straight line $\ln A_n = f(l^2)$ yields H_n. However, the method of Maire and Méring eliminates any error that could be due to the normalization, since the slope of the straight line depends only on the relative values of the coefficients A_n.

3.3.4 Comparison of the Effects of Random Defects and of Position Fluctuations on the Diffraction

When the amplitude of the fluctuations is small, Eq. (3.46) was shown above to take the particular form given in (3.47):

$$q^{(n)} = \left\{\left[1-2\pi^2(h^2\sigma_x^2+k^2\sigma_y^2+Z^2\sigma_z^2)\right] \sum_j p_j \exp\left(2\pi i \vec{s}\cdot\langle\vec{r}_j\rangle\right)\right\}^n$$

and the comparison of this expression with Eq. (3.20) giving the phase difference for a stack in which random translation defects between adjacent layers occur with a probability p_A brings out the fact that, in the former expression, the term $2\pi^2(h^2\sigma_x^2+k^2\sigma_y^2+Z^2\sigma_z^2)$ seems to play the same role as P_A in Eq. (3.20). It can therefore be expected that in the light of the 2D tendency of the structure, the existence of small fluctuations in the positions of the layers will result in the appearance of hk diffraction bands similar to the ones met in systems with random translation defects.

There is, however, a fundamental difference in the modifications of the powder diagrams due to the presence of either of these two types of structural defects. The presence of random translations has an identical effect on all the hkl reflections, whatever the values of h or k (other than $h=k=0$), while the effect of the presence of fluctuations depends on the values of the indices. To illustrate this point, we shall examine different types of fluctuations.

1. **Isotropic distribution of fluctuations coplanar with the layers.**
   From Eq. (3.47), we see that the profiles of all the hkl reflections evolve towards band profiles corresponding to the 2D tendency of the structure, and that this characteristic will be most evident when the value of the sum (h^2+k^2) is high.
2. **Monodimensional fluctuation parallel to the $\vec{a}$ axis ($\sigma_x \neq 0$, $\sigma_y=\sigma_z=0$) or $\vec{b}$ axis ($\sigma_y \neq \sigma_x=\sigma_z=0$).**
   The $0kl$ (or $h0l$) reflections are now quite insensitive to the presence of the fluctuations.
3. **Fluctuations coplanar with the layers and with a vector $\vec{\varepsilon}$ at a constant angle with the directions of the crystallographic axes $\vec{a}$ or $\vec{b}$.**

Let α be the constant angle between the direction of the fluctuation vector $\vec{\varepsilon}$ and a given direction, say, the $\vec{a}$ axis. An actual translation $\vec{t}$ coplanar with a layer will then be described by

$$\vec{t} = \langle\vec{t}\rangle + \vec{a}\,\varepsilon\cos\alpha + \vec{b}\,\varepsilon\,(a/b)\sin\alpha$$

and using now (3.31), we can write

$$q^{(1)} = \exp\,[2\pi i\,\vec{s}\cdot(\langle\vec{t}\rangle + d_{001}\,\vec{u}_z)]\;\overline{\exp\{2\pi i\,[h\varepsilon\cos\alpha + k\varepsilon(a/b)\sin\alpha]\}}$$

$$= \exp\,[2\pi i\,\vec{s}\cdot(\langle\vec{t}\rangle + d_{001}\,\vec{u}_z)]\;\exp\{-2\pi^2\sigma^2[h\cos\alpha + k(a/b)\sin\alpha]^2\}\;. \quad (3.56)$$

This specific type of fluctuation is particularly applicable to lamellar silicates which often have $b=a\sqrt{3}$, so that the second exponential in Eq. (3.56) becomes

$$\exp\{-2\pi^2\sigma^2[h\cos\alpha + (k/\sqrt{3})\sin\alpha]^2\}\;.$$

If the fluctuations occur along a direction corresponding to the diagonal of the unit cell ($\alpha = 60°$), the reflections along the rods $h\bar{h}$ and $\bar{h}h$ will not be affected. Again, if $\alpha = 120°$, or 30°, or 150°, it will be the reflections respectively along the rods hh; $h, \overline{3h}$ or $h, 3h$ which will not be affected.

These various examples show that when the domain of reciprocal space explored is sufficiently extensive so that many diffraction bands and types of reflections can be observed, it is theoretically possible to distinguish unambiguously between a lamellar system with random defects and one with fluctuations in the positions of the layers. In actual practice, and as will be shown below, there are additional complications.

3.3.5 Comparison of the Physical Significances Attached to the Concepts of Random Defects and of Position Fluctuation Defects

Random translation defects were said in Section 1.6.5 to exist between adjacent layers when the direction and, especially, the amplitude of the translation vector between them were random. This concept of stacking defect led us in Section 3.2.1 to express the phase difference q of the waves diffracted by first-nearest neighbors by the relationship in Eq. (3.20), which can be written as

$$q = (1 - P_A) \exp(2\pi i \vec{s} \cdot \vec{r}) \ .$$

Comparing now this concept of random defects with that of fluctuations of position corresponding to a disorder of the second type, we find that the difference between these two models resides in the law of distribution of the deviations shown by the actual translations with respect to the average translation. Thus, if $\langle t_b \rangle$ is the mean translation along the direction b (chosen as example), the concept of random defects entails that the deviations ε_y of the actual translations t_b from $\langle t_b \rangle$ are distributed according to a law whereby any value of ε_y in the range $-b/2$ to $b/2$ is equally probable. This law of distribution is described by the impulse function shown as (a) in Fig. 3.8. In opposition to this, defects due to fluctuations in the position of the layers are such that the probability attached to a given value ε_y of the deviation is largest when ε_y is close to zero. This leads to a bell-shaped or Gaussian-type law of distribution such as (b) in Fig. 3.8.

From the crystallochemical point of view, the notion of fluctuations about a mean value often seems to be physically more realistic than that of random defects. Yet, as far as we know, all the analyses of diffractograms from powders subject to these kinds of defects show profiles characteristic of a more or less pronounced 2D tendency which require, for their interpretation, the use of a mathematical formalism corresponding to Eq. (3.20), as the profiles of all the experimental bands can be recreated by calculation when the terms giving the phase difference q include a constant multiplicative coefficient of the form $(1-P)$, independent of the indices of the reflections. Although this experimental fact might seem to exclude fluctuation defects (since the effects are then functions of the indices), we shall see that it is possible to imagine a structural model with fluctuation defects which also requires a constant multiplicative coefficient.

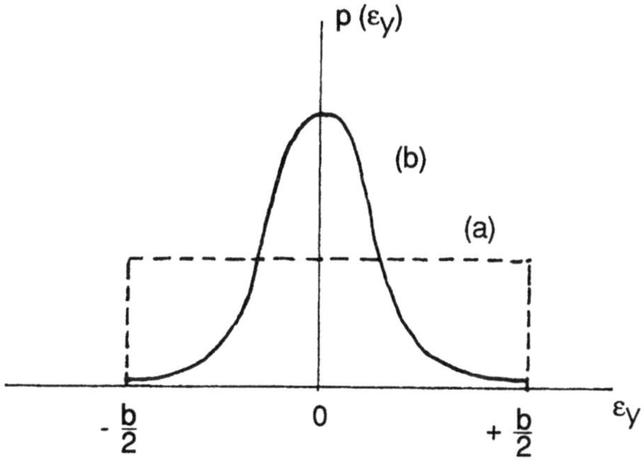

Fig. 3.8. Law of distribution for the deviations ε_y corresponding to random stacking faults (**a**) and to faults due to fluctuation of the positions of the layers (**b**)

Consider the case of stacks which contains a single type of translation $\vec{t}_0$ between adjacent layers. Suppose now that the system is no longer homogeneous, that is the actual translation between **certain pairs** of adjacent layers fluctuates around $\vec{t}_0$ in the *ab* plane by variations described by a Gaussian distribution, and let P′ be the probability of finding such pairs. For the **other pairs** of adjacent layers, the translation is strictly equal to $\vec{t}_0$, and the probability of finding one of these pairs[9] is $(1-P')$. For such a structural model, by neglecting any component of the translation perpendicular to the layer, we find that Eq. (3.41) becomes:

$$q = \exp(2\pi i\vec{s}\cdot\vec{t}_0)\{P'\exp[-2\pi^2(h^2\sigma_x^2+k^2\sigma_y^2)]+(1-P')\} \; . \qquad (3.57)$$

It is evident that, in this expression, the exponential term in the curled bracket becomes negligible when the value of the product $s^2\sigma^2$ is sufficiently large. The relationship in Eq. (3.57) now becomes quite comparable to the form met with random translations, namely

$$q = (1-P')\exp(2\pi i\vec{s}\cdot\vec{t}_0) \qquad (3.58)$$

and the experimental powder patterns showing diffraction bands can therefore also be interpreted on the basis of a structure with fluctuations in the positions of the layers.

We must now find for the product $(\vec{s}\vec{\sigma})^2$ the range of values which legitimate the approximations used and render the diffraction identical (or at least very close) to

[9] This model, which corresponds to a weakening of the bonds between some pairs of adjacent layers, may be envisaged in the case of certain anhydrous mixed-layer minerals like illite-smectite or when bubbles are to be found between some layers [as in kaolinites (Plançon and Tchoubar 1977); or certain micas (Bernard and Pernoux 1953; Rang 1953a, b)]

the one from stacks containing true random translations. We shall take the (02,11) and (20,13) bands of kaolinite as examples, letting P′ = 0.11 and choosing the fluctuations to be coplanar with the layers and isotropic ($\sigma_x = \sigma_y = \sigma$). For $\sigma = 0.5$ Å, computations show that the use of Eq. (3.57) yields a profile for the (02,11) band which is quite different from that corresponding to the one calculated using Eq. (3.20) for a structure with random translations with the probability $P_A = 0.11$ (cf. Figs. 3.9 and 3.10) However, the value $\sigma = 1$ Å, and P′ = 0.11 lead to the same profile for the (02,11) band. Since we compare two models in which the probabilities P′ and P_A have identical values, the exponential term in the curled bracket in Eq. (3.57) is practically equal to zero. Under these conditions, the exponential term will also be practically equal to zero for the (20,13) band since the value of h^2+k^2 increases for this series of reflexions. The profiles of the (20, 13) bands will therefore also be identical for the two models.

In conclusion, the above example shows that it is not always possible to choose between the two models on the basis of X-ray data alone. This must be complemented by further information obtained from crystallochemistry or through the use of other techniques – such as high resolution electron microscopy or the knowledge of other characteristics of the system – to guide in the choice of the structural model.

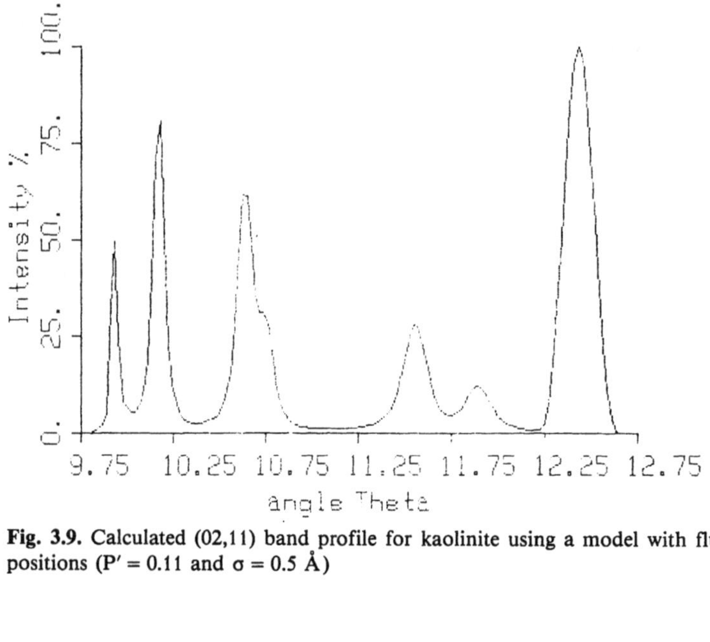

Fig. 3.9. Calculated (02,11) band profile for kaolinite using a model with fluctuations of layer positions (P′ = 0.11 and $\sigma = 0.5$ Å)

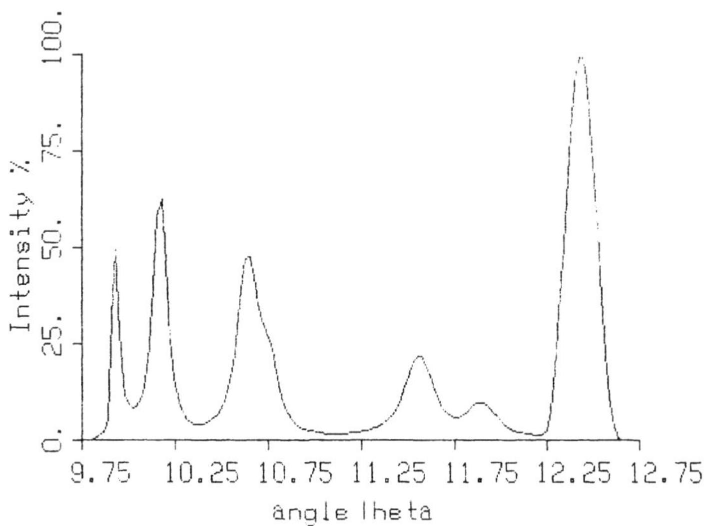

Fig. 3.10. Calculated (02,11) band profile for kaolinite using a model with random layer translations corresponding to a probability $P_A = 0.11$

References

Allegra G (1961 a) A simplified formula for the calculation of the X-ray intensity diffracted by a monodimensionally disordered structure. Acta Crystallogr 14:535

Allegra G (1961 b) Il calcolo dell'intensità di diffrazione dei raggi X da parte di strutture denotanti disordine monodimensionale, pt 1: Criteri matematici. Nuovo Cimento 21:786–801

Allegra G (1961 c) Il calcolo dell'intensità di diffrazione dei raggi X da parte di strutture denotanti disordine monodimensionale, pt 2: Applicazioni della teoria ed alouni modelli struttureli con disordine monodimensionale. Nuovo Cimento 22:661–672

Allegra G (1964) The calculation of the intensity of X-Rays diffracted by a monodimensionally disordered structure. Acta Crystallogr 17:579–586

Bernard R, Pernoux E (1953) Franges d'interférences obtenues par la superposition de deux faisceaux électroniques cohérents. CR Acad Sci Paris 236:187–189

Drits VA, Plançon A, Sakharov BA, Besson G, Tsipurski SI, Tchoubar C (1984) Diffraction effects calculated for structural models of K-saturated montmorillonite containing different types of defects. Clay Mineral 19:541–561

Digonskii VV, Krylov VN (1960) The relation between bonding in graphite and temperature. Zh Prikl Khim 33-12:2638–2643 (in Russian)

Guinier A (1964) Théorie et technique de la radiocristallographie. Chap. 13: Diffraction par les réseaux cristallins imparfaits. Dunod, Paris, pp 490–636

Hendricks SB, Teller E (1942) X-Ray interference in partially ordered layer lattices. J Chem Phys 18:147–167

Houska CR, Warren BE (1954) X-ray study of the graphitization of carbon black. J Appl Phys 25:1503–1509

Kakinoki J, Komura Y (1952) Intensity of X-ray diffraction by one dimensionally disordered crystals. J Phys Soc Jpn 7:30–35

Kakinoki J, Komura Y (1962) Comments on Allegra's paper: a simplified formula for the calculation of the X-ray intensity diffracted by a monodimensionally disordered structure. Acta Crystallogr 15:292

Maire J, Méring J (1960) Croissance des dimensions des domaines cristallins au cours de la graphitation du carbons. Proc 4th Conf Carbon. Pergamon, New York, pp 345–350

Maire J, Méring J (1970) Graphitization of soft carbons. In: Walker PP (ed) Chemistry and physics of carbon, Vol 6. Dekker, New York, pp 125–189

Méring J (1949) L'interférence des rayons X dans les systèmes à stratification désordonnée. Acta Crystallogr 2:371–377

Plançon A, Tchoubar C (1976) Etude des fautes d'empilement dans les kaolinites partiellement désordonnées. II: Modèles d'empilement comportant des fautes par rotation. J Appl Crystallogr 9:279–285

Plançon A, Tchoubar C (1977) Determination of structural defects in phyllosilicates by X-ray diffraction. II. Nature and proportion of defects in natural kaolinites. Clays and Clay Miner 25:436–450

Rang O (1953a) Formbestimmung dünner Einkristall-Lamellen mit Hilfe der durch Kristallgitter-Reflexe hervorgerufenen Scheinstrukturen. Optik 18:90–106

Rang O (1953b) Fern-Interferenzen von Elektronenwellen. Z Phys 136:465–479

Warren BE (1941) X-ray diffraction in random layer lattices. Phys Rev 9:693–698

CHAPTER 4

Statistical Models and Parameters Used to Describe Interstratified Lamellar Systems

Interstratified crystals are good examples of systems with one-dimensional order-disorder phenomena in which layers with different compositions and structures coexist in varying proportions and are superposed with different degrees of order-disorder. Hence, it is not possible to characterize these systems using a single crystal lattice, as is generally the case with crystallized objects[10] and they must be described with the help of statistical concepts calling upon probabilistic parameters.

The problem associated with the quantitative description of disordered interstratified lamellar systems was first tackled by Hendricks and Teller (1942), Wilson (1942), and Méring (1950). In these papers, the statistical parameters were not expressed as a function of the nature of the interaction between layers, since only pure random stacking of the layers was considered in all the models. The concept of interactions among the layers was first introduced by Jagodzinski (1949a, b, c, 1954), and by Kakinoki and Komura (1952, 1954). Random stacking (S = 0, where S is the range of interaction) was clearly differentiated from other cases, in which S was 1 (when the interaction was between first-nearest neighbors) or even as high as 3. The explicit forms of the conditional probabilities were later given by Mac Ewan (1958), then by Cesari et al. (1965) and by Reynolds (1967). These authors however, did not go beyond S = 0 and S = 1. Second-neighbor interaction was studied by Sato (1969). Lastly, the general problem of interaction in stacks of different layer types was tackled by Drits and Sakharov (1976), and it is this approach which will form the basis of the formalism to be given in the present chapter.

4.1 General Parameters Characterizing the Stacking of Different Layers in Interstratified Structures

We first consider an interstratified system in which the mode of stacking follows Markovian statistics (quasi-homogeneous system), and assume that two different

[10] The only interstratified crystals which can be described by a single lattice are those in which the different layers have the same $\vec{a}, \vec{b}$ periodicity and are stacked in strictly periodic sequences; this occurs, for instance, in chlorites, in rectorite, or in the first stage of the graphite intercalation compounds containing a single intercalated metallic cation, as described in Chapter 1.7

types of layers, say A and B, are found in each stack of M layers. If W_A and W_B are the respective probabilities of occurrence of these layers in the mineral, these values are also their relative proportions. If M_A and M_B are the number of layers of types A and B in the stack, we can write:

$$W_A = M_A/M \qquad W_B = M_B/M \qquad \text{and} \qquad W_A + W_B = 1 \ . \tag{4.1}$$

Since there are (M − 1) pairs of nearest neighbors in the stack, we can also count the number M_{AA}, M_{AB}, M_{BA} and M_{BB} of adjacent AA, AB, BA, and BB pairs and express their relative abundance as

$$\begin{aligned} W_{AA} &= M_{AA}/(M-1); \; W_{AB} = M_{AB}/(M-1); \\ W_{BA} &= M_{BA}/(M-1); \; W_{BB} = M_{BB}/(M-1) \end{aligned} \tag{4.2}$$

so that, for a stack with two types of layers.

$$W_{AA} + W_{AB} + W_{BA} + W_{BB} = 1 \ . \tag{4.3}$$

Note that the order of succession of the layers, and therefore the probabilities $W_{AA}, W_{AB}, \ldots$ are defined for a given oriented direction perpendicular to the layers; hereafter, all descriptions of sequences in a stack will be referred to this oriented direction.

Similarly, there are eight possible combinations in sequences of three successive layers composed of types A and B, and the probabilities of occurrence are given by

$$\begin{aligned} W_{AAA} &= M_{AAA}/(M-2); \; W_{AAB} = M_{AAB}/(M-2) \\ W_{ABA} &= M_{ABA}/(M-2); \; W_{BAA} = M_{BAA}/(M-2) \\ W_{BBA} &= M_{BBA}/(M-2); \; W_{BBB} = M_{BBB}/(M-2) \\ W_{BAB} &= M_{BAB}/(M-2); \; W_{ABB} = M_{ABB}/(M-2) \end{aligned} \tag{4.4}$$

with the supplementary relationship:

$$W_{AAA} + W_{AAB} + W_{ABA} + W_{ABB} + W_{BAA} + W_{BAB} + W_{BBA} + W_{BBB} = 1 \ . \tag{4.5}$$

Analogous expressions defining the relative abundances and the link among them can be written for any sequence of n successive layers.

A number of other relationships are inherent in the statistical description of the system. It is evident, for example, that the abundance of pairs of adjacent different layers is $(W_{AB} + W_{BA})$. Again the probability of finding sequences of three successive layers containing two A and one B is $(W_{AAB} + W_{ABA} + W_{BAA})$.

Finally, if we admit that the probability of occurrence of a given type of layer depends on the nature and the mode of stacking of the preceding layers, we are led to introduce the concept of *conditional probability* for a particular type of layer to follow a particular sequence. The occurrence of pairs, of triplets, ... of each combination of kinds of layers can be expressed as a function of these conditional probabilities. For instance, for the pairs, AA, AB, the probabilities of existence are equal to the probabilities of finding a layer A multiplied by the conditional probability that such a layer be followed by a layer A or a layer B; taking into account that the same approach can be used for the description of the pairs BA and BB, we find:

$$W_{AA} = W_A P_{AA}; W_{AB} = W_A P_{AB}; W_{BA} = W_B P_{BA}; \quad W_{BB} = W_B P_{BB} \quad (4.6)$$

in which P_{AA}, P_{AB}, P_{BA} and P_{BB} are, respectively, the conditional probabilities of a layer A following an A layer, a layer B following an A layer, etc. . . . . Obviously, in any two-component (A and B) system, an A layer must be followed either by a layer B or by another layer A, so that

$$P_{AA} + P_{AB} = 1 \quad (4.7)$$

and, similarly,

$$P_{BA} + P_{BB} = 1 \ . \quad (4.8)$$

The computation of the relative proportion of the different sequences of three layers calls upon the conditional probabilities of finding a specific type of layer after the possible pairs of layers. For instance, the relative abundance of AAA triplets is equal to the abundance of AA pairs multiplied by the conditional probability of meeting an A layer after a pair of this type. The relative probabilities associated with each type of triplet are:

$$\begin{aligned} W_{AAA} &= W_{AA} P_{AAA} & W_{AAB} &= W_{AA} P_{AAB} \\ W_{ABA} &= W_{AB} P_{ABA} & W_{ABB} &= W_{AB} P_{ABB} \\ W_{BAA} &= W_{BA} P_{BAA} & W_{BAB} &= W_{BA} P_{BAB} \\ W_{BBA} &= W_{BB} P_{BBA} & W_{BBB} &= W_{BB} P_{BBB} \end{aligned} \quad (4.9)$$

in which $P_{AAA}, P_{AAB}, \ldots P_{BBB}$ are the conditional probabilities of finding an A or a B layer after the given pairs. The conditional probabilities are themselves linked by

$$\begin{aligned} P_{AAA} + P_{AAB} &= 1 & P_{ABA} + P_{ABB} &= 1 \\ P_{BAA} + P_{BAB} &= 1 & P_{BBA} + P_{BBB} &= 1 \ . \end{aligned} \quad (4.10)$$

These expressions can be generalized; thus, for stacks composed of g types of layers, we can write

$$W_i = M_i/M \quad \text{for all} \quad i = A, B, C \ldots$$

$$W_{ij} = M_{ij}/(M-1) \quad \text{for all} \quad i,j = A, B, C \ldots$$

$$W_{ijk} = M_{ijk}/(M-2) \quad \text{for all} \quad i,j,k = A, B, C \ldots$$

$$\vdots$$

etc. ...

with
$$W_{ij} = W_i P_{ij} \; ; \quad W_{ijk} = W_{ij} P_{ijk} \; ; \quad \text{etc.} \ldots$$

and
$$\sum_i W_i = 1 \; ; \quad \sum_i \sum_j W_{ij} = 1 \; ; \quad \text{etc.} \ldots$$

and
$$\sum_j P_{ij} = 1 \; ; \qquad \sum_k P_{ijk} = 1 \; ; \quad \text{etc.} \ldots$$

The determination of the **nature of the interstratification** in a crystal with several types of layers consists in **obtaining the values of the relative proportion W_i of each type of layer, as well as the proportions of the different sequences of pairs, triplets, etc. ...** If the probability of appearance of a given type of layer depends on the n preceding layers (i.e., if, as defined in Chap. 1.7.2, S = n), the number of independent statistical parameters which must be determined for the characterization of the interstratification will be very high when n is large. The next sections will deal with the choice of independent statistical parameters for a given value of S.

4.2 Interstratified Structures with S = 0

In any two-component interstratified structure with S = 0, the sequence of layers A and B is strictly random and the presence of a given type of layer at any point of the stack does not depend on the nature of its neighbors. The conditional probabilities of a layer A following a layer A or a layer B are then both equal to the proportion (relative abundance) of layers A:

$$P_{AA} = P_{BA} = W_A$$

and, similarly,

$$P_{BB} = P_{AB} = W_B \; .$$

The consequence of these relations is that the proportion of given types of pairs of adjacent layers is equal to the product of the proportion of each type of layer:

$$W_{AA} = W_A W_A \; ; \quad W_{AB} = W_{BA} = W_A W_B \; ; \quad W_{BB} = W_B W_B \; .$$

Similarly, the proportions of different possible triplets are given by:

$$W_{AAA} = W_A^3 \; ; \quad W_{AAB} = W_{BAA} = W_{ABA} = W_A^2 W_B$$
$$W_{ABB} = W_{BBA} = W_{BAB} = W_A W_B^2 \; ; \quad W_{BBB} = W_B^3 \; .$$

These expressions can easily be extended to sequences containing any number of different layers.

When S = 0 and the number g of types of layers is 2, and in view of relations (4.1), the determination of a single independent parameter (W_A or W_B) will then be sufficient for the full characterization of the interstratified structure.

In the general case of structures containing g types of layers, there are (g−1) independent parameters and the relative proportions of sequences of 2, 3, ... successive layers may be expressed as

$$W_{ij} = W_i W_j \quad \text{for all} \quad i,j = A, B, C, \ldots$$
$$W_{ijk} = W_i W_j W_k \quad \text{for all} \quad i,j,k = A, B, C, \ldots$$
$$\vdots$$
$$\text{etc.} \ldots .$$

4.3 Interstratified Structures with S = 1

4.3.1 Determination of the Independent Parameters Characterizing Two-Component Structures

In interstratified two-component systems with S = 1, the nature of a layer in the stack depends on the nature of the preceding layer. Hence, if we want to use relationships (4.6) to compute the relative proportions W_{AA}, W_{AB}, W_{BA}, and W_{BB} corresponding to the different types of pairs of layers, we need both the proportions of the A and B layers and the conditional probabilities P_{AA}, P_{AB}, P_{BA}, and P_{BB}. Knowledge of these six parameters is sufficient to yield the relative proportion of any given sequence of n layers. For instance, the abundance of each of the eight possible triplets is given by:

$$W_{AAA} = W_A P_{AA} P_{AA} \qquad W_{AAB} = W_A P_{AA} P_{AB}$$
$$W_{ABA} = W_A P_{AB} P_{BA} \qquad W_{ABB} = W_A P_{AB} P_{BB}$$
$$W_{BAA} = W_B P_{BA} P_{AA} \qquad W_{BAB} = W_B P_{BA} P_{AB}$$
$$W_{BBA} = W_B P_{BB} P_{BA} \qquad W_{BBB} = W_B P_{BB} P_{BB} \; .$$

Furthermore, the six parameters $W_A, W_B, P_{AA}, \ldots P_{BB}$, are not all independent. In order to find the number of independent parameters necessary for full characterization of a two-component interstratified structures with S = 1, we shall determine the relations among them.

Let us first detail the probability of finding an A layer at any given level in the stack. In a two-component system, this layer must form, with the preceding layer, either an AA or a BA pair and the probability of finding this A layer must therefore be the sum of the respective probabilities of meeting an AA pair and a BA pair:

$$W_A = W_{AA} + W_{BA} = W_A P_{AA} + W_B P_{BA} \ . \tag{4.11}$$

Similarly,

$$W_B = W_{AB} + W_{BB} = W_A P_{AB} + W_B P_{BB} \ . \tag{4.12}$$

Writing now Eq. (4-11) as

$$W_A(1 - P_{AA}) = W_B P_{BA}$$

and using Eq. (4.7), we find:

$$W_A P_{AB} = W_B P_{BA} \quad \text{whence} \quad W_{AB} = W_{BA} \ . \tag{4.13}$$

This result – which can also be derived from the relations in (4.12) and (4.8) – shows that, in any two-component system, the number of AB pairs is always equal to the number of BA pairs.

Taking now into account expressions (4.1), (4.7), (4.8) and (4.13), it turns out that for two-component interstratified structures with S = 1, the number of independent parameters that need to be determined is not six but two. It is logical that W_A or W_B be chosen for one of them. Suppose that W_A be the one and that $W_A \geqslant W_B$. From (4.13) it follows that

$$P_{AB} = \frac{W_B}{W_A} P_{BA} = \frac{W_B}{W_A} (1 - P_{BB}) \ . \tag{4.14}$$

It is now convenient to choose, as second independent parameter, one that can vary between 0 and 1, i.e., P_{BA}, or P_{BB}. If P_{BA} be the one chosen, the other four parameters are given by:

$$\begin{aligned} W_B &= 1 - W_A & P_{BB} &= 1 - P_{BA} \\ P_{AB} &= \frac{W_B}{W_A} P_{BA} & P_{AA} &= 1 - P_{AB} \ . \end{aligned} \tag{4.15}$$

Evidently, if $W_B > W_A$, the obvious choice is W_B and P_{AB} (or P_{AA}). In fact, one need only call A the type of layer whose proportion in the stack is greater than 50% for the expressions above to be always applicable.

4.3.2 Classification of Two-Component Structures as a Function of the Degree of Order in the Sequence of Layers

We shall now examine the different degrees of order-disorder which may exist in two-component structures with S = 1.

It is easy to see that, for S = 1, all the possible sequences of layers corresponding to a maximum possible degree of order are obtained when $P_{BB} = 0$ if $W_A \geqslant W_B$ (or $P_{AA} = 0$ if $W_B \geqslant W_A$). In this case, each B layer is between two A layers; this results in a more regular alternation of layers than would be found with a purely random stacking, where $P_{BB} = W_B$. In the following, the former types of configuration will be called *arrangements of layers with maximum possible degree of order* for S = 1, or arrangements of layers with MPDO. Note that, when $0 < P_{BB} < W_B$, the possible configurations will correspond to intermediate degrees of order between MPDO and random layer distribution.

Sato (1965) has suggested that (4.14) be expressed as:

$$P_{BB} = \frac{W_A}{W_B} P_{AA} + 1 - \frac{W_A}{W_B} \tag{4.16}$$

so that all possible two-component structures with S = 1 can be read off a single square diagram having the possible values of P_{AA} as abscissae and P_{BB} as ordinates (Fig. 4.1).

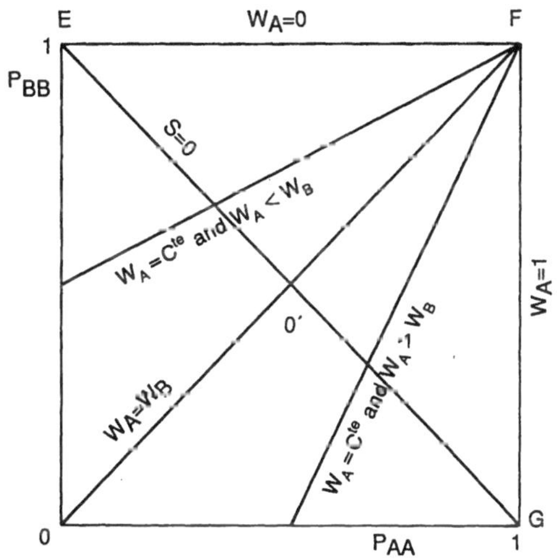

Fig. 4.1. Square diagram for the representation of two-component interstratified structures with S = 1. Each *point* on the diagram corresponds to a structure with a given set of values of W_A, W_B, P_{AA} and P_{BB}. (After Sato 1965)

Expression (4.16) shows that P_{BB} is a linear function of P_{AA} for each value of the ratio W_A/W_B (in other words for each value of W_A), and that this linear variation can be represented, in a diagram, by straight lines passing through the point F, (see Fig. 4.1). The MPDO sequences will be localized on the horizontal (if $W_A \geqslant W_B$ and $P_{BB}=0$) or vertical (if $W_B \geqslant W_A$ and $P_{AA}=0$) axis. It is also clear that a mechanical mixture of A and B layers corresponds to point F of the diagram, at which we have $P_{AA}=1$ and $P_{BB}=1$ simultaneously. Furthermore, all the points along the diagonal EG are such that $P_{AA}+P_{BB}=1$ and therefore that they describe the set of totally disordered (S = 0) two-component structures. Each point on the diagonal corresponds to a possible value of the ratio W_A/W_B. Starting from (4.16) and the fact that $P_{AA}+P_{BB}=1$, it can easily be shown that for each point on this diagonal, $P_{AA}=W_A$ and $P_{BB}=W_B$. Hence, since the conditional probability of A following A is the same as the probability of finding a single A layer, S = 0, and mixed layer two-component systems with S = 0 can be looked upon as a particular case (that of maximum disorder) of systems with S = 1.

Each point of the Sato square diagram represents a particular type of two-component interstratified structure. Those structures with $W_A > W_B$ are within the triangle OGF while those with $W_B > W_A$ are in triangle OEF. In the former case, structures with a different degree of order in the alternation of layers are found within triangle OO′G (in which $0 \leqslant P_{BB} \leqslant W_B$) while those in which the A or B layers tend to be segregated are found within the triangle GO′F ($W_B < P_{BB} \leqslant 1$). Mutatis mutandis, the corresponding areas for $W_A < W_B$ are OO′E ($0 \leqslant P_{AA} \leqslant W_A$) and EO′F ($W_A < P_{AA} \leqslant 1$).

Cesari et al. (1965) have further classified interstratified structures with S = 1 as a function of their degree of order-disorder or of segregation by defining a coefficient of disorder which we shall designate by D_{S1}, such that:

$$D_{S1} = \frac{P_{BB}}{W_B} \tag{4.17}$$

when $W_A > W_B$ and $0 \leqslant P_{BB} \leqslant W_B$. A totally disordered structure ($P_{BB}=W_B$), will have $D_{S1}=1$, while MPDO sequences corresponding to $P_{BB}=0$, will have $D_{S1}=0$ and $0 \leqslant P_{AA} \leqslant W_A$. When $W_B > W_A$ it is convenient to use another coefficient D'_{S1} such that

$$D'_{S1} = \frac{P_{AA}}{W_A} .$$

Referring back to the case when $W_A > W_B$, we can rearrange expression (4.16) to read:

$$W_B = \frac{1-P_{AA}}{2-P_{BB}-P_{AA}} . \tag{4.18}$$

Putting this expression now in Eq. (4.17), we find:

$$P_{AA} = \frac{2P_{BB} - P_{BB}^2 - D_{S1}}{P_{BB} - D_{S1}} , \tag{4.19}$$

which gives the explicit relation between the conditional probabilities P_{AA} and P_{BB} in the different structures having the same degree of disorder in the sequence of layers, i.e., with equal value of D_{S1}. Similarly, when $W_B > W_A$, the conditional probabilities P_{AA} and P_{BB} are linked by:

$$P_{BB} = \frac{2P_{AA} - P_{AA}^2 - D'_{S1}}{P_{AA} - D'_{S1}} \tag{4.20}$$

valid for constant values of D'_{S1}.

Figure 4.2 shows the curves corresponding to the expressions (4.19) and (4.20) for values of D_{S1} or D'_{S1} varying between 0 and 1 by steps of 0.2 units. All the curves (except of course the one corresponding to $D_{S1} = 1$ on the diagonal EG), correspond to mixed-layer systems with S = 1, and the points on each given curve represent structures with equal degree of order-disorder.

The degree of segregation characteristic of a two-component structure has been quantified by Cesari et al. (1965) by a parameter S_{G1} defined as

$$S_{G1} = 1 - \frac{1 - P_{BB}}{1 - W_B} \quad \text{with} \quad W_B < P_{BB} \leqslant 1 . \tag{4.21}$$

The choice of this definition is such that $S_{G1} = 1$ when $P_{BB} = 1$, which corresponds to a mechanical mixture of two phases, each containing a single type of

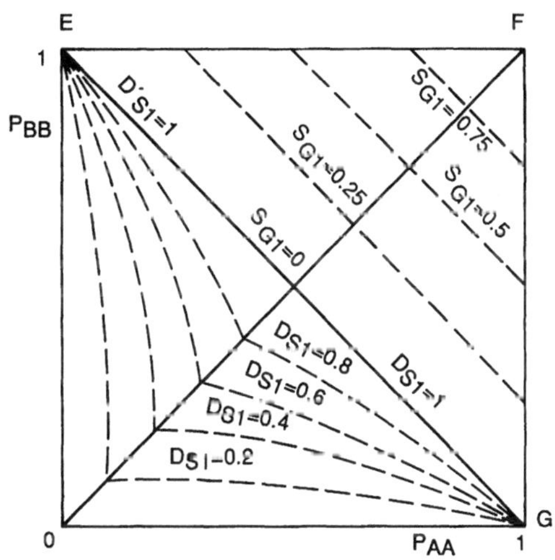

Fig. 4.2. The square diagram of Fig. 4.1 showing the domains corresponding to different degrees of ordering (*below* diagonal EG) and segregation (*above* diagonal EG) of the layers A and B

layer, A or B. Similarly $S_{G1} = 0$ when $P_{BB} = W_B$, namely for totally disordered interstratified structures. Furthermore, if W_B is replaced in (4.21) by its value given by (4.18), we obtain:

$$P_{AA} = S_{G1} - P_{BB} + 1 \ ,$$

which gives the relation between P_{AA} and P_{BB} for different constant values of S_{G1}. This relation is shown in Fig. 4.2 by the dashed lines, each corresponding to equal degrees of segregation found in two-component interstratified structures.

Thus, the square diagram of Fig. 4.2 can be used as a nomogram to compare interstratified systems which differ by their composition and the type of stacking, while still sharing the same interaction parameter $S = 1$.

4.3.3 Interstratified Structures with Three Types of Layers

When an interstratified crystal contains three types of layers, here labeled A, B, and C, with interactions such that $S = 1$, the total number of parameters needed for a full description of the structure is 12, among which the relative proportions are linked by:

$$W_A + W_B + W_C = 1 \ , \tag{4.22}$$

while the nine remaining parameters are the conditional probabilities P_{ij} (with $i, j = A, B, C$) among which the following relationships hold:

$$\begin{aligned} P_{AA} + P_{AB} + P_{AC} &= 1 \\ P_{BA} + P_{BB} + P_{BC} &= 1 \\ P_{CA} + P_{CB} + P_{CC} &= 1 \ . \end{aligned} \tag{4.23}$$

By analogy with the relations in (4.11) and (4.12), we can write:

$$\begin{aligned} W_A &= W_A P_{AA} + W_B P_{BA} + W_C P_{CA} \\ W_B &= W_A P_{AB} + W_B P_{BB} + W_C P_{CB} \\ W_C &= W_A P_{AC} + W_B P_{BC} + W_C P_{CC} \ . \end{aligned} \tag{4.24}$$

Because of relation (4.22), only two of the expressions in (4.24) are independent and the number of independent parameters needed to characterize the three-component structure is, in fact, only six. Furthermore, if the supplementary condition:

$$W_{ij} = W_{ji} \quad \text{or} \quad W_i P_{ij} = W_j P_{ji} \ , \tag{4.25}$$

is respected, we have a structure which was called by Cesari and Allegra (1967) a "symmetrical interstratified structure" and the number of independent parameters is down to five.

We can consider two cases with MPDO:

1. **$W_A > W_B > W_C$ with $W_A \leqslant 0.5$**

Under these conditions, MPDO is obtained when $P_{AA} = P_{BB} = P_{CC} = 0$, and the relations in (4.23) and (4.24) become:

$$P_{AB}+P_{AC}=1 \qquad W_A = W_B P_{BA} + W_C P_{CA}$$
$$P_{BA}+P_{BC}=1 \qquad W_B = W_A P_{AB} + W_C P_{CB}$$
$$P_{CA}+P_{CB}=1 \qquad W_C = W_A P_{AC} + W_B P_{BC} \; .$$

The number of independent parameters is reduced to three. It is convenient to take the relative proportions for two types of layers as independent parameters and P_{CB} for the third. The other parameters are then:

$$P_{AB} = \frac{W_B - W_C P_{CB}}{W_A} \qquad P_{AC} = 1 - P_{AB} \qquad P_{CA} = 1 - P_{CB}$$
$$P_{BA} = \frac{W_A - W_C(1-P_{CB})}{W_B} \qquad P_{BC} = 1 - P_{BA}$$

and the limits for P_{CB}, are found to be 0 and $\left(1 - \frac{W_A - W_B}{W_C}\right)$.

If the structure be such that the relation (4.25) is applicable, all the conditional probabilities P_{ij} (with $i \neq j$) can be expressed as functions of the abundances W_i:

$$W_A P_{AB} = W_B P_{BA} = \frac{W_A + W_B - W_C}{2}$$
$$W_A P_{AC} = W_C P_{CA} = \frac{W_A - W_B + W_C}{2}$$
$$W_B P_{BC} = W_C P_{CB} = \frac{W_B + W_C - W_A}{2} \; .$$

2. **One of the relative proportions, say W_A, is greater than 0.5**

The arrangement with MPDO is obtained when $P_{BB} = P_{CC} = 0$ and when P_{AA} has the smallest possible value, a condition which occurs when $P_{BC} = P_{CB} = 0$, since the presence of adjacent BC or CB pairs increases the number of AA pairs. For this model, the relations in (4.23) and (4.24) work out to be:

$$
\begin{aligned}
P_{AA}+P_{AB}+P_{AC}&=1 \quad & W_A &= W_A P_{AA}+W_B+W_C \\
P_{BA}&=1 \quad & W_B &= W_A P_{AB} \\
P_{CA}&=1 \quad & W_C &= W_A P_{AC} \ ,
\end{aligned}
$$

so that

$$P_{AA}=\frac{W_A-W_B-W_C}{W_A}\ ; \quad P_{AB}=\frac{W_B}{W_A}\ ; \quad P_{AC}=\frac{W_C}{W_A}\ ,$$

and the number of independent parameters is now two.

4.4 Interstratified Structures with S = 2

4.4.1 Relationships Between the Proportions of Different Types of Layers and the Conditional Probabilities

When S = 2, the probability of finding at any point in the stack a layer of a given type depends on that of the two preceding layers. Besides the knowledge of the relative proportions W_i and of the conditional probabilities P_{ij}, full description of a two-component structure with S = 2 requires the knowledge of the conditional probabilities P_{ijk} that a k-type layer will follow an ij-pair. This gives a total of 14 parameters to be determined, but dependences among the parameters decrease this number.

Expressions (4.1), (4.6), (4.7) and (4.8) linking the proportions W_i and the conditional probabilities P_{ij} given in the preceding sections for structures with S = 1 are still valid for S = 2 structures; to them, we can, however, add:

$$
\begin{aligned}
W_{AA} &= W_{AA}P_{AAA}+W_{BA}P_{BAA} \\
W_{AB} &= W_{AA}P_{AAB}+W_{BA}P_{BAB} \\
W_{BA} &= W_{AB}P_{ABA}+W_{BB}P_{BBA} \\
W_{BB} &+ W_{AB}P_{ABB}+W_{BB}P_{BBB}\ .
\end{aligned}
\tag{4.26}
$$

The significance of these expressions can be grasped be examining the first: its right-hand side is composed of two terms which correspond to the probability of finding an A-type layer respectively after an AA or BA pair. Obviously, the sum of these two terms is the proportion of AA pairs within the stack. The homologous three other expressions correspond to the other possible types of pairs.

Expression (4.6), which was used to obtain the relative proportions of pairs composed of different layers when S = 1, can be used for the same purpose when S = 2. Moreover, because of relations (4.3) and (4.13), only two of the expressions in (4.26) are independent. Replacing P_{AAA} and P_{BAB} respectively by $(1-P_{AAB})$ and $(1-P_{BAA})$ in the first two expressions of (4.26) we find:

$$W_{AA}P_{AAB} = W_{BA}P_{BAA} \quad \text{or} \quad W_{AAB} = W_{BAA} \ . \tag{4.27}$$

while replacing P_{ABA} and P_{BBB} by $(1-P_{ABB})$ and $(1-P_{BBA})$ in the last two expressions of (4.26) leads to:

$$W_{BB}P_{BBA} = W_{AB}P_{ABB} \quad \text{or} \quad W_{BBA} = W_{ABB} \ . \tag{4.28}$$

Expressions (4.27) and (4.28) show that in two-component structures with S = 2, the number of AA (or BB) pairs followed by a B (or an A) layer is always equal to the number of these same pairs preceded by a B (or an A) layer. Similarly, it is found that:

$$W_{AAAB} = W_{BAAA} \ ; \qquad W_{AAAAB} = W_{BAAAA} \ ;$$

$$W_{ABBAB} = W_{BABBA} \ ; \quad \text{etc.}$$

Hence, since the 14 parameters required when S = 2 are linked by expressions (4.1), (4.7), (4.8), (4.10), (4.13), (4.27), and (4.28), only four are actually independent. Although we shall indicate, in the next section, the best choice for these independent parameters, let us now summarize the full characterization of two-component systems with S = 2 using the parameters W_i, P_{ij} and P_{ijk}:

- the relative proportions of pairs of layers can be expressed in terms of the conditional probabilities P_{ij} as:

$$W_{ij} = W_i P_{ij} \quad (i,j = A,B) \ ;$$

- the relative proportions of triplets composed of different types of layers are:

$$W_{ijk} = W_{ij}P_{ijk} = W_i P_{ij} P_{ijk} \quad (i,j,k = A,B) \tag{4.29}$$

- the relative proportion of any group of n layers can be expressed using only the proportions W_i and the conditional probabilities attached to each pair for each particular triplet, as:

$$W_{ijkl} = W_{ijk}P_{ijkl} = W_i P_{ij} P_{ijk} P_{ijkl} \ .$$

It follows that a structure with S = 1 can be considered to be a degenerate S = 2 structure in which the conditional probability P_{ijk} would no longer depend on the nature of the ij-pair immediately preceding a k-type layer but only on the presence of the layer j before layer k. In other words, two-component structures with S = 1 are particular cases of S = 2 structures in which:

$$P_{AAA} = P_{BAA} = P_{AA} \qquad P_{ABA} = P_{BBA} = P_{BA}$$

$$P_{AAB} = P_{BAB} = P_{AB} \qquad P_{ABB} = P_{BBB} = P_{BB} \;.$$

4.4.2 Choice of the Independent Parameters

Two of the four independent parameters can be the ones retained for the case S = 1, namely W_A and P_{BA}, corresponding to $W_A \geqslant 0.5$ and $0 \leqslant P_{BA} \leqslant 1$. To obtain the third parameter, let us rewrite the expressions giving W_{ij} as:

$$\begin{aligned} W_{AA} &= W_A P_{AA} = W_A - (1 - W_A) P_{BA} \\ W_{AB} &= W_{BA} = W_A P_{AB} = (1 - W_A) P_{BA} \\ W_{BB} &= W_B P_{BB} = (1 - W_A) P_{BB} \;. \end{aligned} \tag{4.30}$$

When $P_{BA} \leqslant 0.5$ we have, using the last two expressions above:

$$\frac{W_{AB}}{W_{BB}} = \frac{P_{BA}}{1 - P_{BA}} \leqslant 1 \;.$$

In the light of the result given in (4.28), let us now choose, as third independent parameter, P_{ABB}, which is such that $0 \leqslant P_{ABB} \leqslant 1$. We now obtain:

$$P_{BBA} = \frac{W_{AB}}{W_{BB}} P_{ABB} \;; \qquad P_{ABA} = 1 - P_{ABB} \;; \qquad P_{BBB} = 1 - \frac{W_{AB}}{W_{BB}} P_{ABB} \;.$$

– When $P_{BA} \geqslant 0.5$, the last two expressions in (4.30) are used once again to yield

$$\frac{W_{BB}}{W_{AB}} = \frac{1 - P_{BA}}{P_{BA}} \leqslant 1 \;.$$

It is now convenient to choose P_{BBA} for the third independent parameter. As before, we obtain P_{ABB} from (4.28), while P_{BBB} and P_{ABA} are easily calculated from:

$$P_{ABB} = \frac{W_{BB}}{W_{AB}} P_{BBA} \;; \qquad P_{BBB} = 1 - P_{BBA} \;; \qquad P_{ABA} = 1 - \frac{W_{BB}}{W_{AB}} P_{BBA} \;.$$

The last independent parameter can be chosen from among P_{AAA}, P_{BAA}, P_{AAB}, and P_{BAB}. The first two expressions in (4.30) yield:

$$\frac{W_{BA}}{W_{AA}} = \frac{(1-W_A)P_{BA}}{W_A-(1-W_A)P_{BA}}$$

and we shall now determine the conditions associated with $W_{BA}/W_{AA}<1$ or $W_{BA}/W_{AA}>1$.

1. If we require the r.h.s. of this equality to be smaller than 1, we must have:

$$W_A-(1-W_A)P_{BA}>(1-W_A)P_{BA} \ . \tag{4.31}$$

Moreover, since we are considering the case when:

$$0.5 \leqslant W_A \leqslant 1 \quad \text{and} \quad 0 \leqslant P_{BA} \leqslant 1$$

either one of the two sets of inequalities below must be satisfied for the three conditions above to be simultaneously valid:

$$\left[\begin{array}{l} 1/2 \leqslant W_A < 2/3 \\ 0 \leqslant P_{BA} \leqslant \dfrac{W_A}{2(1-W_A)} \end{array}\right. \qquad \left[\begin{array}{l} 2/3 \leqslant W_A \leqslant 1 \\ 0 \leqslant P_{BA} \leqslant 1 \end{array}\right. \tag{4.32}$$

In other words, the ratio W_{BA}/W_{AA} will be less than 1 whenever $W_A>2/3$ and whatever the value of P_{BA}, but for two-component structures with $1/2<W_A<2/3$, the maximum value of the conditional probability P_{BA} is less than 1, except for $W_A=2/3$.

We can now turn back to Eq. (4.27) and take as fourth independent parameter the conditional probability P_{BAA}, whose value will also lie between 0 and 1. The last three (dependent) parameters needed to have full characterization of a two-component structure with S = 2 will therefore be calculated from

$$P_{AAB} = \frac{W_{BA}}{W_{AA}} P_{BAA} \ ; \quad P_{BAB} = 1-P_{BAA} \ ; \quad P_{AAA} = 1-\frac{W_{BA}}{W_{AA}} P_{BAA} \ .$$

2. We now wish to have $W_{BA}/W_{AA}>1$. If we let:

$$1/2 \leqslant W_A \leqslant 2/3 \quad \text{and} \quad W_A/2(1-W_A) \leqslant P_{BA} \leqslant 1 \ ,$$

it is easy to see from the relations (4.30), that $W_{AA}/W_{BA} \leqslant 1$, and one can therefore choose P_{AAB} as fourth independent parameter. The other coefficients follow from

$$P_{BAA} = \frac{W_{AA}}{W_{BA}} P_{AAB} \ ; \quad P_{AAA} = 1-P_{AAB} \ ; \quad P_{BAB} = 1-\frac{W_{AA}}{W_{BA}} P_{AAB} \ .$$

4.4.3 Classification of Structures with S = 2 as a Function of the Degree of Order in the Sequence of Layers

In the particular case described by the conditions $S = 2, W_A > W_B$ and $P_{BB} = 0$, it is possible to use a square diagram analogous to the one proposed by Cesari for $S = 1$. Note however that, under these conditions, the possibility of finding structures with $S = 2$ rapidly decreases as W_A tends towards W_B, and becomes nil when $W_A = W_B$.

With $P_{BB} = 0$, the expressions in (4.27) and (4.30) become:

$$W_{AA} = W_A - W_B \;;\quad W_{AB} = W_{BA} = W_B \;;\quad W_{BB} = 0$$
$$(W_A - W_B) P_{AAB} = W_B P_{BAA} \tag{4.33}$$

The last expression above can be rewritten as

$$(1 - P_{AAA}) = \frac{W_B}{W_A - W_B} (1 - P_{BAB})$$

which can be changed to

$$P_{AAA} = \frac{W_B}{W_A - W_B} P_{BAB} + 1 - \frac{W_B}{W_A - W_B}$$

and, writing now $K_1 = \dfrac{W_B}{W_A - W_B}$, we finally obtain:

$$P_{AAA} = K_1 P_{BAB} + 1 - K_1 \;. \tag{4.34}$$

This expression is similar to the one obtained by Sato and given in (4.16) for $S = 1$ structures, and the corresponding square diagram, much alike that in Fig. 4.1, is shown in Fig. 4.3. The difference lies in the fact that the diagram now only describes those interstratified structures with $S = 2$ which have $W_A > W_B$ and $P_{BB} = 0$ (or, by interchanging the corresponding indices, those with $W_B > W_A$ and $P_{AA} = 0$). The diagram has the probabilities P_{BAB} and P_{AAA} respectively as abscissa and ordinate, and Eq. (4.34) shows that, for any fixed value of W_A, P_{AAA} varies linearly with P_{BAB}. Furthermore, all lines represented by (4.34) must pass through the point (1,1) (point F in the square diagram): thus, $W_A = 2/3$ and $W_A = 1$ correspond respectively to OF and EF, but GF, which should correspond to $W_A = 0.5$, cannot be part of this diagram since the only possible sequence is now ABABAB... owing to the fact that $P_{BB} = 0$. Hence, for this value of W_A, the $S = 2$ structure degenerates to $S = 1$ and Eq. (4.34) is no longer applicable.

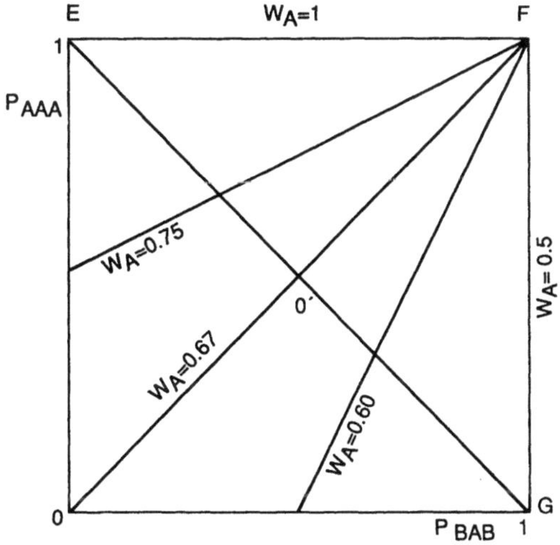

Fig. 4.3. The square diagram describing two-component structures with $S = 2$, $W_A > W_B$ and $P_{BB} = 0$. (After Drits and Sakharov 1976)

Note also that point F of the diagram (where $P_{AAA} = P_{BAB} = 1$) represents total segregation, equivalent to a mechanical mixture of two types of crystals composed respectively of A layers and AB pairs. As for point O, at which $P_{AAA} = P_{BAB} = 0$ and $W_A : W_B = 2:1$, this corresponds to an ordered AABAABAAB ... sequence.

Structures with $S = 2$, MPDO, and different values of W_A, are distributed along the two axes of the square diagram: those with $P_{AAA} = 0$ and $W_A < 2/3$ (along OG) may be described as random arrangements of AB pairs and AAB triplets, while those with $P_{BAB} = 0$ and $W_A > 2/3$ (along OE) as random sequences of AAB triplets and single A layers. Note that structures with $W_A > W_B$ and $P_{BB} = 0$ were classified as arrangements with MPDO when $S = 1$. When $S = 2$, these structures correspond to arrangements with maximum *disorder* and they are distributed along the diagonal EG. For all points along this diagonal, $P_{AAA} + P_{BAB} = 1$, and, replacing P_{BAB} by $(1 - P_{AAA})$ in Eq. (4.34) gives:

$$P_{AAA} - \frac{W_A - W_B}{W_B} .$$

For the structures under consideration, we know furthermore from the relations in (4.33) that:

$$W_A - W_B - W_{AA} = W_A P_{AA}$$

from which:

$$P_{AAA} = P_{AA} ,$$

and, in similar fashion, we can prove that $P_{BAB} = P_{AB}$. These last two relations mean that the probability of finding an AA or AB pair in a triplet does not depend on the nature of the preceding layer.

Lastly, all the points above EG describe S = 2 structures with alternations of zones of A layers and zones of AB pairs with a more or less pronounced tendency to segregation, while points below this diagonal correspond to structures with a varying degree of order in the succession of A layers and AB pairs.

Structures with S = 2, $W_A > W_B$ and $P_{BB} = 0$ can be described by coefficients characterizing the degree of disorder in the alternation of A layers and AB pairs as a function of W_A, with the same approach as that used in the case of S = 1 structure. The two corresponding coefficients, D_{S2} et D'_{S2} are now defined by:

$$D_{S2} = \frac{W_A}{W_B} P_{BAB} \quad \text{when} \quad W_A \geqslant 2/3 \quad \text{and} \quad 0 \leqslant P_{BAB} \leqslant \frac{W_B}{W_A} \; ; \tag{4.35}$$

$$D'_{S2} = \frac{W_A}{W_A - W_B} P_{AAA} \quad \text{when} \quad 1/2 \leqslant W_A \leqslant 2/3 \quad \text{and} \quad 0 \leqslant P_{AAA} \leqslant \frac{W_A - W_B}{W_A} \, . \tag{4.36}$$

Here also, we have $D_{S2} = D'_{S2} = 1$ when $P_{BAB} = P_{AB}$ and $P_{AAA} = P_{AA}$, i.e., when there is maximum disorder for a structure with S = 2 (or a MPDO structure with S = 1, which comes to the same thing). From the definitions, we also find that $D_{S2} = 0$ when $P_{BAB} = 0$, and $D'_{S2} = 0$ when $P_{AAA} = 0$: this corresponds to MPDO in S = 2 structures which then have random sequences of AAB triplets and A layers (for $D_{S2} = 0$), or AAB triplets and AB pairs (for $D'_{S2} = 0$). Hence, these S = 2 structures can also be described as S = 0 structures if one chooses to replace the constitutive individual layers by appropriate combinations of these layers.

Furthermore, we can derive, from Eq. (4.34):

$$\frac{W_B}{W_A} = \frac{1 - P_{AAA}}{2 - P_{AAA} - P_{BAB}} \, , \tag{4.37}$$

which, when substituted in the definition of D_{S2} and D'_{S2}, yields:

$$P_{AAA} = \frac{2P_{BAB} - P_{BAB}^2 - D_{S2}}{P_{BAB} - D_{S2}} \tag{4.38}$$

and

$$P_{BAB} = \frac{2P_{AAA} - P_{AAA}^2 - D'_{S2}}{P_{AAA} - D'_{S2}} \, . \tag{4.39}$$

These two expressions can be used in the graphic representation of the relationship between P_{AAA} and P_{BAB} for different D_{S2} and D'_{S2} values. Structures with the same degree of order in the sequences but different values of W_A (with $W_A > W_B$ and S = 2) are found along each curve within the triangle EOG in Fig. 4.4, in which the value of D_{S2} and D'_{S2} vary by 0.2 units from curve to curve.

It is also possible to describe the tendency to segregation of S = 2 structures by a coefficient analogous to the one used for S = 1 structures and whose definition, in the present case, is

$$S_{G2} = 1 - \frac{1-P_{BAB}}{1-W_B/W_A} \quad \text{with} \quad \frac{W_B}{W_A} \leqslant P_{BAB} \leqslant 1 \ , \tag{4.40}$$

so that

- $S_{G2} = 1$ when $P_{BAB} = 1$, corresponding to total segregation between the ABAB sequences and the stacks exclusively composed of A layers (vertex F of the diagram in Fig. 4.4);
- $S_{G2} = 0$ when $P_{BAB} = W_A/W_B = P_{AB}$, corresponding to total disorder in the sequences of AB pairs and A layers (diagonal EG of the diagram).

Replacing now the ratio W_B/W_A in (4.40) by its equivalent given in Eq. (4.37) yields:

$$P_{BAB} = 1 + S_{G2} - P_{AAA} \ ,$$

which gives us a relation between P_{AAA} and P_{BAB}. This is shown by the dashed straight lines in Fig. 4.4, with S_{G2} as parameter. The whole diagram is therefore a convenient nomogram for the classification of structures with S = 2, $W_A > W_B$ and $P_{BB} = 0$, as a function of their order-disorder relationships.

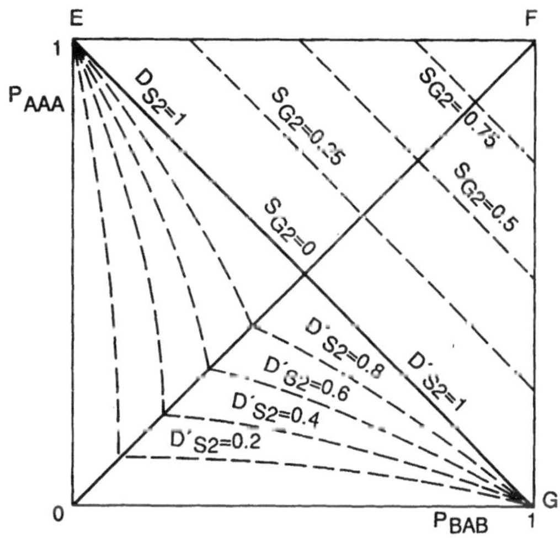

Fig. 4.4. Nomogram for the description of interstratified structures characterized by S = 2, $W_A \geqslant 0.5$ and $P_{BB} = 0$. (After Drits and Sakharov 1976)

4.4.4 Interstratified Structures with S = 2 and g Types of Layers

In the general case of structures containing g types of layers, the characteristic parameters are still the relative proportion of each type of layer and the conditional probabilities of a k-type layer following a given ij pair in the direction perpendicular to the layer planes. The following relationships hold among the parameters:

$$\sum_{i=1}^{g} W_i = 1 \ , \quad \sum_{j=1}^{g} P_{ij} = 1 \qquad \text{for all} \quad i = A, B, C, \ldots$$

$$\sum_{i=1}^{g} W_i P_{ij} = W_j \qquad \text{for all} \quad j = A, B, C, \ldots$$

$$\sum_{k=1}^{g} P_{ijk} = 1 \qquad \text{for all} \quad i, j = A, B, C, \ldots$$

$$\sum_{i=1}^{g} W_{ij} P_{ijk} = W_{jk} \qquad \text{for all} \quad j, k = A, B, C, \ldots .$$

Moreover, the proportions of different sequences of 2, 3, ... layers are linked to the conditional probabilities through:

$$\begin{aligned}
W_{ij} &= W_i P_{ij} && \text{for all} \quad i, j = A, B, C, \ldots \\
W_{ijk} &= W_{ij} P_{ijk} = W_i P_{ij} P_{ijk} && \text{for all} \quad i, j, k = A, B, C, \ldots \\
W_{ijkl} &= W_i P_{ij} P_{ijk} P_{jkl} && \text{for all} \quad i, j, k, l = A, B, C, \ldots \\
&\vdots \\
&\text{etc.} \ldots
\end{aligned}$$

Finally, we must remember that any S = 2 structure in which $P_{ijk} = P_{jk}$ is, in fact, a structure with S = 1.

4.5 Interstratified Structures with S = 3

The characterization of multi-type layer structures in which the layer arrangement is the result of a range of interactions spanning three layers (i.e., in which S = 3) also uses all the conditional probabilities detailed in the description of S = 2 structures, to which are added the conditional probabilities P_{ijkl} (with $i, j, k, l = A, B, C, \ldots$) attached to the possibility of a layer l following an ijk triplet. In the case of two-component structures with S = 3, this leads to a total of 30 parameters, of which 14 are those discussed in connection with the S = 2 structures. The 16 supplementary parameters are the conditional probabilities P_{ijkl} linked by the following relationships:

$$
\begin{aligned}
&P_{AAAA}+P_{AAAB}=1 \quad && P_{BAAA}+P_{BAAB}=1 \\
&P_{AABA}+P_{AABB}=1 \quad && P_{BABA}+P_{BABB}=1 \\
&P_{ABAA}+P_{ABAB}=1 \quad && P_{BBAA}+P_{BBAB}=1 \\
&P_{ABBA}+P_{ABBB}=1 \quad && P_{BBBA}+P_{BBBB}=1 \;.
\end{aligned} \tag{4.41}
$$

By analogy with the relations at (4.26), we also have:

$$
\begin{aligned}
W_{AAA} &= W_{AAA}P_{AAAA}+W_{BAA}P_{BAAA} \\
W_{ABA} &= W_{AAB}P_{AABA}+W_{BAB}P_{BABA} \\
W_{BAA} &= W_{ABA}P_{ABAA}+W_{BBA}P_{BBAA} \\
&\vdots \\
&\text{etc.}
\end{aligned} \tag{4.42}
$$

By combining the appropriate expressions in (4.41) and (4.42), we obtain a number of equalities whose general form is:

$$W_{ijk}P_{ijkl}=W_{lkj}P_{lkji} \;, \tag{4.43}$$

and it can be shown that, due account being taken of (4.41), (4.42), (4.27), and (4.28), only four of the remaining 16 parameters are actually independent. Hence, a two-component S = 3 structure only requires eight independent parameters for full description. This number will be decreased whenever further particular specific constraints exist in the sequence of layers. Thus, assume that for a given system,

$$2/3 \leqslant W_A \leqslant 1 \;; \quad P_{BB}=0 \;; \quad P_{BAB}=0 \;. \tag{4.44}$$

These conditions correspond to MPDO with S = 2. However, since we are dealing with a system in which S = 3, the relations in (4.41) and (4.42) still hold, and we are left with six nonzero conditional probabilities, namely P_{AAAA}, P_{AAAB}, P_{AABA}, P_{ABAA}, P_{BAAA}, and P_{BAAB}. Furthermore, $P_{AABA} = P_{ABAA} = 1$. Also, from (4.43), we know that $W_{AAA}P_{AAAB} = W_{BAA}P_{BAAA}$, and, using (4.29), we find

$$\frac{W_{BAA}}{W_{AAA}}=\frac{W_B}{W_A-2W_B} \;,$$

from which we conclude that $W_{BAA}/W_{AAA} \leqslant 1$ if $3/4 \leqslant W_A \leqslant 1$. If we now choose P_{BAAA} as one of the independent parameters, then:

$$P_{AAAB}=\frac{W_BP_{BAAA}}{(W_A-2W_B)} \;; \quad P_{AAAA}=1-\frac{W_BP_{BAAA}}{(W_A-2W_B)} \quad \text{and} \quad P_{BAAB}=1-P_{BAAA}.$$

On the contrary, if $2/3 \leqslant W_A \leqslant 3/4$, then $W_{AAA}/W_{BAA} \leqslant 1$ and it is more convenient to take, P_{AAAB} as independent parameter.

Hence, the complete description of S = 3 structures constrained by the conditions defined in (4.44) requires only one extra third-neighbor interaction parameter: P_{BAAA} or P_{AAAB}.

In the general case of S = 3 structures containing g types of different layers, we write:

$$\sum_{l=1}^{g} P_{ijkl} = 1 \quad \text{and} \quad W_{jkl} = \sum_{i=1}^{g} W_{ijk} P_{ijkl} \quad \text{for all} \quad i, j, k, l = A, B, C, \ldots$$

but, in general, $W_{ijkl} = W_{lkji}$ is no longer true.

It is obvious that the study of interstratified structures with $S > 3$ can be done using the same approach as the one detailed in the section above. However, full description of these structures becomes quite arduous owing to the increase in the number of independent parameters required, and to the difficulty in the choices of those leading to the simplest expressions.

4.6 Degree of Homogeneity for Powders of Thin Particles with Markovian Interstratification (Quasi-Homogeneous System)

All the relationships obtained in the preceding sections involving the relative proportions of the components and the conditional probabilities of different sequences were established assuming Markovian distributions in "quasi-homogeneous" powders, as defined in Section 1.7.2 and Section 4.1. Inherent in the choice of this model is the consequence, in the case of powders consisting of very thin particles, that the relationships among the relative proportions and the conditional probabilities of the different types of layers are always valid, for each particle and whatever is size, i.e., that they can simultaneously characterize a large number of particles *which may differ not only in the sequence of layer types but also in the respective proportion of each type of layer.* In other words, it means that the powder is inhomogeneous as regards the composition of the particles associated with the type of layers which they contain. The degree of inhomogeneity of such a powder depends on a number of factors and it is this problem which will now be examined.

Let us first consider a thick interstratified crystal. It will be taken as being **statistically homogeneous** if the entire sequence of layers in it follows a single statistical law and if its description requires only a small number of independent parameters for a given value of S.

Suppose now that such a crystal be cleaved into a large number of thin particles, each containing the same number M of layers. The set of thin particles can then be described by the same group of independent probability parameters as were needed for the description of the thick crystal, but the particles in the set may differ in the proportions and in the mutual arrangement of the different types of layers. The following example illustrates this situation (Drits 1987).

Let a thick crystal contain only two types of layers, A and B, with $W_A:W_B = 0.6:0.4$, and let $S = 0$. We now cleave this crystal into particles containing five layers; the composition of each particle will be one among the following possibilities: (5A), (4A+B), (3A+2B), (2A+3B), (A+4B), (5B). The proportion of particles in the set of particles with M_A layers of type A and M_B B layers is:

$$W_{M_A,M_B} = \frac{(W_A)^{M_A}(W_B)^{M_B}(M_A+M_B)!}{M_A!\,M_B!} \quad . \tag{4.45}$$

where $(W_i)^{M_i}$ is the proportion W_i of i-type layers raised to the power M_i. This expression shows that the set will contain, along particles which differ by the proportions and mode of interstratification of A and B layers, particles which contain only A layers ($W_{5A} \approx 0.08$) or B layers ($W_{5B} \approx 0.01$). It is obvious that the proportion W_{M_A,M_B} will be maximum when, for a given value of M, the ratio $M_A:M_B$ is equal to the ratio of the proportions of A and B layers in the thick crystal. In the present case, particles with the composition (3A+2B) are most abundant, with $W_{3A,2B} \approx 0.35$. As the number M of layers in each particle increases, the composition of the particles becomes more and more homogeneous and tends towards the overall value of the A and B proportions ($W_A:W_B = 0.6:0.4$) in the thick crystal. Thus, the proportion of particles in which $W_A:W_B$ lies between 0.7:0.3 and 0.5:0.5 changes from 67 to 75 and 87% as the number of layers in the particles changes from 10 to 20 or 30.

In Fig. 4.5, the histograms illustrate the modifications in the degree of heterogeneity of composition of the particles as M, S, and the $W_A:W_B$ ratio are varied. It is shown, for instance, that with $S = 0$ and $M = 5$, changing W_A from 0.6 to 0.8 induces an increase to 32% of the proportion of particles containing only A layers. On the other hand, with $S = 1$ and $W_A:W_B = 0.6:0.4$, and exclusion of BB pairs ($P_{BB} = 0$; $W_{BB} = 0$), the system of five layer particles is much more homogeneous, since the proportion of (3A+2B) particles is now 65%. Lastly, it can be seen that heterogeneity is not much decreased when the number of layers changes from five to ten. For instance, when $M = 10, S = 0$ and $W_A:W_B = 0.6:0.4$, the total number of particles in which the ratio of the number of A layers to the number of B layers lies between 8:2 and 4:6 is 90%, while it is still 84% for particles with $M = 5$. However, the powder is more homogeneous when $S = 1$ and $W_{BB} = 0$, mainly because of the absence of particles with composition incompatible with the condition $W_{BB} = 0$.

In conclusion, it is important to remember that the degree of heterogeneity of composition of the particles in the powder corresponds to the heterogeneity of the interferential coherence domains determined by X-ray diffraction when the thickness of these domains is of the order of magnitude of the particle thickness.

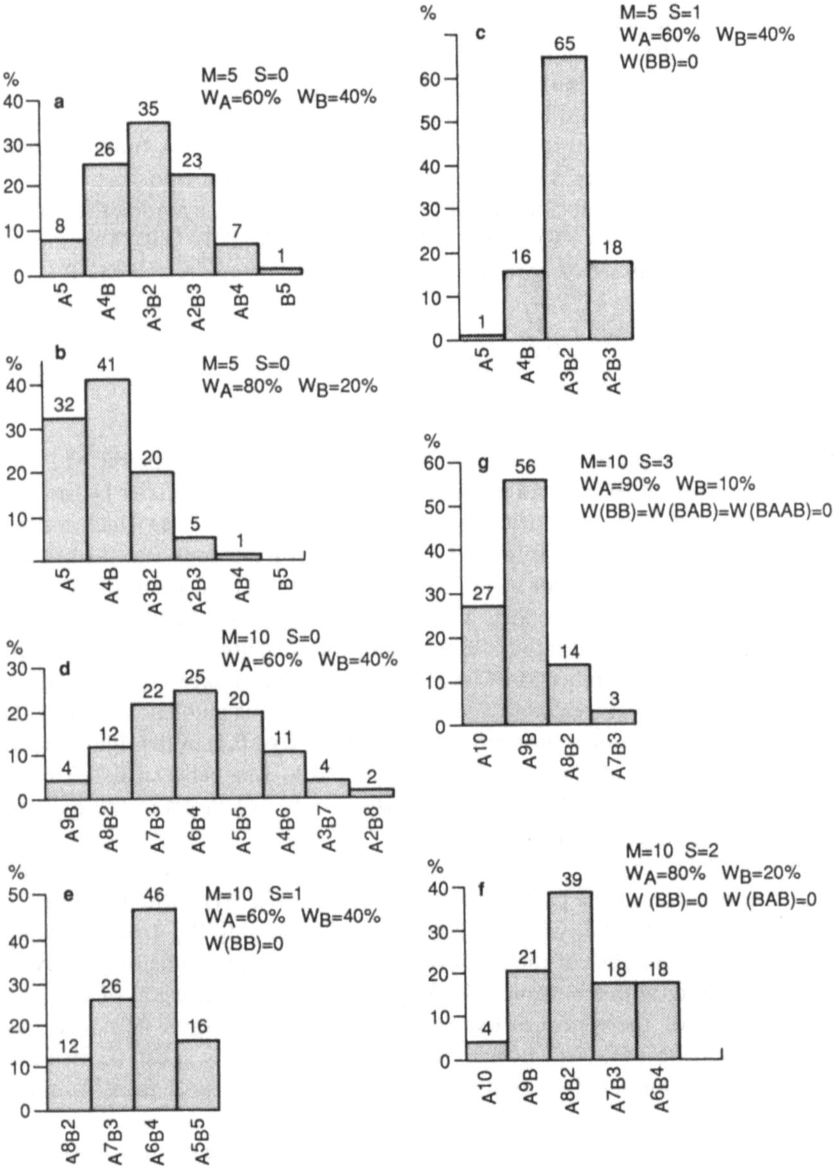

Fig. 4.5. Histograms showing the proportion of particles with a specific composition as a function of the number of layers in the stack, of the $W_A : W_B$ ratio in the powder and of S, the range of interaction factor. (After Drits 1987)

As can be judged by the results of powder diffractometry, many interstratified minerals show coherence domains which correspond to an average thickness of 10 to 15 layers, but this is not sufficient to determine the degree of heterogeneity of the particles themselves. If the mean number of layers in the particles is much larger than the number corresponding to the coherence domains, the degree of heterogeneity of composition of particles is quite low.

4.7 Parameters for the Characterization of Homogeneous Interstratified Systems

Up to this point, the stackings which were described and studied corresponded to sets of "quasi-homogeneous" particles. We shall now detail homogeneous powders in which the proportion of any given type of layer is the same in all the particles, which only differ in the mutual arrangement of the layers (Plançon et al. 1983; Drits et al. 1984).

Let us consider, as general case, a homogeneous powder in which each particle contains a total of M layers of g different types (M_1 layers of type A, M_2 of B, ..., with $M_1+M_2+\ldots+M_g=M$). The proportion W_i of i-type layers is the same in the particles and in the powder. Since the only possible difference in the particles lies in the possible presence of differences in the arrangement of the layers, let E_M be the total number of possible nonidentical arrangements. If the factor S is 0, then

$$E_M = \frac{(M_1+M_2+\ldots+M_g)!}{M_1!M_2!\ldots M_g!} . \tag{4.46}$$

We can note right away that the number of possible arrangements is necessarily smaller in a homogeneous than in a quasi-homogeneous system, since the former excludes, for instance, particles in which the proportion of i-type layers is different from W_i, whereas this possibility can exist in quasi-homogeneous systems.

Consider now the different possible substacks composed of n layers ($n=2,3,\ldots,M-1$) which can be found in the E_M arrangements. In contrast with the latter, the substacks may differ in composition as well as in sequence of layer types (the number of n layers of a given type may be different in two substacks with the same total number of layers). In any particle, the number of different substacks with n layers is $(M-n+1)$ and their total number present in the E_M arrangements in $E_M(M-n+1)$.

We can classify the substacks in categories corresponding to the different compositions, each characterized by its subscript k. Let n_c be the total number of categories with different compositions in the set of substacks containing n layers: k will belong to these categories and we must have $1 \leqslant k \leqslant n_c$ and

$n = n_{1k} + n_{2k} + \dots n_{gk}$, in which $n_{1k}, n_{2k}, \dots n_{gk}$ are the number of A, B, C, ... layers in all the substacks with the composition k. The n_{ik} are such that

$$n_{1k} \leqslant M_1 \; ; \quad n_{2k} \leqslant M_2 \; ; \quad \dots n_{gk} \leqslant M_g \; .$$

The substacks with a given composition k only differ in their sequence of layers. Let a_k^n be one of these sequences and A_k^n the number of possible types of sequences. If $S = 0$, then

$$A_k^n = \frac{(n_{1k} + n_{2k} + \dots + n_{gk})!}{n_{1k}!\, n_{2k}! \dots n_{gk}!} \; . \tag{4.47}$$

For the set E_M of possible arrangements, the total number a_k^n of a given type of sequences (characterized by a specific composition and a specific sequence of layers) is $(a_k^n)_M$, given by

$$(a_k^n)_M = \frac{(M - n + 1)!}{(M_1 - n_{1k})!\,(M_2 - n_{2k})! \dots (M_g - n_{gk})!} \; , \tag{4.48}$$

whereas the total number of different types of sequences with a given composition and number of layers is $A_k^n (a_k^n)_M$.

We shall also let $(A_k^n)_{ij}$ be the group of substacks having an i-type first layer and a j-type last layer, among the set A_k^n of substacks having n layers and composition k. Since there are g types of layers, the set A_k^n will contain g^2 groups $(A_k^n)_{ij}$ differing by the nature of their first and last layer. The relative proportion of each type of subset $(A_k^n)_{ij}$ in each group A_k^n is:

$$f_{(A_k^n)_{ij}} = \frac{n_{ik} n_{jk}}{n(n-1)} \quad \text{if} \quad i \neq j \tag{4.49}$$

and

$$f_{(A_k^n)_{ij}} = \frac{n_{ik}(n_{ik} - 1)}{n(n-1)} \quad \text{if} \quad i = j \; . \tag{4.50}$$

We also have:

$$\sum_{i=1}^{g} \sum_{j=1}^{g} \frac{n_{ik} n_{jk}}{n(n-1)} + \sum_{i=1}^{g} \frac{n_{ik}(n_{ik} - 1)}{n(n-1)} = 1 \; .$$

For the set of substacks with n layers, the proportion of those with a given composition k is:

$$W_{A_k^n} = (C^{n_{1k}})_{M_1} \cdot (C^{n_{2k}})_{M_2} \cdot \dots \cdot (C^{n_{gk}})_{M_g} / (C^n)_M \; , \tag{4.51}$$

in which

$$(C^{n_{ik}})_{M_i} = \frac{M_i!}{n_{ik}!\,(M_i - n_{ik})!} \ ,$$

and this proportion is such that

$$\sum_{k=1}^{n_C} W_{A_k^n} = 1 \ .$$

By taking now into account the expressions given in Eqs. (4.46), (4.47) and (4.48), we modify (4.51) to read

$$W_{A_k^n} = \frac{A_k^n}{E_M} \cdot \frac{(M-n)!}{(M_1 - n_{1k})!\,(M_2 - n_{2k})! \,\ldots\, (M_g - n_{gk})!} = \frac{A_k^n \cdot (a_k^n)_M}{E_M \cdot (M-n+1)} \ . \tag{4.52}$$

Similarly, the proportion in the powder of subsets $(A_k^n)_{ij}$ with fixed i and j is

$$W_{(A_k^n)_{ij}} = W_{A_k^n} \cdot f_{(A_k^n)_{ij}} \tag{4.53}$$

Summing now all the $W_{(A_k^n)_{ij}}$ terms over k, we obtain the proportion of substacks containing n layers, with a first layer i and a last layer j and for all possible compositions k, namely

$$W_{(A^n)_{ij}} = \sum_{k=1}^{n_c} W_{(A_k^n)_{ij}} \tag{4.54}$$

with

$$\sum_{i=1}^{N} \sum_{j=1}^{N} W_{(A^n)_{ij}} = 1 \ .$$

4.7.1 Homogeneous Two-Component (A and B) Systems with S = 0

Before applying the expressions obtained in the preceding sections to the general case of two-component systems, we shall illustrate the meaning of the different definitions in Section 4.7.1 by considering a homogeneous system composed of particles containing ten layers of which eight are type A and two are type B ($M_A = 8$ and $M_B = 2$).

In such a system, there are 45 possible arrangements of the ten layers, or $E_{10} - 45$. Within this number, there are only three possible compositions for substacks of four layers, namely: (4A), (3A+B) and (2A+2B), since the other possible substacks of this group are excluded because in each particle $M_B = 2$. These three compositions will be characterized by the subscript k = 1, 2 or 3, respectively. For any given k, the number of possible sequences is A_k^n: thus, for

$k = 3$, corresponding to the composition (2A+2B) of the substacks, the number of possible sequences in a direction perpendicular to the plane of the layers is $A_3^4 = 6$, namely AABB, BBAA, ABAB, BABA, ABBA und BAAB. In the set E_{10} of possible stackings for the particles, the total number of substacks corresponding to a given sequence for a particular k (such as ABAB for $k = 3$) is $(a_k^n)_M$; in the present case, $(a_3^4)_{10} = 7$. In the E_{10} stackings the total number of substacks with four layers having the composition (2A+2B) is $A_3^4(a_3^4) = 42$. Lastly, the relative proportion of four-layer stacks beginning with an A layer and ending with a B layer is $f_{(A_3^4)_{AB}} = 1/3$ (the sequences are AABB and ABAB), while the relative abundance of four-layer stacks starting and ending with A layers is $f_{(A_3^4)_{AA}} = 1/6$, the only possible sequence here being ABBA.

Let us now return to the general case of a homogeneous powder of particles containing M layers of which M_A are of type A and M_B of type B. In such a system, the proportion of substacks in the powder containing n layers with a specific k is given by a simpler form of Eq. (4.51)

$$W_{A_k^n} = \frac{M_A!}{n_{Ak}!\,(M_A - n_{Ak})!} \cdot \frac{M_B!}{n_{Bk}!\,(M_B - n_{Bk})!} \cdot \frac{n!\,(M-n)!}{M!} \,. \qquad (4.55)$$

It is interesting to compare this expression with the one giving the proportion of this category of substacks in a quasi-homogeneous system with $S = 0$, as read in (4.45)

$$W_{n_{Ak}, n_{Bk}} = \frac{(W_A)^{n_{Ak}} \cdot (W_b)^{n_{Bk}} \cdot (n_{Ak} + n_{Bk})!}{n_{Ak}!\, n_{Bk}!} \,.$$

If both types of powder are composed of particles containing nine layers, with a ratio $W_A : W_B = 2/3 : 1/3$, the proportion of (3A+B) substacks with four layers are 0.476 and 0.395 respectively in the homogeneous and the quasi-homogeneous powders (Drits et al. 1984).

In the general case of the homogeneous powder, the proportion of substacks with a given composition k and a given number n of layers, with the first type A and the last type B, is from (4.49), (4.51) and (4.53):

$$W_{(A_k^n)_{AB}} = \frac{M_A!}{(n_{Ak} - 1)!\,(M_A - n_{Ak})!} \cdot \frac{M_B!}{(n_{Bk} - 1)!\,(M_B - n_{Bk})!} \cdot \frac{(n-2)!\,(M-n)!}{M!} \qquad (4.56)$$

while the proportion of substacks with the first and last layers both of type A is

$$W_{(A_k^n)_{AA}} = \frac{M_A!}{(n_{Ak} - 2)!\,(M_A - n_{Ak})!} \cdot \frac{M_B!}{n_{Bk}!\,(M_B - n_{Bk})!} \cdot \frac{(n-2)!\,(M-n)!}{M!} \qquad (4.57)$$

4.7.2 Homogeneous Two-Component System with $S \neq 0$ and Restrictive Conditions for the Sequence of Layers

We now consider a homogeneous system such that all particles contain M_A layers of type A and M_B layers of type B with $M_A > M_B$ (i.e., $W_A > W_B$), with restrictive conditions on the possible sequences of layers.

1. We shall first suppose that BB pairs are excluded ($P_{BB} = 0$). For quasi-homogeneous systems, this corresponds to $S = 1$ structures with MPDO in the arrangement of the layers. It has already been shown that these structures can be described as totally random sequences of two new components "a" and "b" with "a" = A and "b" = AB, and $S = 0$. By analogy, we shall also describe the homogeneous system with $W_A > W_B$, $P_{BB} = 0$, and $S = 1$, as a system made up of particles containing M' elements such that M_a are A layers and M_b are AB pairs. We then have

$$M_a = M_A - M_B \;;\quad M_b = M_B \;;\quad M' = M_a + M_b = M_A \;.$$

In this case, the substacks of n layers (A and B) can now be described as substacks of n' elements of type "a" and "b", with different compositions k, and

$$n' = n_{ak} + n_{bk} \;,$$

with

$$n_{ak} = n_{Ak} - n_{Bk} \quad \text{and} \quad n_{bk} = n_{Bk}$$

whence

$$n' = n_{Ak} \;.$$

The expression previously given in Eq. (4.49) and describing the proportion of substacks with n-layers of type A and B must be transformed to suit the description of the substacks containing n' elements of type "a" and "b".

$$W_{A_k^{n'}} = \frac{(M_A - M_B)!}{(n_{Ak} - n_{Bk})!\,(M_A - M_B - n_{Ak} + n_{Bk})!} \cdot \frac{M_B!}{n_{Bk}!\,(M_B - n_{Bk})!} \cdot \frac{n_{Ak}!\,(M_A - n_{Ak})!}{M_A!} \;. \tag{4.58}$$

If we examine once again the example given at the end of the preceding section of a homogeneous system of particles containing nine layers with $W_A : W_B = 2/3 : 1/3$ now subject to the restriction $P_{BB} = 0$, the substacks with four layers (3A+B) will become, in the new representation, substacks of three elements and composition "2a+b". The proportion of these "2a+b" substacks is now 0.45, as given by (4.58). This must obviously also be the proportion of (3A+B) substacks.

2. Let us now suppose that the particles in the homogeneous powder are such that $W_A > W_B$, $P_{BB} = 0$, and that the sequence BAB is also excluded ($P_{BAB} = 0$). The result of this new restriction is now that $S = 2$ and that it is those compositions

with $W_A > 2/3$ which show maximum possible order in the arrangement of the layers. In this case also, the homogeneous system can be described as a totally random stacking of new constitutive elements "a" ("a" = A) and "b" ("b" = AAB), with an interaction range factor S = 0, and

$$M_a = M_A - 2M_B \; ; \quad M_n = M_B \; ; \quad M' = M_A - M_B \; .$$

Similarly, in each substack, the number of elements n′ with a given composition k would be given by

$$n' = n_{ak} + n_{bk} \quad \text{with} \quad n_{ak} = n_{Ak} - 2n_{Bk} \quad \text{and} \quad n_{bk} = n_{Bk}$$

so that the proportion of substacks of composition (a, b) is now

$$W_{A_k^{n'}} = \frac{(M_A - 2M_B)!}{(n_{Ak} - 2n_{Bk})!\,(M_A - 2M_B - n_{ak} + 2n_{Bk})!} \cdot \frac{M_B!}{n_{Bk}!\,(M_B - n_{Bk})!} \cdot \frac{(n_{Ak} - n_{Bk})!\,(M_A - M_B - n_{Ak} + n_{Bk})!}{(M_A - M_B)!} \; . \tag{4.59}$$

References

Cesari M, Allegra G (1967) The intensity of X-rays diffracted by monodimensionally disordered structures. Acta Crystallogr 23:200–205

Cesari M, Morelli GL, Favretto L (1965) The determination of the type of stacking in mixed-layer clay minerals. Acta Crystallogr 18:189–196

Drits VA (1987) Mixed layer minerals: diffraction methods and structural features. Proc Int Clay Conf, Denver, vol 1, pp 33–45

Drits VA, Sakharov BA (1976) X-ray structure analysis of interstratified minerals. Nauka, Moscow, 276 pp (in Russian)

Drits VA, Sakharov BA, Plançon A, Ben Brahim J (1984) The distribution of layers in mixed layer crystals of identical composition. Kristallografia 29:350–355 (in Russian)

Hendricks SB, Teller E (1942) X-ray interference in partially ordered layer lattices. J Chem Phys 10:147–167

Jagodzinski H (1949a) Eindimensionale Fehlordnung in Kristallen und ihr Einfluß auf die Röntgeninterferenzen. I. Berechnung des Fehlordnungsgrades aus den Röntgenintensitäten. Acta Crystallogr 2:201–207

Jagodzinski H (1949b) Eindimensionale Fehlordnung in Kristallen und ihr Einfluß auf die Röntgeninterferenzen. II. Berechnung der fehlgeordneten dichtesten Kugelpackungen mit Wechselwirkungen der Reichweite 3. Acta Crytallogr 2:208–214

Jagodzinski H (1949c) Eindimensionale Fehlordnung in Kristallen und ihr Einfluß auf die Röntgeninterferenzen. III. Vergleich der Berechnungen mit experimentellen Ergebnissen. Acta Crystallogr 2:228–304

Jagodzinski H (1954) Der Symmetrieeinfluß auf den allgemeinen Lösungsansatz eindimensionaler Fehlordnungsprobleme. Acta Crystallogr 7:17–25

Kakinoki J, Komura Y (1952) Intensity of X-ray diffraction by one-dimensionally disordered crystal. I. General derivation in the case of the "Reichweite" S = 0 and 1. J Phys Soc Jpn 7:30–35

Kakinoki J, Komura Y (1954) Intensity of X-ray diffraction by one-dimensionally disordered crystal. II. General derivation in the case of the correlation range S ≥ 2. J Phys Soc Jpn 9:169–176

Mac Ewan DMC (1958) Fourier transform methods for studying X-ray scattering from lamellar systems. II. The calculation of X-ray diffraction effects for various types of interstratifications. Kolloid Z 156:61–67

Méring J (1950) Les réflexions des rayons X par les minéraux argileux interstratifiés. Trans 4th Int Congr Soil Science, Amsterdam, vol 1, pp 21–26

Plançon A, Drits VA, Sakharov BA, Gilan ZI, Ben Brahim J (1983) Powder diffraction by layered minerals containing different types of layers and/or stacking defects. J Appl Crystallogr 16:62–64

Reynolds R (1967) Interstratified clay systems: calculation of the total one-dimensional diffraction. Am Mineral 52:661–672

Sato M (1965) Structure of interstratified (mixed-layer) minerals. Nature (London) 208:70–71

Sato M (1969) Interstratified structure with Reichweite g = 2 and its X-ray diffraction patterns. Proc Int Clay Conf, Tokyo, vol 1, pp 207–214

Wilson ALC (1942) Imperfections in the structure of cobalt. II. Mathematical treatment of proposed structure. Proc R Soc London Ser A 180:277–285

CHAPTER 5

Diffraction Methods Adapted to the Structural Analysis of Interstratified Systems

Just like lamellar crystals with stacking faults, irregularly interstratified systems also yield powder diagrams showing important departures from the diffractograms of perfect structures: some of the *hkl* reflections may be broadened, transformed into bands, or show modulations, and their intensities may be weakened. Consequently, determination of d-spacings from such diagrams is meaningless since the values obtained would not correspond to a physical reality.

An important characteristic of the diagrams given by powdered irregular interstratified systems is that the basal reflections constitute an irrational (nonintegral) series, this again being physically unacceptable. One of the main problems associated with the structural analysis of interstratified minerals is therefore the determination from the study of the basal reflections of the nature and number of types of layers present, their proportions and the pattern in the sequence of layer types.

The specific methods which have been elaborated to solve this problem can be divided into two groups:

- in the first, the parameters of the interstratified structure are obtained by the exploitation of the results obtained by a Fourier transformation performed on the diffractometric data giving the intensities and the apparent d-spacings of the reflections (Mac Ewan 1956; D'yakonov 1961, 1962, 1978);
- the second group uses an indirect approach: a model is first drawn up for the structure, and its calculated "theoretical" diffractogram compared to the experimental data (Kakinoki and Komura 1952, 1954a, b, 1965; Mac Ewan 1958; Reynolds 1967, 1980; Cradwick 1975; Weir and Rayner 1976; Drits and Sakharov 1976 etc. ...).

Within this last group, two tendencies are found:

- in the first, the standard expressions and formulas used in the classical methods for triperiodic systems – such as the interference function, etc. – are modified and adapted to the problem on hand (Mac Ewan 1958; Reynolds 1967, 1980; Tettenhorst and Grim 1975a, b).
- in the second, a matrix formalism is used (Hendricks and Teller 1942; Kakinoki and Komura 1952; Sakharov et al. 1982a, etc. ...).

As explained in Chapter 3, the matrix formalism is convenient as it does away with the necessity of repeating tedious calculations when the need arises to compute the intensity of the beam diffracted by substacks of (n+2) layers if the result is known for substacks of (n+1) layers, since all that is needed is multiplying the matrix $[Q]^n$ by the matrix [Q] (cf. Sect. 3.1.2). Another advantage is the possibility of describing the intensity of the diffracted beam at any point in reciprocal space rather than only along the (00) rod, as in the method suggested by Mac Ewan (1958), Reynolds (1967, 1980) and others.

The present chapter is divided into four parts:

- we shall first discuss the advantages and limitations of the direct methods of structural analysis applied to interstratified systems;
- the second part is a description of the indirect methods of analysis based on modifications of the standard methods of crystallography, as applied to basal reflections;
- the third and fourth part, most important of this chapter, give the matrix formalism needed to describe the diffraction phenomena in all of reciprocal space, for interstratified systems with more than one type of layer, or with a single type of layer with correlated faults in the stack (corresponding to $R \geqslant 1$); the formalism will be developed for structural models with increasing complexity in order to arrive at the final description of the general case in which the interstratified system contains any number of different components, with no restriction on the values of S or of R.

5.1 Direct Methods of Structural Analysis

The first direct method of structural analysis of mixed-layer minerals was proposed by Mac Ewan (1956). In this approach, it was supposed that all the layer types have a single, common structure factor (i.e., scattering power) assumed to be known.

D'yakonov (1962) proposed a method free from this limitation and allowing the evaluation of the main probability parameters characteristic of the interstratified structure, with no need for more than a good tracing of the basal reflections. This method will therefore be detailed before being compared to that of Mac Ewan.

5.1.1 The Method of D'yakonov

Let $\varrho(z')$ be the projection of the electronic density of an interstratified crystal at the ordinate z' upon an axis perpendicular to the (001) planes. The amplitude of the wave diffracted by the crystal along the Z axis of the (00) rod is

$$\Phi(Z) = \int \varrho(z') \exp(-2\pi i Z z') dz'$$

and the distribution of the intensity along this axis is given by

$$i_{00}(Z) = \Phi(Z)\Phi^*(Z) = \int \varrho(z')\exp(-2\pi i Z z')dz' \int \varrho(z'')\exp(+2\pi i Z z'')dz''$$

or

$$i_{00}(Z) = \int\int \varrho(z')\varrho(z'')\exp[-2\pi i Z(z'-z'')]\,dz'dz'' \qquad (5.1)$$

in which the variables z′ and z″ cannot both be simultaneously zero.

Note that the integration with respect to z′ is connected with the variation of z′ along the normal to the crystal for a given fixed value of z″. Similarly, the integration with respect to z″ corresponds to the variation of z″ for a fixed given value of z′. The double integration in Eq. (5.1) can however be performed differently.

We can first change simultaneously z′ and z″, keeping their difference z constant, then change the value of z and integrate a second time over z. Mathematically, this is equivalent to writing $z'' = z'+z$ and $dz'' = dz$. Hence Eq. (5.1) becomes

$$i_{00}(Z) = \int_{-\infty}^{+\infty} \varphi'(z)\exp(+2\pi i Z z)\,dz \qquad (5.2)$$

in which

$$\varphi'(z) = \int \varrho(z')\varrho(z+z')\,dz' \;. \qquad (5.3)$$

Although the limits of the integrals in Eq. (5.1) correspond to the boundaries of the crystal, they can be changed to $-\infty$ and $+\infty$ since the electronic density ϱ, and therefore also φ′, both vanish beyond the physical limits of the crystal.

As shown by D'yakonov (1961, 1962), the Fourier transform of Eq. (5.2) is

$$\varphi'(z) = \frac{1}{\pi}\int_0^{\infty} i_{00}(Z)\cos(2\pi Z z)\,dZ \qquad (5.4)$$

and, in the light of Eq. (5.3), it corresponds to the self-convolution of the projection of the crystal electronic density on the axis perpendicular to the (001) plane. The expression for φ′(z) in Eq. (5.4) is therefore equivalent to a monodimensional Patterson function such that its value at each point z is proportional to the product of the electronic charge densities in planes separated by the distance z.

It is moreover known (James 1950; Burger 1959; Guinier 1964) that the monodimensional Patterson function for a lamellar crystal with a single type of layer presents principal and secondary maxima. The principal maxima appear whenever the product $\varrho(z')\varrho(z+z')$ corresponds to two points separated by a value of z equal to a multiple of the period of the function ϱ. The principal peaks in φ′(z) therefore correspond to zero peaks in the Patterson function. The positions and the heights of the secondary maxima depend on the internal structure of the layers.

For mixed-layer crystals, φ′(z) will contain principal peaks if z is equal to the distance between identical layers. For example, for interstratified illite-smectite formed by layers A ($d_A = 10$ Å) and B($d_B = 18$ Å), $\varphi'(z)$ will contain principal peaks at z values 10 Å and 18 Å, as well as at $z = m d_A + n d_B$ corresponding to combinations of mA and nB layers (m and n integers).

The interpretation of $\varphi'(z)$ for mixed-layer structures is complicated by the presence of the secondary peaks that may be superposed on the principal peaks. D'yakonov (1977) proposed a method for the quantitative treatment of $\varphi'(z)$ that diminishes the influence of these secondary peaks.

5.1.2 Computation of the Function $\varphi'(z)$

By combining Eqs. (2.11) and (2.43), we find that $I_{00}(Z)$ is proportional to the intensity $i_{00}(Z)$ of the basal reflections in a powder diagram, and we shall use the latter in our development.

For practical purposes, the integration in Eq. (5.4) is replaced by a summation. Depending on the width and shape of the basal reflections of the diagram, the summation will take different forms (D'yakonov 1977).

1. When the basal reflections are well separated maxima at $Z_1, Z_2, \ldots, Z_k \ldots Z_n$ with individual widths $\Delta Z_k (1 \leqslant k \leqslant n)$, the computation of $\varphi'(z)$ is performed by writing

$$\varphi'(z) = \sum_{k=1}^{n} \int_{Z_k - \Delta Z_k/2}^{Z_k + \Delta Z_k/2} i_{00}(Z) \cos(2\pi Z z) \, dZ \ . \tag{5.5}$$

When the interval of integration is sufficiently small, the integral of the product of a function by a cosine term is equal to the product of the integral of the function by the cosine of the mean value of the argument in the range of integration. Hence, Eq. (5.5) can be changed to

$$\varphi'(z) = \sum_{k=1}^{n} \cos(2\pi Z_k z) \int_{Z_k - \Delta Z_k/2}^{Z_k + \Delta Z_k/2} i_{00}(Z) \, dZ$$

or

$$\varphi'(z) = \sum_{k=1}^{n} I_k \cos(2\pi Z_k z) \tag{5.6}$$

in which I_k is the integrated intensity of the k-th basal reflection found at a distance Z_k of the origin in reciprocal space and corresponding to an apparent interplanar spacing $d_k = 1/Z_k$. This intensity must first be corrected using the Lorentz-polarization factor.

As pointed out by D'yakonov, neglecting the profile of the bands is tantamount to having an infinite thickness for the crystal. Information concerning the sequence of layers, which is reflected in the band profile, is also lost.

2. When $i_{00}(Z)$ is a slowly varying function with badly individualized and broadened basal reflections, the integration in Eq. (5.4) is performed using the

trapezoidal rule after dividing the interval $(Z_{max}-Z_{min})$, in which all the basal reflections are found, into N elementary and equal intervals of width ΔZ such that

$$\Delta Z = \frac{Z_{max}-Z_{min}}{N}\;.$$

If ΔZ is sufficiently small, the integral can be replaced by a sum of products of the mean values of $i_{00}(Z)$ and of $\cos(2\pi Z z)$ over the elementary interval ΔZ, i.e.,

$$\varphi'(z) \approx \frac{\sin[\pi z \Delta Z]}{\pi z \Delta Z} \sum_{k=0}^{N-1} I\left(Z_k + \frac{\Delta Z}{2}\right) \cos\left[2\pi\left(Z_k + \frac{\Delta Z}{2}\right) z\right]\;. \tag{5.7}$$

This relation takes into account the finite dimensions of the crystal, but its correct use implies that

$$\frac{\sin[\pi z(\Delta Z)]}{\pi z(\Delta Z)}$$

be practically equal to 1 throughout the interval $(Z_{max}-Z_{min})$.

Examples of applications of the D'yakonov method will be found in Chapter 9.

5.1.3 Comparison of the Mac Ewan and D'yakonov Direct Methods of Structural Analysis

The method originally proposed by Mac Ewan (1956) has been widely used in the study of interstratified systems. The final expression, similar to Eq. (5.4), is

$$\omega(z) = \frac{1}{\pi} \int_0^\infty G_{00}(Z)\cos(2\pi Z z)\,dZ \tag{5.8}$$

in which $G_{00}(Z)$ is an interference function described by:

$$G_{00}(Z) - \frac{\bar{I}_k(Z)}{|F_{00}(Z)|^2}\;.$$

In this expression, $\bar{I}_k(Z)$ is the integrated intensity of the basal reflexion given by the interstratified crystal, normalized to a single layer and corrected for Lorentz-polarization effects, while $|F_{00}(Z)|^2$ is the structure factor of the layer. Under these conditions, $\omega(z)$ is a function characterizing the probability of finding, in the crystal, interlamellar distances equal to z.

There is a formal analogy between $\omega(z)$ and $\varphi'(z)$ but, as pointed out by D'yakonov (1962), Eq. (5.8) is unsound mathematically, since $G_{00}(Z)$ does not tend to 0 for the upper limit of integration and the integral is therefore divergent. In actual calculations, Eq. (5.8) may be used thanks to the fact that the experimental diffractograms are always recorded over a limited and finite range of 2θ values, so that the integral is then necessarily convergent. Other drawbacks are, however, inherent to the use of Eq. (5.8); the single structure factor implies that the layers are all identical, which is not true for multi-component interstratified crystals; for certain values of Z, the structure factor $|F|^2$ may be very small or even 0, which results in a large uncertainty in the value of $\bar{I}_k(Z)/|F|^2$ whenever $\bar{I}_k(Z)$ is itself small; finally, the indeterminacy in the baseline renders the exploitation of the curve $\omega(z)$ quite difficult. Using a series of examples, Reynolds (1980) showed in detail the kind of errors occurring in the practical use of $\omega(z)$.

5.2 Indirect Methods of Structural Analysis Based on the Computation of the Intensities of Basal Reflections

Indirect methods are so named owing to the fact that they do not yield information on the structure through the direct exploitation of experimental data but rather because they use the intensities computed on the basis of a given structural model. Contributions to the diffraction are provided by all the possible combinations of layers present in the crystals, from single layers to pairs of nearest neighbors, triplets, etc., each combination being associated with its own intrinsic probability. Insofar as only basal reflections are investigated, and in accordance with the lamellar structure described in Chapter 2, an interstratified crystal composed of M layers will be described by a stack along the normal to these layers of M unit cells with an identical set of *a* and *b* parameters. The sequence of unit cells is the same as the sequence of layer types in the stack.

As given by Mac Ewan (1958), the first method of computation involved an interference function corresponding to interstratified crystals with finite thickness and the same scattering power for all types of layers.

Reynolds (1967, 1980) modified this approach to deal with the specific case of two-component interstratified systems by introducing different structure factors, while Weir and Rayner (1974) extended the formalism of Reynolds to three-component systems.

5.2.1 Calculation of an Interference Function Using a Single Structure Factor

The simplification introduced by Mac Ewan (1958) corresponds to layers which can only be differentiated by their thickness, i.e., by the interlayer distances.

Let us compare the intensities of the waves diffracted along the direction of the (00) rod by an interstratified system and by a perfect crystal of M layers identical in all respects. In the latter case, and starting from Eq. (2.6), the intensity $i_{00}(Z)$ is

$$i_{00}(Z) = \sum_{m=1}^{M} |F_m|^2 + 2 \sum_m \sum_n F_m F^*_{m+n} \cos 2\pi Z z_n$$
$$= i_0 + 2 \sum_{n=1}^{M-1} i_n \cos 2\pi Z z_n \tag{5.9}$$

in which z_n is the distance between two n-th nearest-neighbors in the stack, with structure factors $F(Z)$ and $F_n(Z)$. The thickness z_n of any substack of $(n+1)$ layers has a unique value given by $z_n = nd$, where d is the thickness of a single layer. On the contrary, substacks with $(n+1)$ layers in an interstratified system may show different values of z_n between n-th nearest-neighbors. Labeling these different distances z_{nk}, we have

$$z_{nk} = \sum_i n_i d_i \tag{5.10}$$

in which d_i is the thickness of an i-type layer, present n_i times in the substack of $(n+1)$ layers considered. For instance, two types of layers A and B, with $d_A = 10$ Å and $d_B = 18$ Å, can give four different thicknesses for three nearest neighbors in a four-layer substack:

$$z_{31} = 3d_A = 30 \text{ Å}$$
$$z_{32} = 2d_A + d_B = 38 \text{ Å}$$
$$z_{33} = d_A + 2d_B = 46 \text{ Å}$$
$$z_{34} = 3d_B = 54 \text{ Å} \; .$$

Hence, with interstratified systems, Eq. (5.9) must be modified to take into account all the possible values of z_{nk} which exist for the pairs of n-th nearest-neighbors, and we must write

$$i_{00}(Z) = |F(Z)|^2 \left[1 + 2 \sum_{n=1}^{M-1} \frac{M-n}{M} \sum_k \sigma(z_{nk}) \cos(2\pi Z z_{nk}) \right] \tag{5.11}$$

in which $\sigma(z_{nk})$ is the probability of occurrence associated with the distance z_{nk}. By analogy with Eq. (2.10), the interference function $G_{00}(Z)$ is found to be:

$$G_{00}(Z) = 1 + 2 \sum_{n=1}^{M-1} \frac{M-n}{M} \sum_k \sigma(z_{nk}) \cos(2\pi Z z_{nk}) \; , \tag{5.12}$$

which is similar to the expression given by Maç Ewan (1958).

The actual calculation of $G_{00}(Z)$ implies the prior knowledge of the thickness of the individual layers, knowledge of all the distances (z_{nk}) between n-th nearest-neighbors, and of the rule describing the layer subsequences. The probability of occurrence of each z_{nk} is obtained with the help of the relationships developed in Chapter 4, with the proviso that a given z_{nk} may correspond to a number of substacks with the same layer content but with different layer arrangements. With reference to the two-component (A and B) system described above, let us thus suppose that the arrangement of layers corresponds to an order parameter $S = 1$; if the proportion W_A and W_B are known, as well as the probabilities P_{ij} (with $i, j = A, B$), it is for instance possible to calculate the probability σ (38 Å) of finding substacks of thickness 38 Å:

$$\begin{aligned}\sigma(38\,\text{Å}) &= W_{AAB} + W_{BAA} + W_{ABA} \\ &= W_A P_{AA} P_{AB} + W_B P_{BA} P_{AA} + W_A P_{AB} P_{BA}\ .\end{aligned}$$

All the probabilities $\sigma(z_{nk})$ relative to the different distances z_{nk}, for $1 \leqslant n \leqslant M-1$, can be calculated in the same way.

An interesting particular case is that of interstratified systems composed of i different types of layers with thicknesses d_i such that $n_1/d_1 = n_2/d_2 = \ldots = n_i/d_i = Z'$ (a constant), with $n_1, n_2, \ldots, n_i$ equal to integers. The interference function $G_{00}(Z)$ then becomes periodic, with a period Z' and a center of symmetry at $Z'/2$ (Mac Ewan 1958; D'yakonov 1962). Each distance z_{nk} can now be replaced by the product pd', where p is an integer and $d' = 1/Z'$, and Eq. (5.12) now becomes:

$$G_{00}(Z) = \frac{1}{2} + \sum_{p=1}^{M'} \sigma(p) \cos(2\pi Z p d') \qquad (5.13)$$

with $M' = T/d'$, T being the thickness of the interstratified crystal. Also, since $G_{00}(Z)$ is now periodic, the coefficients $\sigma(p)$ of the Fourier series can be expressed as

$$\sigma(p) = \frac{2}{Z'} \int_0^{Z'/2} G_{00}(Z) \cos\left(2\pi p \frac{Z}{Z'}\right) dZ\ . \qquad (5.14)$$

In other words, $\sigma(p)$ is proportional to the number of substacks with the given thickness pd'.

We should also note that though Eq. (5.14) bears a formal resemblance to the expression in Eq. (5.8) given by Mac Ewan, it differs from it by the fact that the upper limit of integration is now finite, so that the integral is always convergent.

5.2.2 Methods of Intensity Calculation Using Different Structure Factors

The simplest case is that of interstratified systems in which the different types of layers all have a center of symmetry in their unit cell, and the imaginary part of each structure factor $F_i(Z)$ then becomes zero.

Let us first examine the case of a two-component (A and B) interstratified stack of layers in which both A and B are centrosymmetric. It is convenient to represent the crystal with the help of a one-dimensional irregular lattice built up as follows:

- the origin is placed at the center of symmetry of the unit cell of the layer at the bottom of the stacks;
- the position of the centers of symmetry of the successive unit cells are determined along the axis Oz (which coincides with the axis of the column); since the thicknesses d_A and d_B are different, the distance between the nodes corresponding to sequences AA, AB or BA, and BB, will respectively be d_A, $(d_A+d_B)/2$ and d_B;
- a weighting coefficient equal to the amplitude of the wave diffracted by the corresponding unit cell (F_A or F_B) is assigned to each node of the lattice thus obtained.

As given by Eq. (5.9) and taking into account the different structure factors, the intensity $i_{00}(Z)$ of the wave diffracted by the stack is

$$i_{00}(Z) = W_A|F_A(Z)|^2 + W_B|F_B(Z)|^2 + 2\sum_{n=1}^{M-1}\frac{M-n}{M}\Bigg[|F_A(Z)|^2\sum_k \sigma(z_{nk}^{AA})\cos 2\pi Z z_{nk}^{AA} + 2F_A(Z)F_B^*(Z)\sum_k \sigma(z_{nk}^{AB})\cos 2\pi Z z_{nk}^{AB} + |F_B(Z)|^2\sum_k \sigma(z_{nk}^{BB})\cos 2\pi Z z_{nk}^{BB}\Bigg] . \tag{5.15}$$

In the above expressions z_{nk}^{AA}, $z_{nk}^{AB} = z_{nk}^{BA}$ and z_{nk}^{BB} are the thicknesses of the substacks, which all contain (n+1) layers (as witnessed by the common index n), but which differ from each other by the nature of the first and last layer. As in Section 5.2.1, the index k is used to differentiate between (n+1) substacks with non-identical compositions but identical A or B layers as first and last layers. The terms $\sigma(z_{nk}^{ij})$ refer to the probability of occurrence of substacks with i-type and j-type first and last layers, respectively, with thickness z_{nk}^{ij} (ij = A, B). We may note that the probabilities are such that $\sigma(z_{nk}^{ij}) = \sigma(z_{nk}^{ji})$. The computation of these probabilities has already been described in Section 5.2.1, but account must now be taken of the distances between nodes along the Oz axis. Thus, for a substack with two nodes of weight F_A and one of weight F_B, the following relations hold:

$$W_{ABA} = W_A P_{AB} P_{BA} ; \quad W_{BAA} = W_B P_{BA} P_{AA} ; \quad W_{AAB} = W_A P_{AA} P_{AB}$$

$$z_{2k}^{AA} = d_A + d_B ; \quad z_{2k}^{AB} = d_A 3/2 + d_B 1/2 = z_{2k}^{BA}$$

$$\sigma(z_{2k}^{AA}) = W_{ABA} ; \quad \sigma(z_{2k}^{AB}) = W_{AAB} + W_{BAA} = \sigma(z_{2k}^{BA}) .$$

Similarly, the proportions (or probabilities of occurrence) of substacks containing five nodes (three of type A and two of B), starting and ending with the same type of layer, are given by:

$$\sigma(z_{4k}^{AA}) = W_{AABBA} + W_{ABBAA} + W_{ABABA} \quad \text{with} \quad z_{4k}^{AA} = 2(d_A + d_B)$$

and

$$\sigma(z_{4k}^{BB}) = W_{BAAAB} \quad \text{with} \quad z_{4k}^{BB} = 3d_A + d_B \ .$$

Reynolds (1967, 1980) noted that the use of Eq. (5.15) in the case of very thin stacks (i.e., with a small number of layers) presents some difficulties, even when there are only two interstratified components as in the systems mica-smectite, mica-chlorite, etc. In such stacks, the center of symmetry must be placed at the center of the octahedral sheet of the 2:1 layers. This choice of origin necessarily neglects the external half of each outermost layer; it does not lead to important errors if the number of layers is large, but this will no longer be true if the stack includes only a small number of layers.

Reynolds (1967, 1980) solved this problem by increasing the number of structural elements taken into account while retaining centrosymmetry. Thus, for interstratified systems with 2:1 layers giving two different types of interlamellar spaces, three structural elements are used to describe the structure: the 2:1 layers themselves (named C) and the interlamellar spaces referred to as types A and B. For each structural element, the origin of its coordinates is placed at a point which creates two equal and symmetrical half-thicknesses. Furthermore, three structure factors, F_C, F_A and F_B, are used, respectively corresponding to the 2:1 layers and to each of the two interlamellar spaces, as well as two distances, d_a and d_b, corresponding to the two pairs of elements a = A+C and b = B+C. Lastly, Reynolds assumes that the number M of layers and the probabilistic parameters attached to the order parameter S are also known. The stack can then be described by an irregular one-dimensional lattice of nodes A, B, and C with respective weights F_A, F_B, and F_C, subject to the condition that two successive structural elements of the same kind never follow each other: each A must be preceded and followed at the same distance by a C, with similar rules for B. In that case, the different possible values of the distances z_{nk} between nth-neighbor layers in a stack of (n+1) layers can be split into the six groups $z_{nk}^{AA}, z_{nk}^{AB}, z_{nk}^{AC}, z_{nk}^{BC}, z_{nk}^{BB}$ and z_{nk}^{CC}, depending on the nature of the two extreme nodes in the one-dimensional lattice. The corresponding probabilities are now obtained as follows:

Consider the set of (m+1) nodes limited at either end by C and corresponding to the substacks of (n+1) layers so that n = m/2. The thickness of these stacks is $z_{nk}^{CC} = (n-p)d_a + pd_b$ where p is the number of b in the stack. The probability of occurrence $\sigma(z_{nk}^{CC})$ of this thickness is obtained using the relationships developed in Chapter 4. Note that $\sigma(z_{nk}^{CC})$ is not only the probability that the distance between the first and last node shall be as given by the expression above, but also the probability attached to the distances between the second and last and between

Fig. 5.1. Scheme of the nodes A, B and C in a substack of seven 2:1 layers separated by interlayer distances A and B, and in a sequence abbaba (with a = A+C and b = B+C)

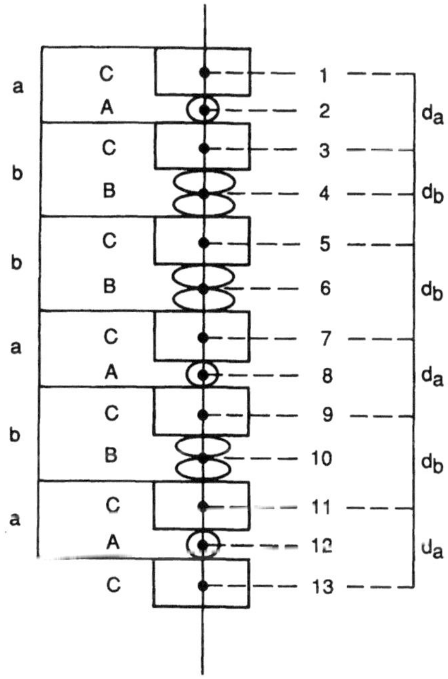

the first and m-th nodes. This is illustrated in Fig. 5.1 showing a stack of seven 2:1 layers separated by interlayer distances A and B and in the sequence abbaba (n = 6 and m = 12). If the order parameter S is 1, the probability of occurrence of this sequence is $W_{abbaba} = W_a P_{ab}^2 P_{bb} P_{ba}^2$. It can also be seen in Fig. 5.1 that this is also the probability attached to the distance $z_{6k}^{CC} = 3d_a + 3d_b$ between the first and last (thirteenth) nodes (both type C) to the distance $z^{AC} = z_{6k}^{CA} = 2.5d_a + 3d_b$ between the second (type A) and the thirteenth (type C) nodes, and to the distance $z_{6k}^{AA} = 2d_a + 3d_b$ between the second and twelfth nodes (both type A). Finally, it should also be noted that there are three other sequences of (a, b)-elements starting and ending with an a, namely ababba, abbbaa and aabbba, with the associated probabilities

$$W_{abbaba} = W_{ababba} = W_1 \quad \text{and} \quad W_{abbbaa} = W_{aabbba} = W_2$$

from which it can be deduced that

$$\sigma(z_{6k}^{AA}) = \sigma(z_{6k}^{CC}) = \sigma(z_{6k}^{AC}) = 2W_1 + 2W_2 \ .$$

Stacks with the same overall composition (3a+3b) but with nonidentical outermost layers have other distances besides $z_{6k}^{CC} = 3d_a + 3d_b$ and $z_{6k}^{AC} = 2.5d_a + 3d_b$ as above, i.e., $z_{6k}^{CB} = 3d_a + 2.5d_b$ and $z_{6k}^{AB} = 2.5d_a + 2.5d_b$. Once again, the probabilities characteristic of these distances are identical; they are given by

$$\sigma = 2(W_{aaabbb}+W_{aababb}+W_{abaabb}+W_{aabbab}+W_{abbaab}+W_{ababab}) \ .$$

In the case of stacks with a low number of layers, starting from Eq. (5.15) and using the structural elements A, B and C coupled as pairs A+C and B+C enabled Reynolds to express the intensity of the electromagnetic wave diffracted by a set of 2:1 layers separated by A and B interlayer spaces as:

$$i_{00}(Z) = F_A^2 L_1 + F_B^2 L_2 + F_A F_B L_3 + F_A F_C L_4 + F_B F_C L_5 + F_C^2 L_6 \ , \qquad (5.16)$$

where $F_i = |F_i|$ and

$$L_1 = W_A \left(\frac{M-1}{M}\right) + 2 \sum_{n=1}^{M-1} \frac{M-n}{M} \sum_k \sigma(z_{nk}^{AA}) \cos 2\pi Z z_{nk}^{AA}$$

$$L_2 = W_B \left(\frac{M-1}{M}\right) + 2 \sum_{n=1}^{M-1} \frac{M-n}{M} \sum_k \sigma(z_{nk}^{BB}) \cos 2\pi Z z_{nk}^{BB}$$

$$L_3 = 2 \sum_{n=1}^{M-1} \frac{M-n}{M} \sum_k \sigma(z_{nk}^{AB}) \cos 2\pi Z z_{nk}^{AB}$$

$$L_4 = 2 \sum_{n=1}^{M-1} \frac{M-n}{M} \sum_k \sigma(z_{nk}^{AC}) \cos 2\pi Z z_{nk}^{AC}$$

$$L_5 = 2 \sum_{n=1}^{M-1} \frac{M-n}{M} \sum_k \sigma(z_{nk}^{BC}) \cos 2\pi Z z_{nk}^{BC} \ .$$

$$L_6 = 1 + 2 \sum_{n=1}^{M-1} \frac{M-n}{M} \sum_k \sigma(z_{nk}^{CC}) \cos 2\pi Z z_{nk}^{CC}$$

To conclude this paragraph, let us apply the formalism developed for thin crystallites to the concrete case of a mica-smectite interstratified system with a range of interaction parameter $S = 1$. Let A be the mica-type interlayer spaces (with nonsolvated K^+ ions) and B be the smectite-type interlayer spaces (containing, for instance, solvated Na^+ ions). When placed in ethylene glycol, this type of mineral species swells: two distances $d_a = 10$ Å and $d_b = 17$ Å are then observed. Let now $W_A = 0.6$, $P_{BB} = 0$ and $M = 4$. The various other parameters needed to compute the intensities of the waves diffracted by the mineral after absorption of the glycol are given in Tables (5.1) and (5.2), in which the results of the computations can also be found.

Table 5.1. Probability (σ) attached to various possible sequences of interlayers A- and B-type spaces when the substack contains a different number of layers

i−j succession	M−n	σ	(M−n)σ
A	3	0.6	1.8
B	3	0.4	1.2
AA	2	0.2	0.4
AB	2	0.4	0.8
BA	2	0.4	0.8
AAA	1	0.067	0.067
AAB	1	0.133	0.133
ABA	1	0.4	0.4
BAA	1	0.133	0.133
BAB	1	0.266	0.266

Table 5.2. Distance (in Å) characteristic of various substacks in mica-smectites stacks containing different number of layers, after immersion in ethylene glycol

i−j succession	z^{AA}	z^{AC}	z^{BB}	z^{BC}	z^{AB}	z^{CC}
A		5				10
B				8.5		17
AA	10	15				20
AB		22		18.5	13.5	27
BA		22		18.5	13.5	27
AAA	20	25				30
AAB		32		28.5	23.5	37
ABA	27	32				37
BAA		32		28.5	23.5	37
BAB			27	35.5		44

5.3 Diffraction by Systems with g Types of Layers, with a Specific Translation r Between the Adjacent i-Type and j-Type Layers, for any Given Value of S

We shall now extend the developments in the preceding sections to the calculation of the intensity diffracted by any given quasi-homogeneous interstratified system, starting from relatively simple systems and proceeding to more general cases. The present section will examine interstratified systems in which the two layers in an (i, j)-pair (i.e., a pair composed of an i-type and a j-type layers) are adjacent and are always related by the same translation $\vec{r}_{ij}$, whatever the value of the order parameter S.

It was shown in Section 3.1.2 that the intensity of the diffraction originating from a system with a disordered structure is most conveniently handled using a matrix formalism in which

- the matrix [W] characterizes the type of layers and their relative proportions;
- the matrix [Φ] regroups the structure factors of the different types of layers;
- the matrix [Q] handles the phase differences among the diffracted waves originating from the different layers in the substack.

Furthermore, Chapter 4 detailed the calculation of the probabilistic parameters characteristic of a given interstratified layer structure, for any given value of the order parameter ($S \geqslant 0$).

Let us now consider an interstratified crystal composed of g types of layers stacked so that two successive layers of type i and j are always connected by the translation vector $\vec{r}_{ij}$. The diffraction produced by such a system may be described by the general expression in Eq. (3.2), normalized to a single layer:

$$i(\vec{s}) = \frac{\Omega}{\sigma}\left\{\overline{|\Phi(\vec{s})|^2} + 2\,\mathrm{Re}\left[\sum_{n=1}^{M-1} \frac{M-n}{M}\, \overline{\Phi(\vec{s})\,\Phi_n^*(\vec{s})\exp(2\pi i\,\vec{s}\cdot\vec{r}_n)}\right]\right\} .$$

In this expression $\overline{|\Phi(\vec{s})|^2}$ corresponds to the intensity of the wave diffracted independently by each of the layers: if we let W_i be the proportion of i-type layers, ($i = 1, 2, \ldots g$), we can change the form of this expression to one identical with Eq. (3.3)

$$\overline{|\Phi(\vec{s})|^2} = \sum_{i=1}^{g} W_i \Phi_i \Phi_i^* . \tag{5.17}$$

The second expression in Eq. (3.2) represents the mean value of the waves diffracted by all the pairs of nth-neighbors related by a translation vector $\vec{r}_n$, itself the sum of elementary translations between successive adjacent layers separating the two layers of the nth-neighbor pair considered. In other words, $\vec{r}_n$ is the translation vector characteristic of the two outermost layers of a given substack containing (n+1) layers. Since the proportions of substacks of 2, 3, 4, ... layers in an interstratified crystal are given by W_{ij}, W_{ijk}, W_{ijkl}, ... (i, j, k, l ... = 1, 2, 3, 4 ... g) as shown in Chapter 4.1, the second mean value can be given a new form

– for n = 1 (corresponding to first-neighbor pairs and substacks of two layers)

$$\sum_{i=1}^{g} \sum_{j=1}^{g} W_{ij} \Phi_i \Phi_j^* \exp(2\pi i \vec{s} \cdot \vec{r}_{ij}) \tag{5.18}$$

– for n = 2 (second-nearest neighbors, substacks of three layers)

$$\sum_{i=1}^{g} \sum_{j=1}^{g} \sum_{k=1}^{g} W_{ijk} \Phi_i \Phi_k^* \exp[2\pi i \vec{s} \cdot (\vec{r}_{ij} + \vec{r}_{jk})] \tag{5.19}$$

– for n = 3 (third-nearest neighbors, substacks of four layers)

$$\sum_{i=1}^{g} \sum_{j=1}^{g} \sum_{k=1}^{g} \sum_{l=1}^{g} W_{ijkl} \Phi_i \Phi_l^* \exp[2\pi i \vec{s} \cdot (\vec{r}_{ij} + \vec{r}_{jk} + \vec{r}_{kl})] \tag{5.20}$$

etc. ...

in which $\vec{r}_{ij}, \vec{r}_{jk}, \vec{r}_{kl}$ ... are the elementary translation vectors between adjacent layers of type ij, jk, kl ... .

Depending on the value of the range parameter S, i.e., on the interactions among the layers, the proportions W_{ij}, W_{ijk}, etc. will take different forms, as shown in Chapter 4. For instance, the proportion of substacks with four layers will be

– when S = 0 $\quad W_{ijkl} = W_i W_j W_k W_l$;

– when S = 1 $\quad W_{ijkl} = W_i P_{ij} P_{jk} P_{kl}$;

– when S = 2 $\quad W_{ijkl} = W_i P_{ij} P_{ijk} P_{jkl}$; etc.

In fact, rather than use Eqs. (5.17), (5.18) etc. ..., it is much more convenient to bring in the matrix formalism recalled at the beginning of this section. Before giving the expressions for the matrices, we should note that two approaches were used to reach [Q]:

– in the first, the result of the calculation of the interference term corresponding to each adjacent (i, j)-pair of layers was set to be the (i, j)-th element of [Q] (cf. Chap. 3.1.2);
– the second grouped the phase differences in the matrix [φ] and the conditional probabilities in a separate matrix [P], so that [Q] was given by the product [φ] [P]. The latter method is not convenient since the size of [Q] is only man-

ageable when **the translation vector $\vec{r}_{ij}$ between two adjacent layers of type i and j is determined by the nature of the layer i.** However, in actual interstratified systems, the translation vector $\vec{r}_{ij}$ may depend on the nature of both layers of the pair and, furthermore, there may exist many values of $\vec{r}_{ij}$ for a given adjacent (i, j)-pair; in either case, a correct description of the interstratified system by $[\varphi]$ and $[P]$ is awkward and, moreover, the matrix $[Q]$ yielded by the product $[\varphi][P]$ rapidly becomes unwieldy. We shall therefore write down directly the expression for the matrix $[Q]$.

5.3.1 Expressions for the Matrices [W], [Φ] and [Q], when S = 0 or 1

Following the definitions given in Section 3.1.2, and under the prevalent restrictions, the matrices are

$$[W] = \begin{bmatrix} W_1 & 0 & \dots & 0 \\ 0 & W_2 & \dots & 0 \\ \vdots & \vdots & & \vdots \\ 0 & 0 & \dots & W_g \end{bmatrix} \qquad [\Phi] = \begin{bmatrix} \Phi_1^*\Phi_1 & \Phi_1^*\Phi_2 & \dots & \Phi_1^*\Phi_g \\ \Phi_2^*\Phi_1 & \Phi_2^*\Phi_2 & \dots & \Phi_2^*\Phi_g \\ \vdots & \vdots & & \vdots \\ \Phi_g^*\Phi_1 & \Phi_g^*\Phi_2 & \dots & \Phi_g^*\Phi_g \end{bmatrix}$$

$$\text{and} \quad [Q] = \begin{bmatrix} P_{11}\varphi_{11} & P_{12}\varphi_{12} & \dots & P_{1g}\varphi_{1g} \\ P_{21}\varphi_{21} & P_{22}\varphi_{22} & \dots & P_{2g}\varphi_{2g} \\ \vdots & \vdots & & \vdots \\ P_{g1}\varphi_{g1} & P_{g2}\varphi_{g2} & \dots & P_{gg}\varphi_{gg} \end{bmatrix} \tag{5.21}$$

in which $\varphi_{ij} = \exp(2\pi i \vec{s}\vec{r}_{ij})$; when $S = 0, P_{ij} = W_j$, but when $S = 1$, P_{ij} are the conditional probabilities of finding a j-type layer following a layer of type i.

The intensity of the wave diffracted by the interstratified system is given by Eq. (3.11), which is the matrix form of Eq. (3.2)

$$i(\vec{s}) = \frac{\Omega}{\sigma}\left[\mathrm{Tr}\{[\Phi][W]\} + 2\,\mathrm{Re}\sum_{n=1}^{M-1}\frac{M-n}{M}\,\mathrm{Tr}\{[\Phi][W][Q]^n\}\right]. \tag{5.22}$$

5.3.2 Expressions for [W], [Φ] and [Q] when S = 2

Kakinoki and Komura (1954a, b) showed that it is possible to retain the matrix formulation of Eq. (3.11) for the intensity of the diffracted wave for disordered systems with $S > 1$, provided $[W]$, $[\Phi]$ and $[Q]$ are constructed in such a way that each element itself be a submatrix, with its expression formally that of the appropriate matrix corresponding to $S = 1$. Using the forms suggested by Sakharov et al. (1982a, 1983), we shall therefore write out $[W]$, $[\Phi]$ and $[Q]$ for $S = 2$, remembering that such an interstratified system can be completely described using the abundance ratios W_i and the conditional probabilities P_{ij} and P_{ijk}.

Consider now the (i, j)-pairs in the stack. For each of them, the type of the next layer can be found among the g different types present in the stack, and the probability of occurrence of the sequence ijk is $W_{ijk} = W_{ij}P_{ijk}$ where $1 \leqslant k \leqslant g$ for any (i, j)-pair with fixed i and j values. Hence, when S = 2, each (i, j)-term of [W], [Q] and [Φ] obtained for S = 1 must be replaced by a submatrix consisting of (g×g) elements. As a result, the matrices [W], [Q] and [Φ] for S = 2 will be of order g^2 and are:

$$[W] = \begin{bmatrix} [W]_1 & 0 & \dots & 0 \\ 0 & [W]_2 & \dots & 0 \\ \vdots & \vdots & & \vdots \\ 0 & 0 & \dots & [W]_g \end{bmatrix} \qquad [\Phi] = \begin{bmatrix} [\Phi]_{11} & [\Phi]_{12} & \dots & [\Phi]_{1g} \\ [\Phi]_{21} & [\Phi]_{22} & \dots & [\Phi]_{2g} \\ \vdots & \vdots & & \vdots \\ [\Phi]_{g1} & [\Phi]_{g2} & \dots & [\Phi]_{gg} \end{bmatrix}$$

$$\text{and} \quad [Q] = \begin{bmatrix} [Q]_{11} & [Q]_{12} & \dots & [Q]_{1g} \\ [Q]_{21} & [Q]_{22} & \dots & [Q]_{2g} \\ \vdots & \vdots & & \vdots \\ [Q]_{g1} & [Q]_{g2} & \dots & [Q]_{gg} \end{bmatrix} \tag{5.23}$$

The submatrices $[W]_i$, $[\Phi]_{ij}$ and $[Q]_{ij}$ can now be detailed.

a) For any given index i, each submatrix $[W]_i$ is a diagonal matrix

$$[W]_i = \begin{bmatrix} W_{i1} & 0 & \dots & 0 \\ 0 & W_{i2} & \dots & 0 \\ \vdots & \vdots & & \vdots \\ 0 & 0 & \dots & W_{ig} \end{bmatrix} \tag{5.24}$$

in which $W_{i1}, W_{i2}, \dots W_{ig}$ are the relative proportions of adjacent (i, 1)-, (i, 2)-, ... (i, g)-pairs in the stack, with the possibility of writing also $W_{ij} = W_i P_{ij}$.

b) Each submatrix $[\Phi]_{ij}$ is a square matrix of order (g×g) in which all elements are identical and equal to $(\Phi_i^* \Phi_j)$.

$$[\Phi]_{ij} = \begin{bmatrix} \Phi_i^* \Phi_i & \Phi_i^* \Psi_j & \dots & \Phi_i^* \Phi_j \\ \Phi_i^* \Phi_j & \Phi_i^* \Phi_j & \dots & \Phi_i^* \Phi_i \\ \vdots & \vdots & & \vdots \\ \Phi_i^* \Phi_j & \Phi_i^* \Phi_j & \dots & \Phi_i^* \Phi_j \end{bmatrix} \tag{5.25}$$

which can also be expressed as $[\Phi]_{ij} = \Phi_i^* \Phi_j [U]$ where [U] is a matrix of order (g×g) in which all the elements are unity.

c) Prior to giving the expression for the submatrix $[Q]_{ij}$, we shall examine the elements of the matrices $[Q]^2$, $[Q]^3, \dots,$ when S = 1: their general expressions are, respectively, $P_{ij}P_{jk}\varphi_{ij}\varphi_{jk}$, $P_{ij}P_{jk}P_{kl}\varphi_{ij}\varphi_{jk}\varphi_{kl}\dots$. In other words, the indices of the factors in the group of conditional probabilities or in the group of exponential phaseshift factors are such that the second index of any factor is the same as the first index of the next factor. It is evident that this property must be retained in system with $S = 2$, both for the P_{ijk} and for the φ_{ij} factors. This implies that the $P_{ijk}\varphi_{ij}$ terms in the submatrix $[Q]_{ij}$ will be found in the j-th row and the k-th columns, all the other elements of $[Q]_{ij}$ being zero. This submatrix is therefore

$$[Q]_{ij} = \begin{array}{c} \\ 1 \\ 2 \\ \vdots \\ j \\ \vdots \\ g \end{array} \begin{array}{c} \begin{array}{ccccccc} 1 & 2 & \dots & k & \dots & g \end{array} \\ \left[\begin{array}{cc:c:c:c:c} 0 & 0 & \dots & 0 & \dots & 0 \\ 0 & 0 & \dots & 0 & \dots & 0 \\ \hdashline P_{ij1}\varphi_{ij} & P_{ij2}\varphi_{ij} & \dots & P_{ijk}\varphi_{ij} & \dots & P_{ijg}\varphi_{ij} \\ \hdashline 0 & 0 & \dots & 0 & \dots & 0 \end{array}\right] \end{array} \tag{5.26}$$

in which P_{ijk} is the conditional probability of a k-type layer following an (i, j)-pair and in which φ_{ij}, as in Eq. (5.21), is

$$\varphi_{ij} = \exp(2\pi i \vec{s}\cdot\vec{r}_{ij}) \ .$$

As an illustration, we give [W], [Φ] and [Q] for an interstratified system of two components (A and B), with $S = 2$. Eqs. (5.23), (5.24), (5.25) to (5.26), yield

$$[W] = \left[\begin{array}{cc:cc} W_{AA} & 0 & 0 & 0 \\ 0 & W_{AB} & 0 & 0 \\ \hdashline 0 & 0 & W_{BA} & 0 \\ 0 & 0 & 0 & W_{BB} \end{array}\right] \qquad [\Phi] = \left[\begin{array}{cc:cc} \Phi_A^*\Phi_A & \Phi_A^*\Phi_A & \Phi_A^*\Phi_B & \Phi_A^*\Phi_B \\ \Phi_A^*\Phi_A & \Phi_A^*\Phi_A & \Phi_A^*\Phi_B & \Phi_A^*\Phi_B \\ \hdashline \Phi_B^*\Phi_A & \Phi_B^*\Phi_A & \Phi_B^*\Phi_B & \Phi_B^*\Phi_B \\ \Phi_B^*\Phi_A & \Phi_B^*\Phi_A & \Phi_B^*\Phi_B & \Phi_B^*\Phi_B \end{array}\right]$$

$$[Q] = \left[\begin{array}{cc:cc} P_{AAA}\varphi_{AA} & P_{AAB}\varphi_{AA} & 0 & 0 \\ 0 & 0 & P_{ABA}\varphi_{AB} & P_{ABB}\varphi_{AB} \\ \hdashline P_{BAA}\varphi_{BA} & P_{BAB}\varphi_{BA} & 0 & 0 \\ 0 & 0 & P_{BBA}\varphi_{BB} & P_{BBB}\varphi_{BB} \end{array}\right]$$

in which the dotted lines partition the matrices into the submatrices. For $n = 0, 1, 2, 3$, the matrix formulation will lead to Eqs. (5.17)–(5.20). For example,

$$\mathrm{Tr}\{[\Phi][W]\} = \sum_{i=1}^{g} W_i \Phi_i \Phi_i^* \ .$$

Since

$$[\Phi][W] = \begin{bmatrix} W_{AA}\Phi_A^*\Phi_A & W_{AB}\Phi_A^*\Phi_A & W_{BA}\Phi_A^*\Phi_B & W_{BB}\Phi_A^*\Phi_B \\ W_{AA}\Phi_A^*\Phi_A & W_{AB}\Phi_A^*\Phi_A & W_{BA}\Phi_A^*\Phi_B & W_{BB}\Phi_A^*\Phi_B \\ W_{AA}\Phi_B^*\Phi_A & W_{AB}\Phi_B^*\Phi_A & W_{BA}\Phi_B^*\Phi_B & W_{BB}\Phi_B^*\Phi_B \\ W_{AA}\Phi_B^*\Phi_A & W_{AB}\Phi_B^*\Phi_A & W_{BA}\Phi_B^*\Phi_B & W_{BB}\Phi_B^*\Phi_B \end{bmatrix}$$

then

$$\begin{aligned} \mathrm{Tr}\{[\Phi][W]\} &= W_{AA}\Phi_A^*\Phi_A + W_{AB}\Phi_A^*\Phi_A + W_{BA}\Phi_B^*\Phi_B + W_{BB}\Phi_B^*\Phi_B \\ &= (W_{AA}+W_{AB})\Phi_A^*\Phi_A + (W_{BA}+W_{BB})\Phi_B^*\Phi_B \ . \end{aligned}$$

From Eqs. (4.11) and (4.12),

$$W_{AA}+W_{AB} = W_A \quad \text{and} \quad W_{BA}+W_{BB} = W_B$$

hence $$\mathrm{Tr}\{[\Phi][W]\} = \sum_i W_i\Phi_i\Phi_i^* = W_A\Phi_A^*\Phi_A + W_B\Phi_B^*\Phi_B \ .$$

5.3.3 Expressions for [W], [Φ] and [Q] when S = 3

In a stack containing g types of layers, let us consider groups of four successive layers and, for one of these groups, let the first three layers be an (i, j, k)-triplet. The conditional probability that the fourth layer shall be of type l, where l is one of the g types present in the stack, is P_{ijkl}. Under these conditions, when S = 3, each one of the elements of the submatrices $[W]_i$, $[\Phi]_{ij}$ and $[Q]_{ij}$ of [W], [Φ] and [Q] for S = 2 will itself be a submatrix $[W]'_{ij}$, $[\Phi]'_{ij}$ and $[Q]'_{ij}$, of order (g×g).

The matrix $[W]'_{ij}$ is:

$$[W]'_{ij} = \begin{bmatrix} W_{ij1} & 0 & \dots & 0 \\ 0 & W_{ij2} & \dots & 0 \\ \vdots & \vdots & & \vdots \\ 0 & 0 & \dots & W_{ijg} \end{bmatrix} \tag{5.27}$$

in which $W_{ij1}, W_{ij2}, \dots W_{ijg}$ are the proportions of the (i, j, 1)-, (i, j, 2)-, … (i, j, g)-triplets in the stack.

The matrix $[\Phi]'_{ij}$ is

$$[\Phi'_{ij}] = \Phi_i^*\Phi_j[U] \tag{5.28}$$

in which [U] is the unity square matrix of order (g×g), already defined.

Finally, the matrix $[Q]'_{ij}$ is:

$$[Q]'_{ij} = \begin{array}{c} \\ 1 \\ \vdots \\ k \\ \vdots \\ g \end{array} \begin{array}{c} \begin{array}{ccccccc} 1 & 2 & \dots & l & \dots & g \end{array} \\ \left[\begin{array}{ccccccc} 0 & 0 & \dots & 0 & \dots & 0 \\ \vdots & \vdots & & \vdots & & \vdots \\ P_{ijk1}\varphi_{ij} & P_{ijk2}\varphi_{ij} & \dots & P_{ijkl}\varphi_{ij} & \dots & P_{ijkg}\varphi_{ij} \\ \vdots & \vdots & & \vdots & & \vdots \\ 0 & 0 & \dots & 0 & \dots & 0 \end{array} \right] \end{array} \tag{5.29}$$

in which the P_{ijkl} factors are conditional probabilities as defined above and φ_{ij} is consistent with the definitions in Eqs. (5.21) and (5.26).

Once again, we shall detail [W], [Φ] and [Q] to illustrate the case of a two-component system (A and B) with S = 3,

$$[W] = \left[\begin{array}{cc|cc|cc|cc} W_{AAA} & 0 & 0 & 0 & 0 & 0 & 0 & 0 \\ 0 & W_{AAB} & 0 & 0 & 0 & 0 & 0 & 0 \\ \hline 0 & 0 & W_{ABA} & 0 & 0 & 0 & 0 & 0 \\ 0 & 0 & 0 & W_{ABB} & 0 & 0 & 0 & 0 \\ \hline 0 & 0 & 0 & 0 & W_{BAA} & 0 & 0 & 0 \\ 0 & 0 & 0 & 0 & 0 & W_{BAB} & 0 & 0 \\ \hline 0 & 0 & 0 & 0 & 0 & 0 & W_{BBA} & 0 \\ 0 & 0 & 0 & 0 & 0 & 0 & 0 & W_{BBB} \end{array} \right]$$

$$[\Phi] = \left[\begin{array}{cccc|cccc} \Phi_A^*\Phi_A & \Phi_A^*\Phi_A & \Phi_A^*\Phi_A & \Phi_A^*\Phi_A & \Phi_B^*\Phi_A & \Phi_B^*\Phi_A & \Phi_B^*\Phi_A & \Phi_B^*\Phi_A \\ \Phi_A^*\Phi_A & \Phi_A^*\Phi_A & \Phi_A^*\Phi_A & \Phi_A^*\Phi_A & \Phi_B^*\Phi_A & \Phi_B^*\Phi_A & \Phi_B^*\Phi_A & \Phi_B^*\Phi_A \\ \Phi_A^*\Phi_A & \Phi_A^*\Phi_A & \Phi_A^*\Phi_A & \Phi_A^*\Phi_A & \Phi_B^*\Phi_A & \Phi_B^*\Phi_A & \Phi_B^*\Phi_A & \Phi_B^*\Phi_A \\ \Phi_A^*\Phi_A & \Phi_A^*\Phi_A & \Phi_A^*\Phi_A & \Phi_A^*\Phi_A & \Phi_B^*\Phi_A & \Phi_B^*\Phi_A & \Phi_B^*\Phi_A & \Phi_B^*\Phi_A \\ \hline \Phi_A^*\Phi_B & \Phi_A^*\Phi_B & \Phi_A^*\Phi_B & \Phi_A^*\Phi_B & \Phi_B^*\Phi_B & \Phi_B^*\Phi_B & \Phi_B^*\Phi_B & \Phi_B^*\Phi_B \\ \Phi_A^*\Phi_B & \Phi_A^*\Phi_B & \Phi_A^*\Phi_B & \Phi_A^*\Phi_B & \Phi_B^*\Phi_B & \Phi_B^*\Phi_B & \Phi_B^*\Phi_B & \Phi_B^*\Phi_B \\ \Phi_A^*\Phi_B & \Phi_A^*\Phi_B & \Phi_A^*\Phi_B & \Phi_A^*\Phi_B & \Phi_B^*\Phi_B & \Phi_B^*\Phi_B & \Phi_B^*\Phi_B & \Phi_B^*\Phi_B \\ \Phi_A^*\Phi_B & \Phi_A^*\Phi_B & \Phi_A^*\Phi_B & \Phi_A^*\Phi_B & \Phi_B^*\Phi_B & \Phi_B^*\Phi_B & \Phi_B^*\Phi_B & \Phi_B^*\Phi_B \end{array} \right]$$

$$[Q] = \left[\begin{array}{cc|cc|cc|cc} P_{AAAA}\varphi_{AA} & P_{AAAB}\varphi_{AA} & 0 & 0 & 0 & 0 & 0 & 0 \\ 0 & 0 & P_{AABA}\varphi_{AA} & P_{AABB}\varphi_{AA} & 0 & 0 & 0 & 0 \\ \hline 0 & 0 & 0 & 0 & P_{ABAA}\varphi_{AB} & P_{ABAB}\varphi_{AB} & 0 & 0 \\ 0 & 0 & 0 & 0 & 0 & 0 & P_{ABBA}\varphi_{AB} & P_{ABBB}\varphi_{AB} \\ \hline P_{BAAA}\varphi_{BA} & P_{BAAB}\varphi_{BA} & 0 & 0 & 0 & 0 & 0 & 0 \\ 0 & 0 & P_{BABA}\varphi_{BA} & P_{BABB}\varphi_{BA} & 0 & 0 & 0 & 0 \\ \hline 0 & 0 & 0 & 0 & P_{BBAA}\varphi_{BB} & P_{BBAB}\varphi_{BB} & 0 & 0 \\ 0 & 0 & 0 & 0 & 0 & 0 & P_{BBBA}\varphi_{BB} & P_{BBBB}\varphi_{BB} \end{array} \right]$$

5.3.4 The Matrices [W], [Φ] and [Q] in the Case of Interstratified Systems with g Components, for any Given Value of S

The expressions for the matrices [W], [Φ] and [Q] given in the preceding paragraphs and corresponding to $S = 0, 1, 2$ or 3 show that changing S to S+1 modifies both the order of the matrices and the number and position of the elements which they contain. Each element of the matrices corresponding to $S = 1$ becomes a $(g \times g)$ matrix when $S = 2$ and, similarly, each of the elements in the new matrices also becomes a $(g \times g)$ matrix when $S = 3$. In general, for any value of S, the replacement of the matrix elements by submatrices must be performed g^{s-1} times, resulting in the creation of matrices of order g^s, while the total number of submatrices will be $g^{2(s-1)}$. For any given values of g and S the following features of the matrices should be noted:

– the elements on the principal diagonal of [W] correspond to the relative proportion of each possible composition of the substack containing S layers; the number of elements in the diagonal is g^s and the number of indices in each element of [W] is S;
– the matrix [Φ] can be written as in Eq. (5.25), where the submatrices are expressed as $[\Phi]_{ij} = \Phi_i^* \Phi_j [U]$, where [U] is a matrix of order g^{s-1} in which all the elements are unity;
– each nonzero element of [Q] is the product of two terms; the first is the probability of one of the g types of layers following a substack of S given layers; the second is a phase-shift term corresponding to the translation between the first two layers in the substack; there are g^{s+1} nonzero elements in [Q] and the number of indices in the probability term is (S+1), whereas the phase-shift term is always limited to two indices.

This "extrapolation by analogy" approach has been used by a number of authors (Kakinoki and Komura 1954a; Allegra 1964; Plançon 1981) to describe the matrices corresponding to $S > 1$. **Practically**, the method is, however, insufficient to give full control of the matrix formalism for all combination of values of g and S, unless the numerical rules given below (Sakharov et al. 1982a, 1983) are followed, and provided the structure factor of each type of layer is known, as well as the translation vectors between adjacent layers and the corresponding probability coefficients.

In the general case, it is convenient to use the numbers 1, 2, 3, ... to identify the different types of layers instead of the letters A, B, C, ... We then observe that:

a) In the matrix [W], the indices i j k ... of the successive nonzero elements in the principal diagonal form an ordered set of numbers, with each number larger than the one preceding it. Thus, when $g = 3$ (A = 1, B = 2, and C = 3) and $S = 2$, the successive indices of W_{ij} are 11, 12, 13, 21, 22, 23, 31, 32, and 33. Similarly, the combination $g = 2$ and $S = 3$ gives, for the indices of W_{ijk}: 111, 112, 121, 122, 211, 212, 221, and 222.

b) The matrix [Q] is constructed by placing horizontally and vertically, in ascending order, the numbers corresponding to the indices of the elements P_{ijk}. Each nonzero element of [Q] is at the intersection of a row and a column such that the first $(n-1)$ digits of the n-digit column-suffix are identical to the last $(n-1)$ digits of the n-digit suffix of the row. The indices (i, j, k) of the corresponding $P_{ijk}\varphi_{ij}$ element are then formed by the suffix of the row followed by the last digit of the column-suffix. Thus for an interstratified system with $g = 3$ and $S = 2$, we would write

$$[Q] = \begin{array}{c} \\ \mathit{11} \\ \mathit{12} \\ \mathit{13} \\ \mathit{21} \\ \mathit{22} \\ \mathit{23} \\ \mathit{31} \\ \mathit{32} \\ \mathit{33} \end{array} \begin{array}{c} \begin{array}{ccccccccc} \mathit{11} & \mathit{12} & \mathit{13} & \mathit{21} & \mathit{22} & \mathit{23} & \mathit{31} & \mathit{32} & \mathit{33} \end{array} \\ \left[\begin{array}{ccccccccc} 111 & 112 & 113 & 0 & 0 & 0 & 0 & 0 & 0 \\ 0 & 0 & 0 & 121 & 122 & 123 & 0 & 0 & 0 \\ 0 & 0 & 0 & 0 & 0 & 0 & 131 & 132 & 133 \\ 211 & 212 & 213 & 0 & 0 & 0 & 0 & 0 & 0 \\ 0 & 0 & 0 & 221 & 222 & 223 & 0 & 0 & 0 \\ 0 & 0 & 0 & 0 & 0 & 0 & 231 & 232 & 233 \\ 311 & 312 & 313 & 0 & 0 & 0 & 0 & 0 & 0 \\ 0 & 0 & 0 & 321 & 322 & 323 & 0 & 0 & 0 \\ 0 & 0 & 0 & 0 & 0 & 0 & 331 & 332 & 333 \end{array}\right] \end{array}$$

in which, for instance, the index 121 corresponds to the product $P_{121}\exp(2\pi i \vec{s}\cdot\vec{r}_{12})$.

c) Lastly, for given S and g values, $[\Phi]$ is composed of submatrices $[\Phi_{ij}]$ of order g^{s-1} which are at the intersection of row i and column j.

5.4 Intensity of the Wave Diffracted by Systems with g Types of Layers, for any Value of S and R

5.4.1 Matrix Formalism for Systems with Identical Layers in the Same Azimuthal Orientation, with Translational Defects and an Interaction Parameter R ≥ 1

The starting point for the calculation of the intensity of the diffracted waves is Eq. (3.2). Since all the layer structure factors are identical, the relationship becomes

$$i(\vec{s}) = \frac{\Omega}{\sigma}\,|\Phi(\vec{s})|^2\left\{1+2\,\mathrm{Re}\sum_{n=1}^{M-1}\frac{M-n}{M}\,\overline{\exp(2\pi i\vec{s}\vec{r}_n)}\right\}$$

or, using the matrix formalism

$$i(\vec{s}) = \frac{\Omega}{\sigma} |\Phi(\vec{s})|^2 \left\{ Tr\,[U]\,[W] + 2\,Re \sum_{n=1}^{M-1} \frac{M-n}{M}\, Tr\,[U]\,[W]\,[Q]^n \right\} .$$

Since Tr [U] [W] = 1, the last equality yields

$$i(\vec{s}) = \frac{\Omega}{\sigma} |\Phi(\vec{s})|^2 \left\{ 1 + 2\,Re \sum_{n=1}^{M-1} \frac{M-n}{M}\, Tr\,[U]\,[W]\,[Q]^n \right\} . \qquad (5.30)$$

The description of lamellar structures composed of identical layers involving translation defects with a range of interaction $R \geq 1$ is best done without reference to the layers themselves but rather using the succession of translations. If $\vec{r}_1$, $\vec{r}_2$, $\vec{r}_3, \ldots, \vec{r}_N$ be the N different translation vectors present in the stack, each can be associated with its probability of occurrence W_1, W_2, $W_3, \ldots, W_N$ with

$$\sum_{i=1}^{N} W_i = 1 .$$

For a given value of R, the coefficients P_{ij}, P_{ijk}, $P_{ijkl}, \ldots$ (with $i, j, k, l, \ldots = 1, 2, 3, 4 \ldots N$) represent the conditional probabilities for different possible sequences of translations. These conditional probabilities are inter-related and connected with the relative proportions W_i in the same fashion as in interstratified structures presented in Chapter 4 and they follow the same rules. Matrices [W] and [Q] given in Section 5.3 for various values of S ($S = 0, 1, 2, 3, \ldots$) can then be used in the same form in Eq. (5.30), on condition that R and N play here the same roles as S and g.

5.4.2 Matrix Formalism for Interstratified Structures with any Number of Translations Without Mutual Interaction (R = 0)

In all the examples examined in the preceding paragraphs, the translation vector $\vec{r}_{ij}$ ($i, j = 1, 2, 3, \ldots g$) between two adjacent layers of a given type was assumed to be always the same. Now we shall consider structures with any S in which layer pairs of a given type may be related by several translations distributed with $R = 0$.

Evidently, the description of this type of structure cannot be limited to that of the probability parameters characteristic of the sequence of the layer types in a direction perpendicular to the layers. We also need to know the various possible relative positions of adjacent layers of different types. We are therefore led to introduce new probabilistic coefficients W_k^{ij} to characterize the proportions of the different translations $\vec{r}_k^{\,ij}$ found between two adjacent layers respectively of type i and j. If N_{ij} is the total number of translations possible between the two adjacent layers of an (i, j)-pair, then $1 \leqslant k \leqslant N_{ij}$, and

$$\sum_{k=1}^{N_{ij}} W_k^{ij} = 1 \quad (\text{with } i,j = 1,2,\ldots g) \ .$$

Consider now an interstratified system of g different types of layers such that the relative position of the two adjacent layers of a given (i, j)-pair is one among N_{ij} possibilities. The probability attached to a given substack is equal to the product of the probability coefficients characteristic of each type of subsequence of different layers multiplied by the product of the probability coefficients characteristic of the proportion of each type of relative position. Taking, as an example, $g = 2$ for a two-component (A and B) system, and supposing that $N_{ij} = 2$ for any adjacent (i, j)-pair, we see that the possible two-layer combinations are A_1A, A_2A, A_1B, $A_2B\ldots$, or that the corresponding translation vectors are $\vec{r}_1^{AA}$, $\vec{r}_2^{AA}$, $\vec{r}_1^{AB}$, $\vec{r}_2^{AB}$, etc.... Hence, the probability of finding a substack such as $A_1A_2B_1A_1B_2B$ is:

if $S = 0$: $\quad W_A^3 W_B^3 W_1^{AA} W_2^{AB} W_1^{BA} W_1^{AB} W_2^{BB}$;

if $S = 1$: $\quad W_A P_{AA} P_{AB}^2 P_{BA} P_{BB} W_1^{AA} W_2^{AB} W_1^{BA} W_1^{AB} W_2^{BB}$;

if $S = 2$: $\quad W_A P_{AA} P_{AAB} P_{ABA} P_{BAB} P_{ABB} W_1^{AA} W_2^{AB} W_1^{BA} W_1^{AB} W_2^{BB}$;

etc.

We can now rewrite Eqs. (5.18), (5.19) and (5.20), which give the intensity of the wave diffracted by the outermost of the (n+1) layers in a substack, as

$$\sum_{i=1}^{g}\sum_{j=1}^{g} W_{ij}\Phi_i^*\Phi_j \sum_{m=1}^{N_{ij}} W_m^{ij} \exp(2\pi i\vec{s}\cdot\vec{r}_m^{ij}) \ ; \quad \text{for} \quad n = 1 \tag{5.31}$$

$$\sum_{i=1}^{g}\sum_{j=1}^{g}\sum_{k=1}^{g} W_{ijk}\Phi_i^*\Phi_k \sum_{m=1}^{N_{ij}}\sum_{n=1}^{N_{jk}} W_{mn}^{ijk} \exp[2\pi i\vec{s}\cdot(\vec{r}_m^{ij}+\vec{r}_n^{jk})] \ ; \ \text{for } n = 2 \tag{5.32}$$

$$\sum_{i=1}^{g}\sum_{j=1}^{g}\sum_{k=1}^{g}\sum_{l=1}^{g} W_{ijkl}\Phi_i^*\Phi_l \sum_{m=1}^{N_{ij}}\sum_{n=1}^{N_{jk}}\sum_{p=1}^{N_{kl}} W_{mnp}^{ijkl} \exp[2\pi i\vec{s}\cdot(\vec{r}_m^{ij}+\vec{r}_n^{jk}+\vec{r}_p^{kl})]$$

$$\text{for} \quad n = 3 \tag{5.33}$$

in which W_{ij}, W_{ijk}, W_{ijkl} are modified according to the value of the parameter S and in which $W_{mn}^{ijk} = W_m^{ij} W_n^{jk}$ and $W_{mnp}^{ijkl} = W_m^{ij} W_n^{jk} W_p^{kl}$.

It is possible to verify that the matrix formalism for the intensity in (5.22) is still valid provided the phase-shifts φ, in [Q], be expressed as:

$$\varphi_{jk} = \sum_{m=1}^{N_{jk}} W_m^{jk} \exp(2\pi i\vec{s}\cdot\vec{r}_m^{jk}) \ . \tag{5.34}$$

5.4.3 Matrix Formalism in the General Case of Interstratified Systems

In general, the sequence of layer types does not depend on the relative mutual positions of the successive layers. In other words, for any given value of S, the parameter R which describes the interactions between translations may have any value, and vice versa (Sakharov et al. 1982b, 1983).

Let us then consider a system with different types of layers. Between any two given layer i and j, there may be M_{ij} different translations noted $\vec{r}_m^{ij}$ (with $m = 1, 2, \ldots, M_{ij}$), with equivalent notations for the pairs j and k ($\vec{r}_n^{jk}$ and M_{jk}), k and l ($\vec{r}_q^{kl}$ and M_{kl}), etc. We let W_m^{ij} be the proportion or relative abundance of the translation $\vec{r}_m^{ij}$, P_{mn}^{ijk} be the conditional probability of having the translation $\vec{r}_n^{jk}$ after a translation $\vec{r}_m^{ij}$ in an (i, j, k)-triplet, P_{mnq}^{ijkl} be the conditional probability of having the translation $\vec{r}_q^{kl}$ after the sequence of translations $(\vec{r}_m^{ij}+\vec{r}_n^{k})$ in a substack of four layers i, j, k, l, etc. Among the different probability coefficients, a number of relationships exist:

$$\sum_{m=1}^{M_{ij}} W_m^{ij} = 1 \; ; \quad \sum_{n=1}^{M_{jk}} P_{mn}^{ijk} = 1 \; ; \quad \sum_{q=1}^{M_{kl}} P_{mnq}^{ijkl} = 1 \tag{5.35}$$

$$\sum_{m=1}^{M_{ij}} W_m^{ij} P_{mn}^{ijk} = W_n^{jk} \; ; \quad \sum_{m=1}^{M_{ij}} W_{mn}^{ijk} P_{mnq}^{ijkl} = W_{nq}^{jkl} \tag{5.36}$$

in which $W_{mn}^{ijk} = W_m^{ij} P_{mn}^{ijk}$.

These relationships give the possibility of describing quantitatively any substack of different layers, with their actual positions, when the parameters R and S are known. For instance, the probability of having a substack $i_m j_n k_q l$ for given i, j, k, l types and m, n, q indices will be

for S = 1 and R = 1: $W_i P_{ij} P_{jk} P_{kl} W_m^{ij} P_{mn}^{ijk} P_{nq}^{jkl}$;

for S = 1 and R = 2: $W_i P_{ij} P_{jk} P_{kl} W_m^{ij} P_{mn}^{ijk} P_{mnq}^{ijkl}$;

for S = 2 and R = 2: $W_i P_{ij} P_{ijk} P_{jkl} W_m^{ij} P_{mn}^{ijk} P_{mnq}^{ijkl}$;

etc.

It is obvious that in the case examined, as well as in the general case, the conditional probabilities included in the factors W_{mn}^{ijk} and W_{mnq}^{ijkl} of Eqs. (5.35) and (5.36) must be compatible with the chosen values of R.

The intensity of the wave diffracted by the interstratified system can be computed, once again, using the matrix formalism of Eq. (5.22) provided the matrices [W], [Φ], and [Q] are expressed in a new way. This will be detailed using, as an example, an interstratified system containing two types of layers labeled 1 and 2, with range parameters S = 1 and R = 1. Let us assume furthermore that the number of possible translations between the (1, 1)-, (1, 2)-, (2, 1)-, and (2, 2)-pairs of layers are respectively $M_{11} = 1$, $M_{12} = 2$, $M_{21} = 3$ and $M_{22} = 1$. Matrix [W] is now

$$[W] = \left[\begin{array}{ccc|cccc} W_1 W_1^{11} & & & & & & \\ & W_1 W_1^{12} & & & & & \\ & & W_1 W_2^{12} & & & & \\ \hline & & & W_2 W_1^{21} & & & \\ & & & & W_2 W_2^{21} & & \\ & & & & & W_2 W_3^{21} & \\ & & & & & & W_2 W_1^{22} \end{array}\right] \tag{5.37}$$

in which W_1 and W_2 are the proportions of type 1 and type 2 layers respectively and W_1^{11}, W_m^{12}, W_n^{21}, and W_1^{22} are the relative abundances of the different possible types of translations between the pairs (1, 1), (1, 2), (2, 1) and (2, 2).

Similarly, matrix [Φ] is expressed as

$$[\Phi] = \left[\begin{array}{ccc|cccc} \Phi_1^*\Phi_1 & \Phi_1^*\Phi_1 & \Phi_1^*\Phi_1 & \Phi_2^*\Phi_1 & \Phi_2^*\Phi_1 & \Phi_2^*\Phi_1 & \Phi_2^*\Phi_1 \\ \Phi_1^*\Phi_1 & \Phi_1^*\Phi_1 & \Phi_1^*\Phi_1 & \Phi_2^*\Phi_1 & \Phi_2^*\Phi_1 & \Phi_2^*\Phi_1 & \Phi_2^*\Phi_1 \\ \Phi_1^*\Phi_1 & \Phi_1^*\Phi_1 & \Phi_1^*\Phi_1 & \Phi_2^*\Phi_1 & \Phi_2^*\Phi_1 & \Phi_2^*\Phi_1 & \Phi_2^*\Phi_1 \\ \hline \Phi_1^*\Phi_2 & \Phi_1^*\Phi_2 & \Phi_1^*\Phi_2 & \Phi_2^*\Phi_2 & \Phi_2^*\Phi_2 & \Phi_2^*\Phi_2 & \Phi_2^*\Phi_2 \\ \Phi_1^*\Phi_2 & \Phi_1^*\Phi_2 & \Phi_1^*\Phi_2 & \Phi_2^*\Phi_2 & \Phi_2^*\Phi_2 & \Phi_2^*\Phi_2 & \Phi_2^*\Phi_2 \\ \Phi_1^*\Phi_2 & \Phi_1^*\Phi_2 & \Phi_1^*\Phi_2 & \Phi_2^*\Phi_2 & \Phi_2^*\Phi_2 & \Phi_2^*\Phi_2 & \Phi_2^*\Phi_2 \\ \Phi_1^*\Phi_2 & \Phi_1^*\Phi_2 & \Phi_1^*\Phi_2 & \Phi_2^*\Phi_2 & \Phi_2^*\Phi_2 & \Phi_2^*\Phi_2 & \Phi_2^*\Phi_2 \end{array}\right] \tag{5.38}$$

Lastly, using a conventional symbolism which will be detailed below, matrix [Q] can be written as

$$[Q] = \begin{array}{c} \\ 1_1 1 \\ 1_1 2 \\ 1_2 2 \\ 2_1 1 \\ 2_2 1 \\ 2_3 1 \\ 2_1 2 \end{array} \begin{array}{ccc|cccc} 1_1 1 & 1_1 2 & 1_2 2 & 2_1 1 & 2_2 1 & 2_3 1 & 2_1 2 \\ \hline 1_1 1_1 1 & 1_1 1_1 2 & 1_1 1_2 2 & 0 & 0 & 0 & 0 \\ 0 & 0 & 0 & 1_1 2_1 1 & 1_1 2_2 1 & 1_1 2_3 1 & 1_1 2_1 2 \\ 0 & 0 & 0 & 1_2 2_1 1 & 1_2 2_2 1 & 1_2 2_3 1 & 1_2 2_1 2 \\ \hline 2_1 1_1 1 & 2_1 1_1 2 & 2_1 1_2 2 & 0 & 0 & 0 & 0 \\ 2_2 1_1 1 & 2_2 1_1 2 & 2_2 1_2 2 & 0 & 0 & 0 & 0 \\ 2_3 1_1 1 & 2_3 1_1 2 & 2_3 1_2 2 & 0 & 0 & 0 & 0 \\ 0 & 0 & 0 & 2_1 2_1 1 & 2_1 2_2 1 & 2_1 2_3 1 & 2_1 2_1 2 \end{array} \tag{5.39}$$

where the upper case (UC) digits in a notation such as $2_3 1_1 1$ are understood to represent the succession of layer types, whereas the lower case (LC) indices identify the translations between the pairs corresponding to the UC digits. In other words, the substack 211 considered has a translation $\vec{r}_3^{21}$ between the type 2 and the type 1 layer, and a translation $\vec{r}_1^{11}$ between the two type 1 layers. The symbol $2_3 1_1 1$ is therefore a shorthand notation for

$$P_{21}\ P_{31}^{211} \exp(2\pi i \vec{s}\cdot\vec{r}_3^{21})\ .$$

Once again, and as detailed in Sections 5.3.3 and 5.3.4, the matrices can be partitioned in a number of submatrices, but the latter may be no longer square.

In the most general case, i.e., when $S, R, g, M_{ij}, M_{ijk}\ldots$ have any value, the matrices [W], [Φ] and [Q] can be written down by applying the following rules:

1. [W] **is a diagonal matrix.** Each element in the matrix [W] corresponds to a product of two probabilities: on one hand, the probability $W_{ijk}\ldots$ (in which the number of subscript indices $ijk\ldots$ is equal to S) of having a substack of S layers, and, on the other hand, the probability $W_{mn\ldots}^{ijk\ldots}$ (in which there are (R+1) superscript and R subscript indices) attached to the sequence of translations. In order to construct this matrix the value of Sup [S, (R+1)], i.e., which, of S or of (R+1), is the largest, is first determined. If $(R+1) \geqslant S$ the sequence of the superscripts for elements reading from the top left corner to the bottom right corner, is given by the superscripts of the coefficients W_m^{ij}, W_{mn}^{ijk}, etc. depending on the value of (R+1). With reference to the above example, each element in the matrix [W] is represented by the symbolic notation (ij_m) where ij represents the types of layers (from among the g types present in the system) in an adjacent (i, j)-pair and m is one of the M_{ij} possible translation between them. The order of the elements in the diagonal is such that, reading from the top left corner to the bottom right corner, the successive values of the symbolic number ij are either equal or increase, as do also the values of m corresponding to a given (i, j)-pair. This therefore leads to the results given in Eq. (5.37) corresponding to the successive values of ijm: 11_1, 12_1, 12_2, 21_1, 21_2, 21_3, and 22_1.

If $S > (R+1)$, the sequence of the superscripts for elements is given by the subscripts of the coefficients W_{ij}, W_{ijk}, etc. .... The number of indices depends on the value of S. As an example, let $S = 3$, $R = 1$, $g = 2$, $M_{11} = 1$, $M_{12} = 2$, $M_{21} = 3$ and $M_{22} = 1$. The sequence of the indices in the diagonal elements of matrix [W] is such that, reading from the top left corner to the bottom right corner, the successive values of the symbolic number ijk are either equal or increase as do also the values of m corresponding to a given (j, k)-pair. This leads to the following sequence: 111_1, 112_1, 112_2, 121_1, 121_2, 121_3, 122_1, 211_1, 212_1, 212_2, 221_1, 221_2, 221_3, and 222_1. The symbol ijk_m is a shorter notation for $W_{ijk}W_m^{jk}$.

2. **The matrix [Φ] is square.** None of its elements is zero, and each is a product of the structure factor of each of two layer types among all the g layer types. In this matrix, the square submatrices, in which each element is equal to $\Phi_i^*\Phi_i$, may be separated along the principal diagonal. The sequence of these submatrices along the diagonal is determined by the increasing value of i. The order of each of these submatrices is

$$g^{S-R-1}\times L_i \quad \text{if} \quad S > R+1$$

or

$$L_i \quad \text{if} \quad S \leqslant R+1$$

in which

$$L_i = \sum_{j=1}^{g} \sum_{k=1}^{g} \cdots \sum_{l=1}^{g} \sum_{m=1}^{g} \underbrace{M_{ij} M_{jk} \cdots M_{lm}}_{\text{R terms}}$$

so that the number of factors in each term of the sum is equal to R.

After writing down the submatrices of the diagonal, [Φ] is completed with rectangular submatrices, each containing a single type of element, namely $\Phi_i^* \Phi_j$ (with $i \neq j$) where i is the row and j the column of the square submatrices bordering the rectangular submatrices.

3. **The elements of matrix** [Q] correspond to the product of a number of factors, namely (i) the conditional probabilities of having substacks containing (S+1) layers; (ii) the conditional probabilities characteristic of the different translations in the substacks with (R+2) layers; (iii) a phase-shift factor corresponding to the translation between the first two layers of these latter substacks. In order to express the elements of this matrix using the symbolic notation mentioned before, a "guide-row" and an identical "guide-column" are written down respectively above and to the left of the matrix [Q] to be filled, composed of sequences of UC digits (for the types of layers) forming numbers which are either equal to or greater than the preceding one, and LC digits appearing as subscript to identify the translations between layers of different types, also in increasing order. In the symbolic notation which characterizes each element of the matrix, the number of UC digits describing the sequence of layers in the substack is

$$\{1 + \mathrm{Sup}\,[S, (R+1)]\} \ .$$

The number of LC digits is equal to (R+1).

In the example examined above, we found Sup [S, (R+1)] = 2, hence the number of UC digits must be 3, while the number of LC subscripts is 2. Each nonzero element of [Q] is found at an intersection of a column and a row from the guides such that the last UC digit of the row is the same as the first UC digit of the column, and the symbolic notation is then formed by the number of the column following the first UC digit and the first LC subscript of the row.

Lastly, it may be useful to point out that the order of the matrices [W], [Φ] and [Q] is

$$g^{S-R-1} \sum_{i=1}^{g} L_i \quad \text{when} \quad S > R+1$$

and

$$\sum_{i=1}^{g} L_i \quad \text{when} \quad S \leq R+1 \ .$$

5.5 General Remarks

The mathematical formalism described above in a general form involves a large number of independent variables. For this reason there is always a probability that the modelized diffractive effects, close to the experimental XRD curve, may be calculated using different sets of values of these independent variables. In other words, an unambiguous result can be hardly be expected in the general case practically; the study of complex defective structures is feasible only if part of the variables involved can be determined using independent considerations. Some examples of the application of such an approach are given in the subsequent chapters.

References

Allegra G (1964) The calculation of the intensity of X-rays diffracted by monodimensionally disordered structures. Acta Crystallogr 17:172–178
Burger J (1959) Vector space and its application in crystal structure investigation. John Wiley & Sons, New York
Cradwick PD (1975) On the calculation of one-dimensional X-ray scattering from interstratified material. Clay Mineral 10:347–356
Drits VA, Sakharov BA (1976) X-ray structure analysis of interstratified minerals. Nauka, Moscow, 255 pp (in Russian)
D'yakonov YuS (1961) Application of Fourier transform methods for interpretation of mixed-layer mineral diffraction patterns. Kristallografia 6-4:624–625 (in Russian)
D'yakonov YuS (1962) Interpretation of mixed-layer mineral diffraction patterns by direct methods of Fourier transform. In: Sidorenko GA (ed) X-ray study of mineral raw materials. Nedra-Moscow 1:pp 34–42 (in Russian)
D'yakonov YuS (1974) Development of direct methods of Fourier transform for the interpretation of mixed-layer mineral diffraction patterns. In: Frank-Kamenetskii VA (ed) Crystallochemistry and structure of minerals. Nauka, Leningrad, pp 33–43 (in Russian)
D'yakonov YuS (1977) About techniques for calculation of layer distribution curves in the direct method for the interpretation of mixed-layer structures. In: Sidorenko GA (ed) X-ray study of mineral raw materials, vol. 11. Nedra, Moscow, pp 76–80 (in Russian)
D'yakonov YuS (1978) Interpretation of layer distribution curves in the direct method of interpretation of mixed layer curves. In: Sidorenko GA (ed) X-ray study of mineral raw materials. Nedra, Moscow, pp 7–14 (in Russian)
Guinier A (1964) Théorie et technique de la radiocristallographie. Dunod, Paris, 740 pp
Hendricks SB, Teller E (1942) X-ray interference in partially ordered layer lattices. J Chem Phys 10:147–167
James RW (1950) The optical principles of the diffraction of X-rays. Bell, London
Kakinoki J, Komura Y (1952) Intensity of X-ray diffraction by one-dimensionally disordered crystals. I: General deviation in the case of the "Reichweite" S = 0 and 1. J Phys Soc Jpn 7:30–35
Kakinoki J, Komura Y (1954a) Intensity of X-ray diffraction by one-dimensionally disordered crystals II: General derivation in the case of the correlation range S≥2. J Phys Soc Jpn 9:169–176
Kakinoki J, Komura Y (1954b) Intensity of X-ray diffraction by one dimensionally disordered crystal. The close packed structure. J Phys Soc Jpn 9:177–183

Kakinoki J, Komura Y (1965) Diffraction by a one-dimensionally disordered crystal. I: The intensity equation. Acta Crystallogr 19:137–147

Mac Ewan DMC (1956) Fourier transform methods for studying scattering from lamellar system I: A direct method for analysing interstratified mixtures. Kolloid Z 149:96–108

Mac Ewan DMC (1958) Fourrier transform methods for studying X-ray scattering from lamellar systems. II: The calculation of X-ray diffraction effects for various types of interstratifications. Kolloid Z 156:1:61–67

Plançon A (1981) Diffraction by layer structures containing different kinds of layers and stacking faults. J Appl Crystallgr 14:300–304

Reynolds RC (1967) Interstratified clay systems: calculation of total one-dimensional diffraction functions. Am Mineral 52:661–672

Reynolds RC (1980) Interstratified clay minerals. In: Brindley GW, Brown G (eds) Crystal structures of clay minerals and their X-ray identification. Mineral Soc, London, pp 229–304

Reynolds RC, Hower J (1970) The nature of interlaying in mixed-layer illite-montmorillonite. Clays Clay Mineral 18:25–36

Sakharov BA, Naumov AS, Drits VA (1982a) X-ray diffraction by mixed-layer structures with random distribution of stacking faults. Dokl Akad Nauk SSSR 265:339–343 (in Russian)

Sakharov BA, Naumov AS, Drits VA (1982b) X-ray intensities scattered by layer structure with short range ordering parameters $S \geqslant 1$ and $G \geqslant 1$. Dokl Akad Nauk SSSR 265:871–874 (in Russian)

Sakharov BA, Naumov AS, Drits VA (1983) X-ray scattering by defect layer structures. Kristallografia 28:951–958 (in Russian)

Tettenhorst R, Grim RE (1975a) Interstratified clays. I: Theoretical. Am Mineral 60:49–59

Tettenhorst R, Grim RE (1975b) Interstratified clays. II: Some experimental results. Am Mineral 60:60–65

Weir AH, Rayner JH (1974) An interstratified illite-smectite from Denchworth Series Soil in Weathered Oxford Clay. Clay Mineral 10:173–187

CHAPTER 6

Experimental Techniques Adapted to the Study of Microdivided Lamellar Systems

The method of structural analysis by modelization of X-ray diffraction patterns and comparison with the experimental diagrams is only efficient if all the experimental functions which can perturb the intensity distributions in XRD pattern are perfectly controlled. Perturbations may be of two kinds:

- some may have an instrumental origin and be connected with the size and divergence of the X-ray beam, its monochromaticity, the size or shape of the sample, etc.
- others may be connected with the sample itself, its absorption of the incident beam, or preferred orientation of the particles in the powder, etc.

A careful control of the experimental conditions may minimize some of these perturbations (especially those of the first category) to a level many orders of magnitude smaller than the deformations produced by the presence of structural defects in the specimen. However, other perturbations may not be minimized or even reduced. Accurate mathematical expressions to represent them must then be sought and introduced in the computation of the theoretical diffractogram, as was done in Chapter 2.4 with the functions describing the effect of orientation of the particles in a powder.

In the present chapter, we shall first briefly describe the principles underlying the various experimental mountings used in powder diffractometry in order to obtain a proper understanding of the causes responsible for the perturbations. We shall then define some of the experimental conditions well adapted to the quantitative study of the real structure of disordered lamellar systems.

6.1 Survey of the Techniques Most Frequently Used in Powder Diffractometry

6.1.1 The Powder Diagram

If a powder of triperiodic perfect microcrystals is isotropic, it will contain crystallites oriented so that their set of crystallographic planes characterized by the indices $(h_1 k_1 l_1)$ and separated by the interplanar distance d_1 will be at the Bragg angle θ_1, with respect to an incident X-ray beam of wavelength λ. Diffraction occurs along a direction at an angle $2\theta_1$ with respect to the incident beam, with $2d_1 \sin\theta_1 = \lambda$. If the number of such crystallites in the powder is large, the set of diffracted beams will constitute a continuous conical sheet with half apical angle $2\theta_1$. Other crystallites in the powder will, for instance, be oriented so that the set of planes characterized by the indices $(h_2 k_2 l_2)$ and the distance d_2 will diffract, giving rise to a second conical sheet of diffraction with half apical angle $2\theta_2$, etc. Thus, globally, the interaction of the X-ray beam with the isotropic powder results in the formation of a number of coaxial cones, each characterized by the half angle $2\theta_i$ at its apex (Fig. 6.1).

6.1.2 The Debye-Scherrer-Hull Mountings

Many experimental determinations of powder diagrams use instrumental arrangements directly traceable to that devised almost simultaneously by Debye and Scherrer (1916) and by Hull (1917), shown in Fig. 6.2. The beam issuing from the X-ray tube is first passed through the filter (F) to eliminate the Kβ radiation. Before falling on the specimen (S), the remaining Kα radiation is rendered nearly parallel by passage through the collimator (C). The trace of the conical sheets of diffraction form a series of rings (or portions thereof) with diameter $4R\theta_i$ in the median plane of a photographic film (P) placed cylindrically on the circumference of a circle of radius R. Depending on the type of X-ray camera, the axis of the cylinder, which coincides with the center of the specimen, may be horizontal or vertical, as in Fig. 6.2.

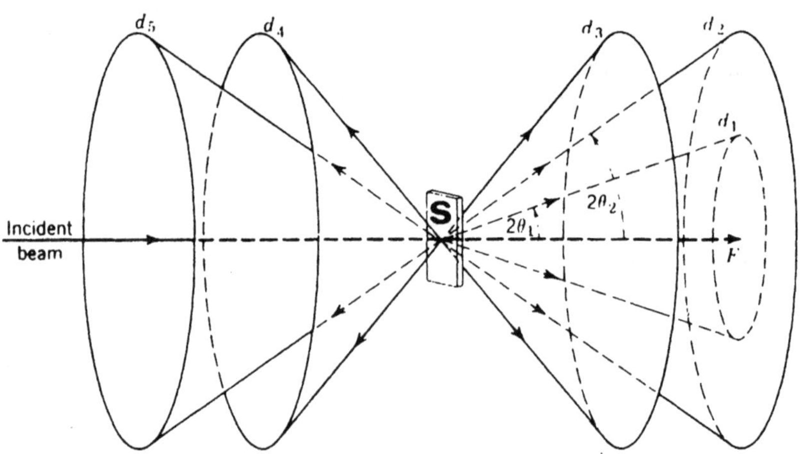

Fig. 6.1. Diffraction cones produced by a powder sample S

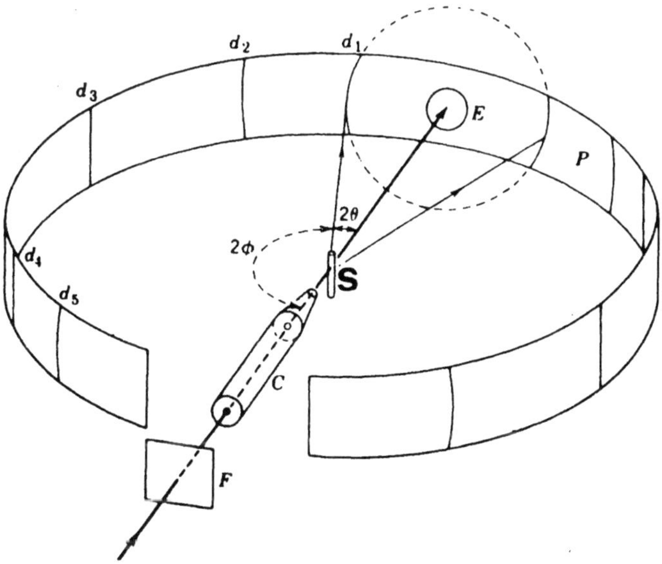

Fig. 6.2. Geometry of the Debye-Scherrer arrangement for powder diffractometry

This instrumental arrangement has a number of drawbacks:

1. Collimation of the beam in height and in width results in a serious loss of available energy particularly detrimental to the study of disordered structures since, in that case, the diffracted beam is not localized in narrow domains close to the Bragg angles $2\theta_i$ but is rather spread over practically all space with more or less important local variations whose importance depends on the nature and amount of structural defects in the crystals. Hence, the intensity corresponding to a given angle is usually too small to yield good photographic records.
2. In spite of the collimation, the beam is slightly divergent and the accurate measurement of the angle $2\theta_i$ is marred by the consequent broadening of the diffraction lines.
3. A background often appears on the film because of (1) short or long wavelength radiations which are not eliminated by the filter and (2) parasitic diffusions due to L radiation from tungsten evaporated by the filament and condensed on the target.
4. Obviously, the $K\alpha_1$ and $K\alpha_2$ radiations cannot be separated by the filter.
5. Although the use of a photographic film permits the simultaneous recording of all the lines in the diffractogram, it suffers from a number of intrinsic handicaps: narrow range of linearity of blackening vs. integrated exposure, granularity of the emulsion, necessity of further photometry to determine the intensity and the line profile, etc.

6.1.3 Use of a Recording Counter and of a Monochromator

A first improvement is the replacement of the photographic film by a scintillation or a proportional counter geared to move uniformly along the goniometric circle, and permitting the direct and continuous recording of the diffraction effects. However, even with low speeds of rotation of the counter, the number of photons captured at each position is small and, once again, this technique is not quite adapted to the study of disordered systems since the counting statistics do not yield sufficient accuracy in measurements of the intensity. Moreover, the other drawbacks mentioned above remain.

Elimination of the $K\beta$ radiation and separation of the $K\alpha_1 - K\alpha_2$ doublet in the individual lines is best obtained by passing the X-ray beam through a monochromator. [For general information concerning monochromator geometries, choice of the crystal, etc., see Guinier (1964) or Klug and Alexander (1974)].

The Johann-Guinier monochromator is the type most often used (Johann 1931; Cauchois 1933). The active part is a crystal (quartz, graphite ...) bent elastically so that the normal to a specific set of hkl reflections converge onto the center of curvature of the crystal platelets. After reflection, rays of wavelength λ all seem to converge to the point F (Fig. 6.3) but do not actually pass through a common point since this monochromator is astigmatic. If the source is a line, the diffracted beam converges to a line rather than to a point focus.

The monochromator can be used either on the incident beam, and one then speaks of an "upstream" mounting, or on the beam diffracted by the specimen, when the mounting is said to be of the "downstream" type. Each has merits and disadvantages:

- The downstream monochromator stops all fluorescence radiation emitted by the sample, as well as incoherent diffusion and parasitic reflections. It is, how-

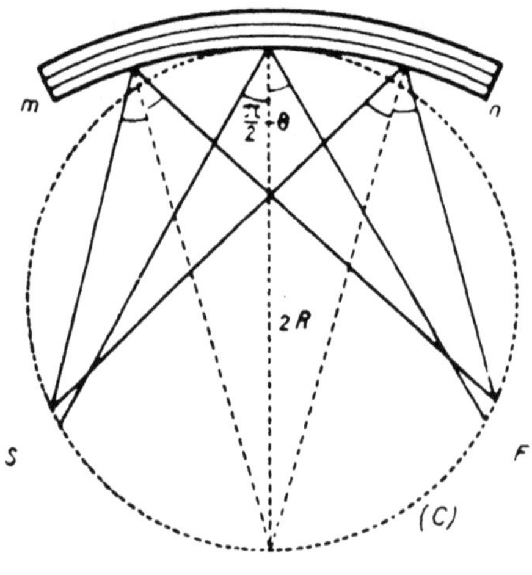

Fig. 6.3. Johann-type bent-crystal monochromator

ever, not adapted to recording techniques which use linear or curved localization detectors (cf. Chap. 7.2.5).

- Upstream monochromators are well suited to instrumentation involving localized detectors; furthermore, the specimen under examination receives only a minimum amount of radiation, and this amount can easily be monitored quantitatively. On the other hand, fluorescence radiation, incoherent diffusion, etc. are not eliminated and it is necessary to include some means of energy discrimination in the instrumentation. In spite of these shortcomings, the present development in the field of localization detectors had led to a marked preference for upstream mountings of monochromators, with a radius of curvature specifically adapted to the wavelength of the incident radiation and to the dimension of the focalization circle necessary to give good spatial resolution of the reflections.

Figure 6.4 sketches the principal features of an installation using an upstream monochromator to be used in reflection (Fig. 6.4a) or transmission mountings (Fig. 6.4b).

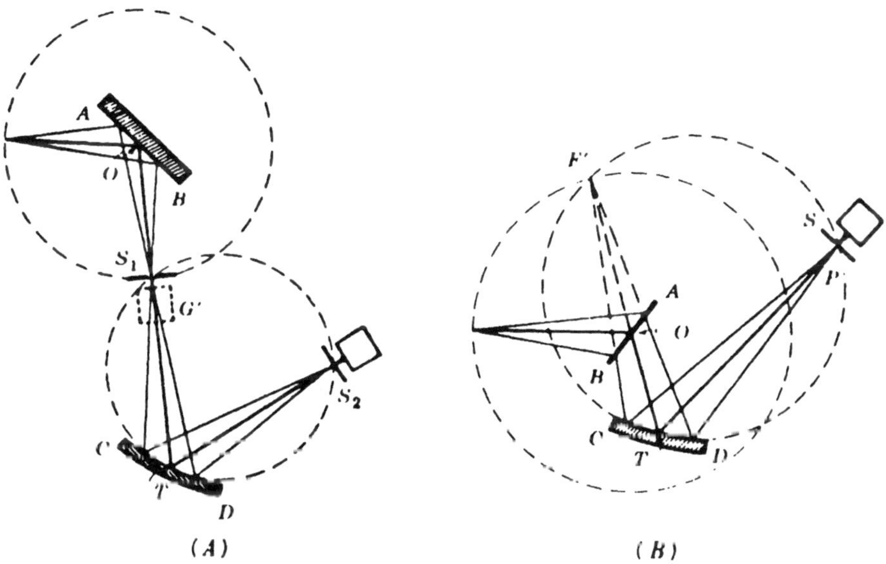

Fig. 6.4. Diffractometric mountings used with bent monochromators; **A** reflection geometry; **B** transmission geometry

6.1.4 Advantages and Drawbacks of the Reflection and Transmission Mountings

In almost all classical set-ups for powder diffractometry, and particularly those in which the determination of the position of the reflections is the principal goal, the specimen is placed in a semi-capillary Lindemann glass tube completely immersed in the incident X-ray beam. Another possibility consists in preparing a flat, platelet-like deposit of powder which can then be used to obtain reflection or transmission diffractograms.

1. The Reflection Mounting

As shown in Fig. 6.4a, the diffracted beam emerges out of the specimen from the same face as the one which receives the incident beam. The mean advantage of this geometry lies in the ease of preparation of the specimen: the powder can be simply deposited on a horizontal support, or suspended in an appropriate liquid and filtered onto a Millipore filter which can be used as sample holder. These experimental conditions are particularly well suited to the study of lamellar systems when only the $00l$ reflections are required: in this case it is necessary to prepare the deposit so that most particles will show a **preferred orientation** with their (001) plane parallel to the horizontal support[11]. When the diffractogram is scanned in the symmetrical $\theta-2\theta$ reflection conditions[12], an enhancement of intensity of the $00l$ bands is obtained from this type of oriented preparation.

In spite of these advantages, the reflection mounting is beset by a number of major disadvantages:

- as shown by Guinier (1964) and confirmed by Klug and Alexander (1974), it is not easy to measure the correct coefficient of absorption of the specimen; this is particularly true when the latter has a thin flat shape because in that case the primary beam is not completely absorbed;
- in a symmetrical $\theta-2\theta$ mounting, the incident beam is at the Bragg angle θ with respect to the midplane of the specimen, so that for small values of θ, the path of the beam within the sample is long, and convolution effects on band profile cannot be neglected.

2. The Transmission Mounting

In this type of mounting (Fig. 6.4b), the diffracted beam emerges from the face opposite the one receiving the incident beam. Sample preparation strives to minimize preferred orientation in the particles, with the drawback that, in a $\theta-2\theta$ symmetrical mounting, the diffractogram now includes reflections corresponding

[11] One of the methods used to prepare oriented deposits is the slow evaporation of a dilute suspension of the particles in a flat-bottomed container.

[12] A sample is said to be in a symmetrical configuration when the angle between the primary beam and the surface of the sample is equal $(\pi/2-\theta)$ or θ (where θ is the Bragg angle) for the transmission and reflection mountings, respectively.

to all allowed values of h, k, l besides the $00l$ series. On the other hand, the determination of the absorption coefficient is now easy (cf. Sect. 6.4.2) and it is also a simple matter to adjust the thickness of the sample for convolution effects (due to sample thickness) to become negligible (cf. Sect. 6.3.2).

6.2 Adaptation of Transmission Techniques to the Study of Microdivided Lamellar Systems

Figure 6.5 gives the general layout of the instrumentation required for transmission diffractometry especially adapted to the study of disordered lamellar structures. The principal characteristics of the various components and the optimum conditions for their use will now be detailed.

6.2.1 The X-Ray Source

Since many powder diffractometers use X-ray tubes with a so-called "line focus" coupled to a crystal monochromator, the focused beam is a line segment. The attainable angular resolution depends on the width of this segment and it is important that the source itself be narrow. Preference is therefore given to sources about $0.04 \times 8.0\,mm^2$ from fine-focus tubes.

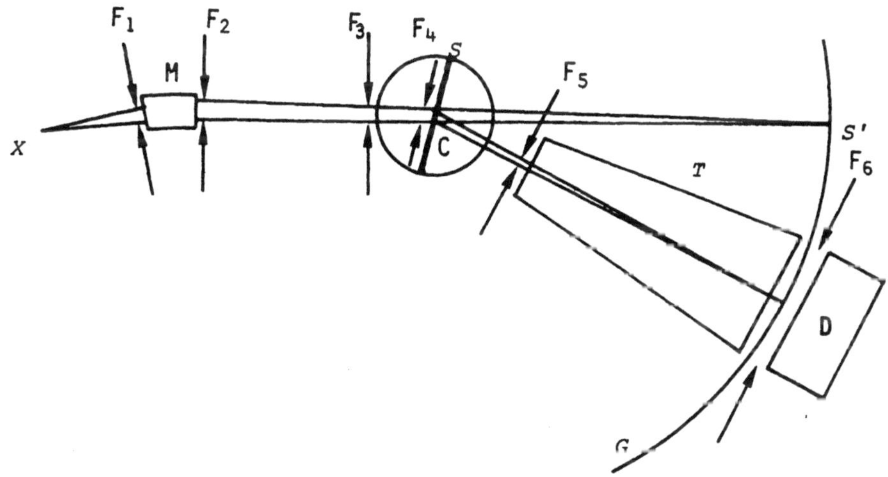

Fig. 6.5. Transmission diffractometry; X-ray source; S' focal point of the beam after monochromatization; G trace of the goniometric circle; M monochromator; S sample; D localization detector; C chamber containing the sample-holder; T evacuated conical tube; F_1 to F_6 slits with adjustable height and width

The choice of anticathode, i.e., of the wavelength to be used, depends on the accuracy required in the domain of reciprocal space explored, with the proviso that the final resolution is linked to the characteristics of the recording system and of the monochromator. Moreover, if the instrumentation does not include an energy-discriminating system, the choice of anticathode must take into account the necessity of avoiding sample fluorescence. Thus, the $CuK\alpha_1$ radiation ($\lambda = 0.15404$ nm) is well adapted to the study of lamellar silicates; however, iron-rich samples are best studied with $MoK\alpha_1$ radiation ($\lambda = 0.070930$ nm) since the difference in energy with the fluorescence radiation is sufficiently important that a 0.1-mm-thick aluminum filter will stop the latter. For the same reason, the filter can be replaced by an electronic discriminator.

6.2.2 The Monochromator

The Johann-Guinier bent crystal monochromator is the type most often used, in an upstream configuration. The crystal is given a constant radius of curvature by the preformed jaws of the vise in which it is held. Large values of the radius are best since they yield a better angular resolution for the peaks. Similarly, the breadth of the beam at the focus decreases as the focal length increases; since this breadth is one of the instrumental functions likely to perturb the intrinsic profile of the reflections, it is important that it be as small as possible. Consider, for instance, a quartz monochromator with a radius of curvature equal to 1.4 m, focusing the $Cu\,K\alpha_1$ radiation at about 51 cm. If the source is a fine-focus tube, the breadth of the beam at the focus is about 100 μm, corresponding to an angular width $\Delta(2\theta) = 0.02°$ for a goniometer of radius 250 mm, and peaks of intrinsic width smaller than $7\Delta(2\theta)$ will be broadened by more than 1%.

Monochromators with variable radius of curvature exist[13] which enable a better adaptation of the system geometry to the nature of the specimen being studied. Their principle (Lemonier et al. 1978) is shown in Fig. 6.6, drawn with reference to synchroton radiation issuing from a beam of orbiting electrons. Horizontal focusing is obtained by reflection of the incident X-ray beam on the (111) plane of a germanium single crystal cut at an angle between 5 and 10° and curved cylindrically. The monochromator is used in the spectral domain ranging from 0.7 to 3.5 Å, but the best resolution is obtained when $\lambda = 1.4$ Å. The resolution is approximately given by:

$$\frac{\Delta\lambda}{\lambda} \approx (l_1^2 + l_2^2)^{1/2}$$

in which l_1, the spectral width of the reflected beam, is

[13] ANVAR licence no. 14598.

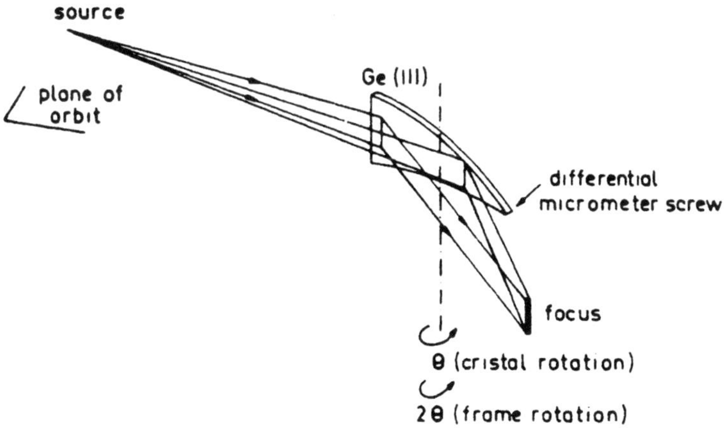

Fig. 6.6. Principle of the curved crystal monochromator

$$l_1 = \frac{4a}{\pi V} d_h^2 |F_h| \, |\cos 2\theta| \left(\frac{\sin r}{\sin i}\right)^{1/2}$$

and F_h is the structure factor of the Bragg reflection, d_h is the spacing between the reflecting planes, V is the unit cell volume, a is the radius of the electron orbit giving off the synchrotron radiation used, θ is the Bragg angle, the other parameters are as defined in the caption of Fig. 6.7, and l_2 is given by

$$l_2 = \delta\theta \operatorname{ctg} \theta \ ,$$

where $\delta\theta$ is the variation of the Bragg angle along the crystal depending on the horizontal extension of the source through

$$\delta\theta = \frac{h}{P} + \frac{L}{2} \left| \frac{\sin r}{P'} - \frac{\sin i}{P} \right| \ .$$

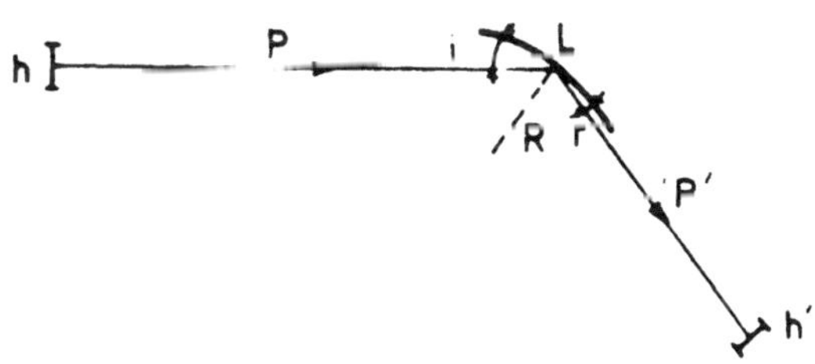

Fig. 6.7. Relevant parameters in the description of the characteristics of the monochromator; h and h': horizontal extensions of source and focus; P and P' source to crystal and crystal to sample distances; R radius of curvature of the crystal; $i = \theta + \alpha$; $r = \theta - \alpha$; L length of the crystal

The best resolution is obtained when

$$\frac{\sin r}{P'} = \frac{\sin i}{P} .$$

Lastly, the monochromator must be equipped with the proper devices for centering and orienting it with respect to the primary beam, and with a set of adjustable slits (F_1 and F_2, Fig. 6.5) used to limit the height and breadth of the beam and to select the $K\alpha_1$ radiation.

6.2.3 Particular Features of the Specimen

A thin flat form for the sample is generally obtained by placing a small amount of powder between two thin films of plastic (polystyrene, mylar, etc.) stretched on a rigid frame. The thickness of the sample – and therefore of the frame – is chosen on the basis of the following criteria:

1. The intensity of the diffracted beam must be sufficient to yield good photon counting statistics. Too thin samples mean an insufficient quantity of matter, while absorption becomes excessive if the sample is too thick. Calculations show that the best conditions are obtained when the intensities of the transmitted and incident radiations are in the ratio $1/e$, with $e = 2.7183$ (Guinier 1964).
2. Since the sample thickness is never negligible, convolution effects always exist which influence the band profile. For a given thickness, broadening of the band depends on the angle of incidence of the primary beam, the diffraction angle 2θ and the intrinsic width of the diffracted beam. Thus, if broadening is to be kept below 1% of the intrinsic width, the maximum thickness of the sample must be calculated, using the method of Sect. 6.3.2, for the reflections in the range to be explored.

Furthermore, in most microdivided lamellar systems, the particles in the powder show some preferential orientation, and the function characteristic of this orientation must be determined for each sample. This is generally done by the method (described in greater detail in Sect. 6.5.2) in which the intensity of the first $00l$ reflection is determined, with the beam incident at a grazing angle on the sample. Because of this geometry, the sample must be given the shape of a thin rectangle, many centimeters long.

6.2.4 The Goniometer

The use of a monochromator in a transmission mounting with focal length about 510 mm implies that the radius of the goniometer be about 250 mm, with two advantages accruing: a good angular resolution for the tracing, and the possibility of fitting voluminous attachments (such as cryostats, controlled atmosphere chambers, ovens, etc.) in the sample area.

The axis of the goniometer may be vertical or horizontal, but the specific requirements are:

1. A step-scan control acting on the counter and on the sample. Correct band profiles can only be obtained if the counter is kept at each position for a time sufficient to count a number of photons yielding correct statistics. This is particularly necessary if the system is not equipped with a localization detector.
2. The possibility of decoupling the rotation of the specimen from that of the counter. This will, for instance, be needed in the experimental determination of the particle orientation function.

6.2.5 The Detector and Counting Equipment

Powder diagrams in which both the profile and the intensity of the bands can be quantitatively exploited are obtained using either proportional counters or localization detectors.

1. Proportional counters

The physical principles at the basis of proportional counters have been fully described (Guinier 1964; Klug and Alexander 1974) and therefore need no further discussion.

The counter is placed on the movable arm of the goniometer with the entrance slit at the focal point of the X-ray beam. When step (rather than continuous) scanning is used, the number n of photons counted at each position must be compatible with the accuracy required: for an accuracy of 1%, at least 10000 photons must be counted since the accuracy is given by $1/\sqrt{n}$. Two counting techniques can be used: fixed time or fixed count, this latter method having the advantage of constant accuracy for all points of the diagram.

Here also, convolution effects must be minimized by careful control of the entrance slit-width, usually equal to the width of the X-ray beam at the focus. Somewhat larger entrance slits can be used when the reflections themselves are broad, but the ratio 1/7 between the slit width and the intrinsic breadth of the band must be maintained if broadening due to convolution must be kept below 1%.

An advantage of the proportional counter is the possibility of placing an antiscattering tube between the sample and the entrance slit of the counter to eliminate parasitic reflections and/or diffuse scattering due, for instance, to the presence of air. The tube is fixed on the arm of the goniometer and turns with the counter. On the contrary, one of the disadvantages attached to the use of step scanning with a proportional counter is the time needed for a recording, which may require many days; this method therefore cannot be used for structures which change in time.

2. Localization detectors

This type of detector is composed of a special unit and of the electronic circuitry needed to shape, encode, and count the pulses created by the incoming photons.

The detector consists of two half-cathodes behind a gold-plated molybdenum wire anode (see Fig. 6.8). The interpenetration of the two comb-shaped half-cathodes, which are photoetched on an insulating substrate, eliminates any possible splitting of the bands. The anode is biased to a high potential and both half-cathodes are connected to entrance pre-amplifiers. The electrodes are in an atmosphere of inert gases adapted to the energy of the incident photons: argon or methane for the lower and most commonly used energies, krypton or xenon for more energetic beams. The system is used in the proportional range.

Under the influence of the incoming photon, a column of gas is ionized along the path of the photon. The electronic circuitry detects the ions and localizes the photon by computing the center of the charges induced on each half-cathode by the ionization radiation. After adequate shaping, the digitalized information (which consists of the angular abscissa of a point on the anode and the number of photons absorbed at this point) is encoded and memorized in a multichannel analyzer or a microcomputer.

In the linear localization detectors presently available commercially[14], the active and useful length of the wire anode is 5cm and the "electronic resolution" can differentiate between photons with points of impact separated by 0.1 mm. When such a detector is fitted tangentially on a goniometric circle with radius 250 mm, this corresponds to an angular resolution of about 0.01° in θ. The total length of the anode wire permits a simultaneous scan corresponding to about 10° in 2θ.

Curved localization detectors have recently been offered which cover an angular range of 120° in 2θ. However, since their radius of curvature is fixed, they can only be used with a monochromator with focal length double the radius of curvature of the detector, in contrast to the linear detectors which are of more general use.

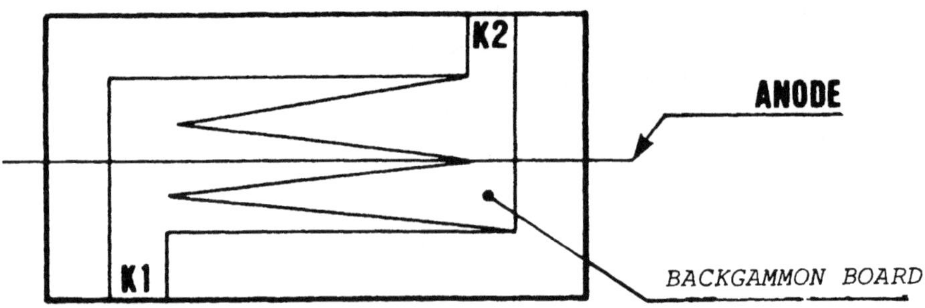

Fig. 6.8. Diagram of a linear localization detector

[14] Linear localization detectors are manufactured: in France, by INEL; in the German Federal Republic, by SIEMENS; in Japan, by RIGAKU.

It should also be noted that antiscattering tubes cannot be used with localization detectors, and that the only possibility of eliminating the diffusion of X-rays by air is to intercalate an evacuated chamber between the sample and the detector. Furthermore, other parasitic diffusions are recorded and it is therefore necessary to substract their contribution from the overall X-ray diffractogram.

6.3 Perturbing Factors which can be Minimized

As mentioned at the beginning of this chapter, some of the factors which can modify the intrinsic shape or intensity of a diffraction band can be controlled and brought to such a low level that their effect can be neglected in the subsequent quantitative exploitation of the diffractogram. Their origin is connected, on one hand, with the height and breadth of the primary X-ray beam and of the different slits through which it passes and, on the other hand, with the thickness of the sample.

Prior to showing how this minimization can be obtained, let us first point out the simple relation between the widths at half height (WHH) x_1 and x_2 of two functions and the WHH x_3 of their convolution product whenever the functions are symmetrical (for example Gaussian, Lorentzian or pulse shaped), namely:

$$x_3^2 = x_1^2 + x_2^2 \ . \tag{6.1}$$

If x_1 is the intrinsic WHH of a symmetrical reflection, we can determine by Eq. (6.1) the proportion of broadening $(x_3 - x_1)/x_1$ of this reflection when it is convoluted with an instrumental function of known WHH x_2. The numerical limiting values given in Sect. 6.2.2 and 6.2.5 were obtained using the above relationship: the reflection of intrinsic WHH x_1 will only be broadened by 1% if the WHH of the perturbing function x_2 is such that $x_2 \approx x_1/7$. In general, convolution effects of perturbing functions on symmetrical reflections may be neglected in quantitative interpretations of experimental diagrams when broadening of the intrinsic WHH is less than 1 to 3%.

6.3.1 Choice of the Slit-Widths in the Path of the X-Ray Beam

The widths of the various slits can be set as detailed above, but, the minimum angular width is determined by that of the beam at the focus. However, samples in which the interferential coherence domains are large may yield diffraction bands sufficiently narrow that the width of the primary X-ray beam can no longer be considered to be negligible. The apparent size L of the coherence domain is linked to a number of other parameters by

$$L = \lambda R / 7 \Delta l_0 \cos \theta_0$$

in which λ is the wavelength of the incident radiation, R is the radius of the goniometric circle, Δl_0 is the width of the beam at the focus and θ_0 is the Bragg angle. Convolution effects can be important or negligible depending on the value of the apparent and actual size of the coherent domains. For example, if the actual size L of the coherent domains is 500 Å, and using the relationships given above, neglect of convolution effects will be possible if Δl_0 is less than 100 µm when $d_{001} = 10$ Å, $\lambda = 1.5404$ Å and $R = 250$ mm.

In actual practice, when the width of the primary beam cannot be reduced, it is necessary to obtain a recording of the beam itself and to calculate the spectrum by convoluting the function representing the incident beam profile with the beam diffracted by a given structure before comparing the result with the experimental data[15].

6.3.2 Optimal Slit Height

When a linear source is used, the diffraction bands are deformed by the "umbrella effect": the peaks are not symmetrical and are broadened on the low-angle side, beyond the maximums. This is due to the fact that the apices of the cones of diffraction from a given (hkl) family of planes lie on a line of length h_s equal to the length of the source. Although the axes of the cones are parallel, they do not coincide and therefore broaden the trace (cf. Fig. 6.9). Computation shows that the umbrella effect will broaden the reflection at the angle θ by less than 1% of the intrinsic width of the beam if h_s (in mm) and θ (in degrees) are linked by $h_s < 0.62\,\theta$.

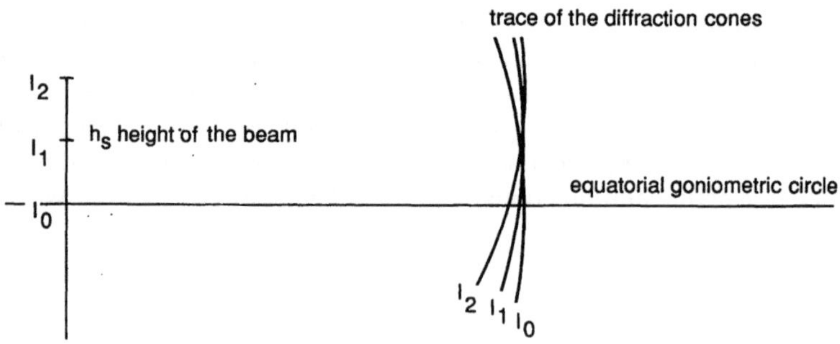

Fig. 6.9. The "umbrella effect"

[15] Although mathematical methods for deconvolution exist, theory and practice show that it is easier to perform the convolution and compare the result with the experimental data than go through the deconvolution.

6.3.3 Choice of Sample Thickness

As already stated in Sections 6.1.4 and 6.2.3, the sample thickness is extremely important. The calculation of the optimum thickness must therefore involve μ, the coefficient of absorption, as well as ϱ, the density of the sample. With reference to Fig. 6.10 let S be the normal cross-section of the incident beam and i its intensity per unit volume. If a is the thickness of the sample, the contribution dI of a slice dx of sample at the depth x to the intensity of the beam diffracted at the angle 2θ is given by

$$dI = iS\,dx \exp(-\mu\varrho x) \exp[-\mu\varrho(a-x)/\cos 2\theta] \; .$$

If θ is small, the total intensity of the diffracted beam is

$$I = iSa \exp(-\mu\varrho a)$$

and the maximum value of I will be found by setting $dI/da = 0$, or

$$dI/da = iS[\exp(-\mu\varrho a) - \mu\varrho a \exp(-\mu\varrho a)] = 0$$

whence

$$a = 1/\mu\varrho \; .$$

With this value of a, corresponding to optimal sample thickness, the intensity of the primary beam is reduced, after diffraction and absorption, by the factor 1/e.

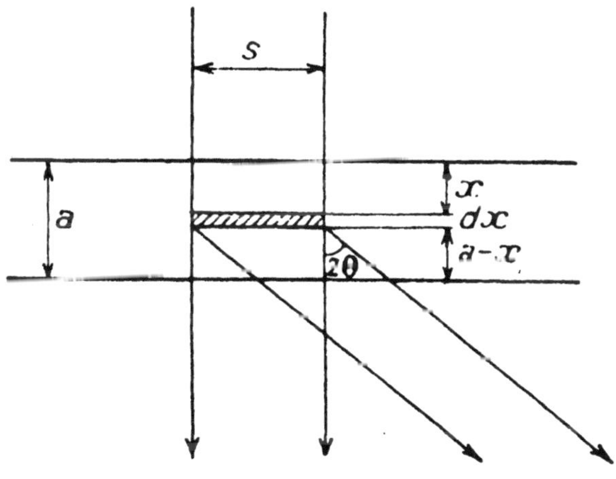

Fig. 6.10. Calculation of the absorption factor for a platelet mounted for transmission diffractometry

In principle, the optimized thickness therefore depends on the wavelength and on the nature of the sample. Practical experience and calculations both show that samples of lamellar silicates 0.5 mm thick (for ϱ about 0.4 g/cm^3) give diffractograms in which energy losses are minimal and aberration negligible, provided Cu Kα_1 is used for alumino-magnesian silicates and Mo Kα_1 for iron-rich clays.

One important point should be noted: optimizing the sample thickness with respect to the absorption does not eliminate the necessity of taking into account the convolution effects due to the same sample thickness. As shown in Fig. 6.11, the rays converging on the entrance slit or on a point of the linear detector are distributed in the angular range $2\theta-\alpha$ to $2\theta+\alpha$. The aberration due to the thickness of the sample will manifest itself by a convolution of the profile with a pulse-function of width:

$$\Delta l = 2\mathrm{a} \sin\theta \ .$$

This function is continuously variable, and its width increases with the Bragg angle θ.

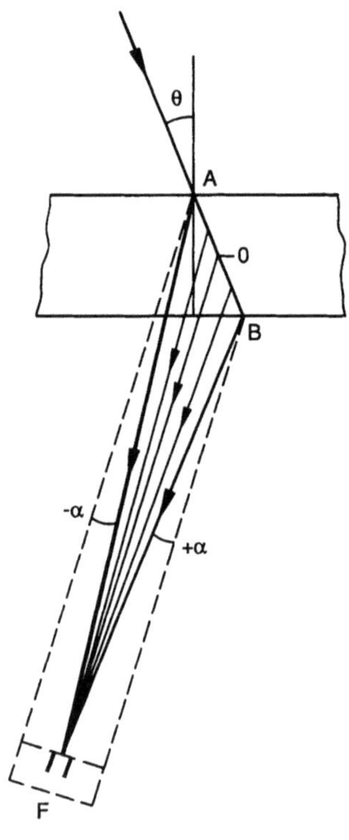

Fig. 6.11. Effect of sample thickness in transmission diffractometry

6.4 Principal Corrections on the Diffraction Patterns

The effects of some perturbing factors can only be eliminated by correcting a number of parameters read off the diagram. They are mainly connected with polarization of the diffracted beam by the sample, partial absorption by the specimen, and lack of linearity in the response of the localization detector if one is used.

6.4.1 Correction of Effects Due to Polarization of the X-Ray Beams

The rays emitted by the anode of the X-ray tube are not polarized in any particular direction, but since they undergo two reflections, namely on the bent monochromator crystal and on the specimen before being collected by the detector, some of the initial intensity is lost and the diffracted beam must be multiplied, at each point, by a depolarization factor P (Guinier 1964) equal to

$$P = \frac{1+\cos^2 2\theta_0}{1+\cos^2 2\theta \cos^2 2\theta_0} \tag{6.2}$$

in which θ_0 is the angle of reflection of the primary beam with respect to the monochromator crystal and θ is the Bragg angle of reflection.

The "Lorentz-polarization factor" quoted in a number of source-books and tables – and different from the one given here – takes into account both the polarization of the X-ray beam and the fact that the tracing corresponds to only a part of the total intensity scattered into space by the sample. The Lorentz factor depends both on the geometry of the experimental set-up and on the conditions of diffraction and recording. Inasmuch as our main concern here is with structural analysis based on the comparison of experimental data with calculated diffractograms, we shall not make use of the Lorentz factor in correcting experimental powder patterns since its effect is automatically included in the integration of the intensity along each (hk) rod [cf. Eq. (2.14) and, more generally, Chap. 2.3.1 and 2.3.2].

6.4.2 Correction of Effects Due to Sample Absorption

1. Experimental determination of the product $\mu\varrho x$ in a forward transmission diffraction set-up

As shown in Section 6.3.3, the reduction of the initial intensity I_0, after passing through a sample of thickness x, is

$$I = I_0 \exp(-\mu\varrho x) \ . \tag{6.3}$$

One of the classical methods used to determine the absorption coefficient of a powder is to shape it as a platelet of thickness x_0, and to compare the integrated intensity I' of a reference reflection when the platelet is placed perpendicularly to the beam with the value I determined in absence of the platelet. The product $\mu \varrho x_0$ is then given by

$$\mu \varrho x_0 = \ln(I'/I) \ . \tag{6.4}$$

Evidently, in these measurements, parasitic absorptions (due to the presence of air, of the plastic sheets between which the sample may be placed, etc.) which should be eliminated are also involved. Furthermore, monitored changes in the sample which necessitate keeping it in a closed chamber (controlled temperature or humidity atmosphere, cryostat, etc.) may prevent its removal from the chamber during the run. The method of in situ measurements of Ben Brahim et al. (1986) can then be used advantageously.

Working first on the empty chamber without the sample but in conditions rigorously identical to the ones used in the subsequent experiments, a blank series of measurements are performed in which the intensity I_1 of the beam transmitted without deviation when it is perpendicular to the sample holder (Fig. 6.12a) is compared to the intensity I_1' obtained when the incident beam is at an angle α to the surface of the sample holder (Fig. 6.12b).

A second series of measurements is then performed, with the sample in place, and again with the beam perpendicular and at the angle a with respect to the surface of the sample, yielding two new values I_2 and I_2'.

Letting now K_m be the coefficient of absorption of the plastic films etc. in the first series of measurements, we can write

$$\begin{aligned} I_1 &= I_0 \exp(-K_m) \\ I_1' &= I_0 \exp(-K_m/\sin a) \ . \end{aligned} \tag{6.5}$$

Similarly, if K_{a+m} be the coefficient of absorption of the sample and of the plastic films in the second series of measurements.

$$\begin{aligned} I_2 &= I_0 \exp(-K_{a+m}) \\ I_2' &= I_0 \exp(-K_{a+m}/\sin a) \ . \end{aligned} \tag{6.6}$$

From these last four expressions, we derive

$$K_m = \frac{\sin a}{1-\sin a} \ln \frac{I_2}{I_2'} \quad \text{and} \quad K_{a+m} = \frac{\sin a}{1-\sin a} \ln \frac{I_1}{I_1'} \ .$$

The ratio

$$\exp(-K_{a+m})/\exp(-K_m) = \exp(K_m - K_{a+m})$$

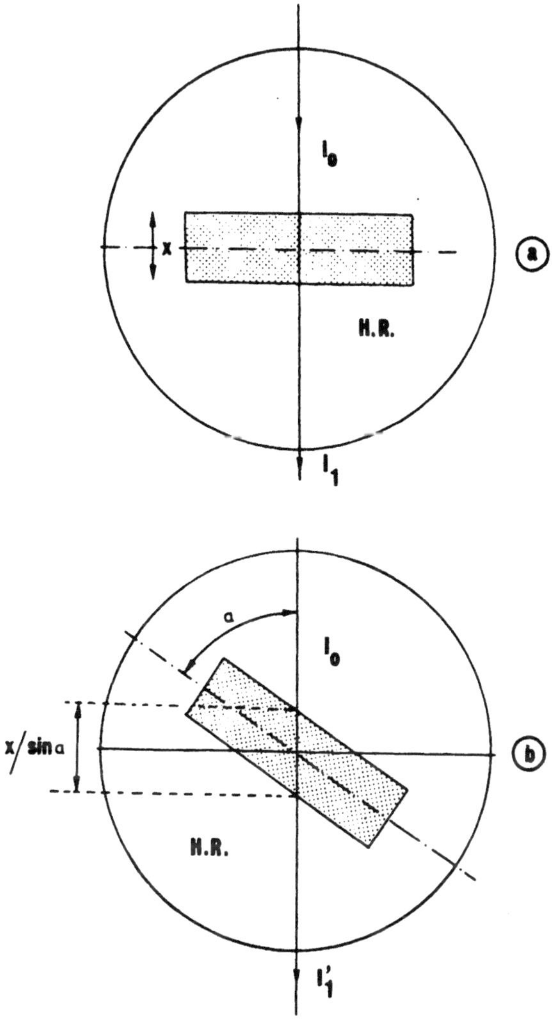

Fig. 6.12. In situ measurement of the absorption coefficient of a hydrated clay

is now a measure of the absorption by the powder sample by itself.

2. Absorption correction

The recorded diffractogram is the sum of the pattern given by the sample itself and of parasitic diffusions which must be determined under rigorously identical conditions and substracted from the total. If now I_g and I_b respectively be the intensities of the global spectrum and of the "blank" spectrum at the angle 2θ, the intensity at this point due to the sample by itself is given by the following expression valid in the general case

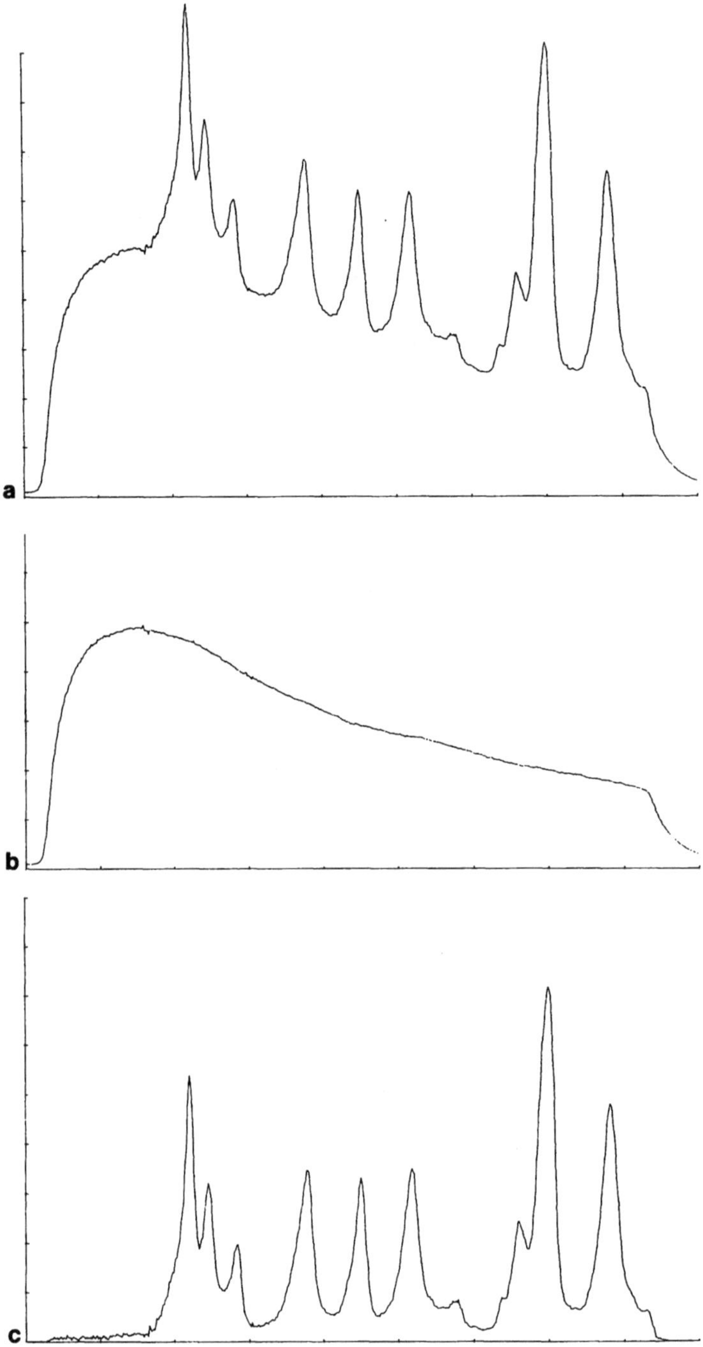

Fig. 6.13. Diffraction from a sample of celadonite; **a** raw intensity distribution Ig; **b** corrected diffusion background, $I_b \exp[(K_m - K_{a+m})/\cos\theta]$; **c** corrected experimental diffractogram $I(2\theta)$

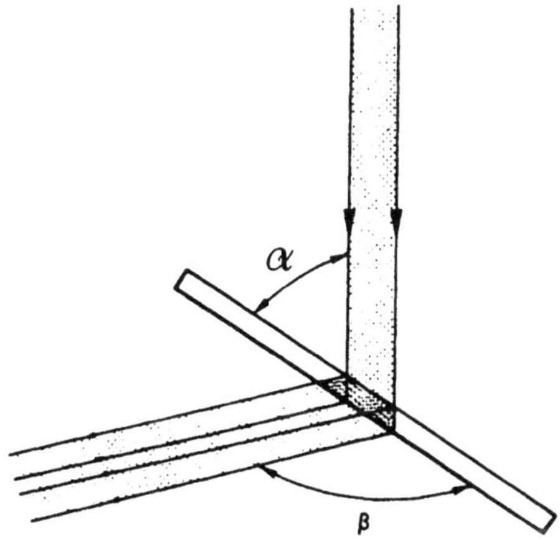

Fig. 6.14. Schematic representation of the variable α and β involved in the elimination of absorption effects in a diffractogram obtained in an asymmetrical transmission configuration

$$I(2\theta) = I_g - I_b \exp[(K_m - K_{a+m})/\cos\theta]$$

as shown in Fig. 6.13.

If the diffractogram has been obtained using a transmission non symmetrical $\theta-2\theta$ mounting, the intensity $I(2\theta)$ is corrected for absorption effects by multiplying it by the coefficient A given by

$$A = \frac{K_a}{x_0} \cdot \frac{\sin\alpha - \sin\beta}{\sin\beta} \cdot \frac{1}{\exp(-K_a/\sin\alpha) - \exp(-K_a/\sin\beta)} \tag{6.7}$$

in which K_a is the product $\mu \varrho x_0$ for the sample, while α and β are the angles between the surface of the sample and the incident and diffracted beam respectively (cf. Fig. 6.14). The intensity to be compared with the intensity of the theoretical (computed) spectra is the corrected value $I_C(2\theta) = A\,I(2\theta)$.

For a symmetrical $\theta-2\theta$ transmission mounting, the correction factor A_{sym} is

$$A_{sym} = \frac{\cos\theta}{x_0} \exp(K_a/\cos\theta) \; . \tag{6.8}$$

6.4.3 Correction for the Nonlinear Response of the Localization Detector

Local impedance variations of the anode wire can result in differences in the number of counts caused by an equal number of pulses received at different points of a localization detector. This lack of linearity is detected by recording, in a transmission set-up and on either side of the directly transmitted beam, the wide-angle diffusion (corresponding to $s \approx 10\,\text{nm}^{-1}$) produced by a platelet of amor-

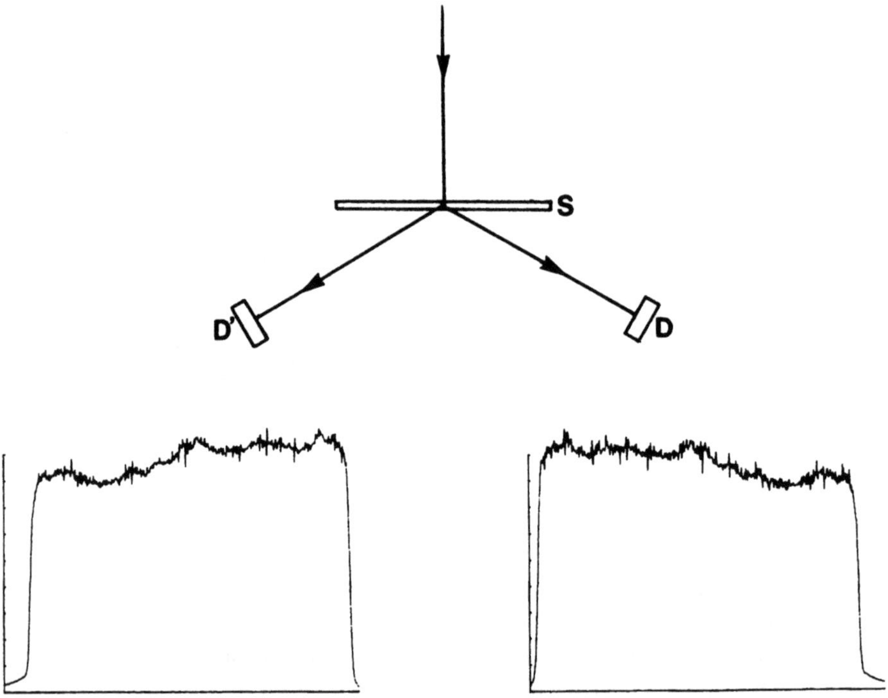

Fig. 6.15. Recording geometry and response curve of a localization detector

phous glass placed perpendicularly to the incident beam. Two nearly straight lines of extremely small slope (about 1%) are obtained, as shown in Fig. 6.15. If the detector were perfectly linear, the point-by-point addition of the lines would yield a constant diffusion background, whereas actual (nonlinear) detectors yield a nonhorizontal line with some fairly sudden variations in the contents of some channels. The ratio R_i (where i is the number of the channel) of the mean intensity of the curve to the actual intensity at the given channel yields a measure of the local lack of linearity. The correction then consists in multiplying the number of counts of each channel by the corresponding value of $1/R_i$.

Obviously, this correction must be performed before the one involving polarization effects or adsorption of the beam.

6.5 Perturbing Factors Introduced in the Computation of the Theoretical Diffractograms

6.5.1 Lorentz Factor

We already mentioned in Section 6.4.1 that, in order to calculate the intensity $I_{hk}(Z)$, it was necessary to take into account a geometric term known as the Lorentz factor whose value depends on the type of XRD technique used as well as on the physical state of the sample: thus, when the diffractometer is used with a powder, the Lorentz factor is equal to $1/\sin^2\theta$, but it is equal to $1/\sin\theta$ if equatorial plane reflections from a mosaic crystal are investigated.

In order to analyze the more frequent case of a partially oriented (textured) powder sample, as considered in Section 2.4.2, we can rewrite Eq. (2.37) as:

$$I_{00}(Z) = \frac{1}{\Omega\sigma} \iint |F_{00}(Z)|^2 G_{00}(Z) G(\varepsilon_1, \varepsilon_2) \bar{p}(\varphi) \frac{dA}{s^2} \, .$$

If the orientation function $\bar{p}(\varphi)$ can be considered to be constant in the range of integration, it can be taken out of the integral, together with $F_{00}(Z)$ and $G_{00}(Z)$, so that the expression above now becomes:

$$I_{00}(Z) = \frac{1}{\Omega\sigma s^2} |F_{00}(Z)|^2 G_{00}(Z) \bar{p}(\varphi) \int G(\varepsilon_1, \varepsilon_2) \, dA \, .$$

As the integral now depends only on the cross-section of the (00) rod, the intensities of the basal reflections from partially oriented samples change with the angle of diffraction as $1/s^2$, which corresponds to the $1/\sin^2\theta$ Lorentz factor for an isotropically oriented powder.

Figure 2.21 can be used to estimate the degree of anisotropy of the texture when the assumption $\bar{p}(\varphi) \approx \bar{p}(\pi/2)$ is not valid. The maximum deviation of angle φ from the axis of the rod is $1/Rs$ where R is the radius of the coherent domain. Even for values of s relatively as small as $s = 0.1$ Å^{-1}, as for a 001 reflection from mica, and with $R = 100$ Å, we find $\Delta\varphi \approx 6°$, and it will be even smaller for the next and following orders of reflections. The routine technique of sample preparation by sedimentation is not likely to give such textural perfection and the Lorentz factor for powders must be used for most powder samples.

In the particular case of a powdered solid with a highly perfect texture, the intensity calculations should be done using the set of equations in Section 2.4.2, except for those containing T(U), as Eq. (2.38), since the definition of T(U) did not take the orientation function into account. The actual geometric term depends on the texture and changes from $1/\sin^2\theta$ to $1/\sin\theta$. The same variation of the Lorentz factor is found in the procedure for intensity calculations developed in detail by

Reynolds (1976) for partially oriented samples in terms of a real-space coordinate system. However, this work, which revealed the influence of a number of parameters on the Lorentz factor, did not insist adequately on their importance.

The range of integration over the cone of the diffracted beams depends on the ratio of the height h_d and the width w of the detector aparture. At low values of h_d (h_1 in Fig. 6.16), the range has a constant value equal to the slit height, but for high values of h_d (h_2, in Fig. 6.16), the limits of integration are defined by the intersection of the diffraction cone with the inner plate of the slit. This results in a decrease of intensity for the low-angle reflections since, in the last case, the limits of integration are closer.

Besides the effect of aperture size, the recorded intensity depends on the illuminated area of the sample. Figure 6.17 shows the propagation of the beam through the set of Soller slits. The geometrical term (GT) includes two integrations:

$$GT = \int_0^{h_d/2} \int_{\beta_1}^{\beta_2} p(\varphi)\,d\varphi\,dx = \int_{h_d/2}^{x_{lim}} \int_{\beta_3}^{\beta_4} p(\varphi)\,d\varphi\,dx \ .$$

Figure 6.18 shows some results of numerical calculations of GT for a sample with an isotropic orientation function in different experimental conditions, together with values of the conventional Lorentz factor for isotropic powders (curve 1) and for ideal textures (curve 2). The parameters used in the calculations were those of the DRON-UM diffractometer, namely: goniometer radius $R_0 = 180$ mm, Soller slit height and length 1 mm and 40 mm respectively. The angular size of a single Soller slit ($\beta_1 + \beta_2$) was 0.3°, but, because of the presence of the second term in the expression above for GT, the actual vertical divergence was increased by a factor of 4.5. In Fig. 6.18, curves 3 and 4 were obtained for aperture widths $e = 0.5$ mm and $e = 0.1$ mm, respectively. All the curves were normalized for equal value at $\theta = 20°$. Figure 6.18 shows that for $e = 0.1$ mm (curve 4), the intensities of the low-angle reflections are reduced, similarly to what could have been observed with a high quality texture. Thus, the calculated GT for the 001 reflexions of a glycol-treated smectite ($\theta = 2.6°$) is half the powder Lorentz factor, but it is 4.5 times larger than that for an ideal texture. For a slit with $e = 0.5$ mm and at the same value of θ, the difference with the powder Lorentz factor is only 2%.

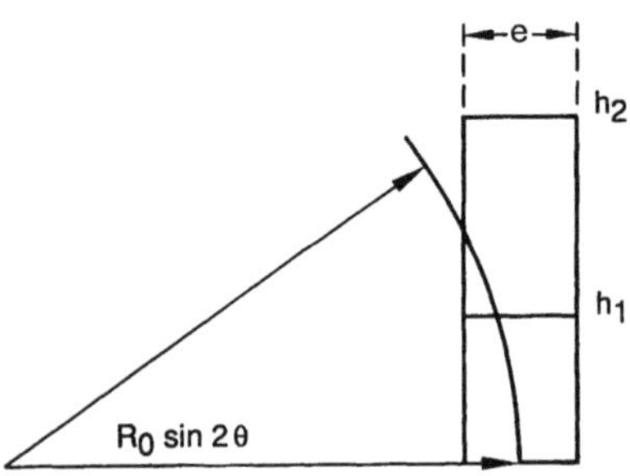

Fig. 6.16. Intersection of a diffracted X-ray beam cone with the detector aperture in the case of high (h_2) and low (h_1) slits: R_0 is the goniometer radius; θ is the diffraction angle; e is the width of the slit

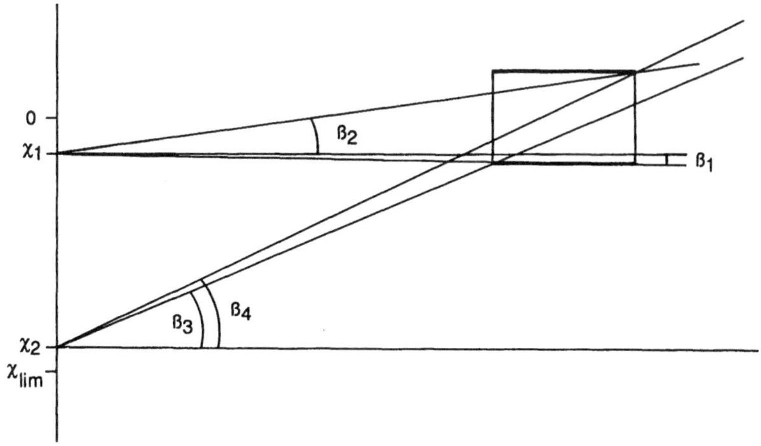

Fig. 6.17. Propagation of X-ray beams from the part OX_{lim} of an elongated sample through a single Soller slit fragment; β are integration limits over the diffraction cone

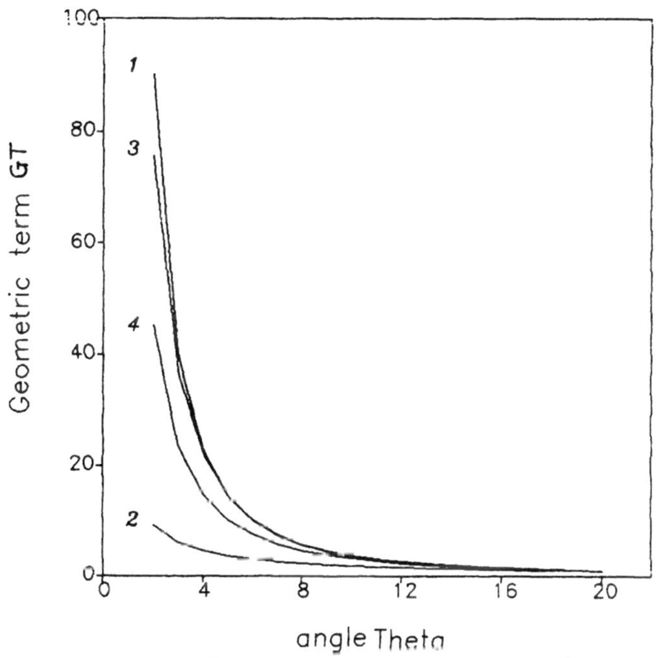

Fig. 6.18. Plot of the geometrical term versus the diffraction angle θ. Curves *1, 3, 4* correspond to an isotropic powder sample: *1* infinitely wide slit (isotropic powder Lorentz factor). *3* e = 0.5 mm slit; *4* e = 0.1 mm slit. Curve *2* corresponds to an ideal texture Lorentz factor

It is evident, from these values, that striving to obtain the best possible resolution from a diffractometer by using excessively narrow slits would not allow a proper representation of the intensities if the powder Lorentz factor were used for the diffraction at low angles. The problem of correct calculation of intensities for low angles becomes more complex when the orientation function $p(\varphi)$ is not constant and, especially, when the reflecting planes are not parallel to the textural planes.

6.5.2 Orientation Function for the Particles in a Powder

It was shown in Section 2.4 that, from a theoretical viewpoint, the existence of preferred orientation of the particles in the powder is revealed by modifications of the intensity distribution in the diagrams, and that the intensity along an (hk) rod, as given in Eq. (2.38), was to be computed from

$$I_{hk}(s) = \frac{1}{\Omega \sigma s} \int |F_{hk}(Z)|^2 G_{hk}(Z) T(U) \bar{p}(\varphi) d\varphi$$

in which φ is the angle between the diffusion vector $\vec{s}$ and the (001) plane of each particle (cf. Fig. 2.6). The function $\bar{p}(\varphi)$ itself is linked to the normalized function $p(\gamma)$ which represents the probability of finding in the powder a particle in which the (001) plane is at an angle γ with the surface of the sample platelet. The relationship between $\bar{p}(\varphi)$ and $p(\gamma)$ is expressed in Eq. (2.34), which is the integral of $p(\gamma)$ between the limits $\gamma_{min} = \varphi$ and $\gamma_{max} = \pi - \varphi$, for each value of φ. This relation is given with reference to a symmetrical $\theta - 2\theta$ transmission mounting (Sect. 2.4.2); the relation for the nonsymmetrical $\theta - 2\theta$ mounting is given in Section 2.4.3.

The effect of preferential orientation in the particles of the powder is illustrated in Fig. 6.19a in which the (02, 11) diffraction band is shown, as calculated using Eq. (2.38) for a smectite with interlayer Cs^+ cations and with the orientation function $p(\gamma)$ given in Fig. 6.19b. The presence of preferential orientation exaggerates the intensity of the leading edge of the band (an effect which could erroneously be attributed to the structure factor) and simultaneously thins out this part of the band (which could entail an overestimation of the size of the interferential coherence domains in the $(00l)$ plane). It is important to know whether the powder sample has preferential orientation if one is to reach a correct interpretation of the relative intensity variations and of the band profiles. The $p(\gamma)$ function must therefore be determined experimentally for each sample examined.

It was shown in Section 2.4.4 that the intensities of the $00l$ reflections from a lamellar system in a transmission mounting are directly related to $p(\gamma)$ owing to the fact that, for this series of reflections, the coefficient $\bar{p}(\varphi)$ in Eq. (2.38) is reduced to the value of the function $p(\gamma)$ at a single point: for any given orientation of the platelet, characterized by the angle α between the incident beam and the surface, the (001) planes which contribute to the intensity of $00l$ reflections

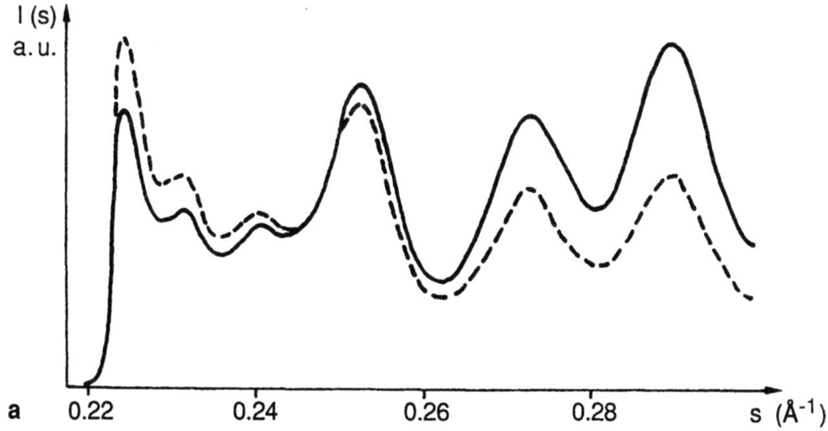

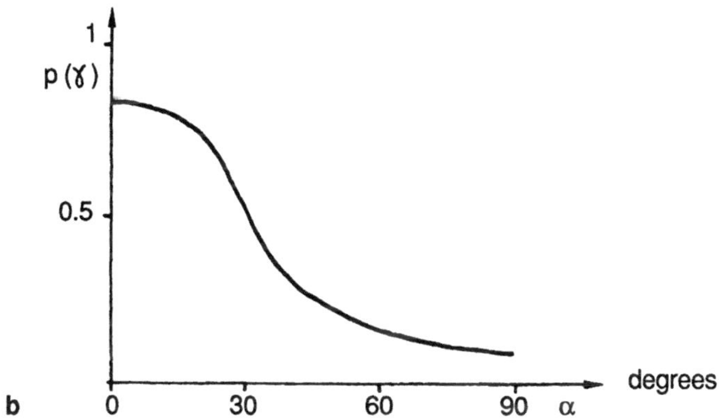

Fig. 6.19. **a** Modification of the theoretical profile of the (02, 11) band from a well-crystallized Cs^+-treated smectite: without preferential orientation (*full line*), or in the presence of the preferential orientation (*dashed curve*) shown in part **b** of the figure; **b** preferential orientation curve of the particles in the powder sample

are at an angle $\gamma = \theta + \alpha$ with respect to the midplane of the platelet, where θ is the Bragg angle for the $00l$ reflection considered. In other words, the integrated intensities of the basal $00l$ reflections are directly proportional to the proportion $p(\gamma)$ of the particles. It is this property which is used for determining the orientation function $p(\gamma)$.

Let $(I_{00l})_\alpha$ be the experimentally determined value of the integrated intensity for a certain $00l$ reflection with the sample in a position characterized by the angle α with respect to the incident beam, and let $(I_{00l})_{\alpha s}$ be the integrated intensity of the same reflection when the sample is in the symmetrical $\theta - 2\theta$ position. The relation $\alpha_s = \pi/2 - \theta$ then holds and the (001) planes of the particles which participate in the diffraction are at an angle $\gamma = 90°$ with respect to the surface of the platelet.

The intensities $(I_{001})_\alpha$ and $(I_{001})_{\alpha s}$ obtained in the positions characterized by the angles α and α_s must first be corrected for sample absorption. Using Eq. (6.7) and (6.8), we obtain:

$$(I'_{001})_\alpha = A \cdot (I_{001})_\alpha \quad \text{and} \quad (I'_{001})_{\alpha s} = A_{sym}(I_{001})_{\alpha s}$$

in which the primes on the intensities indicate corrected values.

If the powder were isotropic, the corrected intensity $(I'_{001})_\alpha$ would always be equal to $(I'_{001})_{\alpha s}$, for all values of α since, in that case, the density $p(\gamma)$ of particles at the angle θ with (001) planes at the Bragg angle $\gamma = \alpha + \theta$ with respect to the surface of the sample would be independent of the orientation of the sample and, for all values of α, we would have $p(\gamma) = \text{constant}$. However, with preferential orientation of the particles, $(I'_{001})_\alpha$ is proportional to the value of $p(\gamma)$ specific to each value of α. Writing then $\omega_0 = \pi/2 - \gamma$, we find

$$(I'_{001})_\alpha = p(\omega_0)(I'_{001})_{\alpha s}$$

and its equivalent

$$A(I_{001})_\alpha / A_{sym}(I_{001})_{\alpha s} = p(\omega_0) \ .$$

Note that $(I_{001})_{\alpha s}$ is a single value whereas $(I_{001})_\alpha$ varies with each value of α, as also does the factor A. Hence, by measuring $(I_{001})_\alpha$ for different positions of the sample on either side of the symmetrical $\theta - 2\theta$ position, the graphs of $(I_{001})_\alpha/(I_{001})_{\alpha s}$ and of A_{sym}/A as a function of α can be drawn. The function $p(\alpha)$ is obtained by computing, point by point, the ratio of the two curves. Taking then into account that $\alpha = \pi/2 - \theta - \omega_0$, we transform $p(\alpha)$ into $p(\omega_0)$ and, finally, dividing each value of $p(\omega_0)$ by the value $p(90°)$ corresponding to $\omega_0 = 90°$ yields the orientation function $p(\gamma) = p(90° - \omega_0)$ which is then normalized by

$$4\pi \int_0^{\pi/2} p(\gamma)\,d\gamma = 1 \ .$$

Figure 6.20 shows $(I_{001})_\alpha/(I_{001})_{\alpha s}$ and A_{sym}/A as functions of α for a partially oriented graphite powder (de Courville et al. 1979), as well as the normalized $p(\gamma)$. Instrumental constraints usually limit α to the range $0° \leqslant \alpha \leqslant 70°$, and it is therefore necessary to extrapolate the curves to $\omega_0 = 0°$. This can best be done using an analytic function with a good fit to the experimental points.

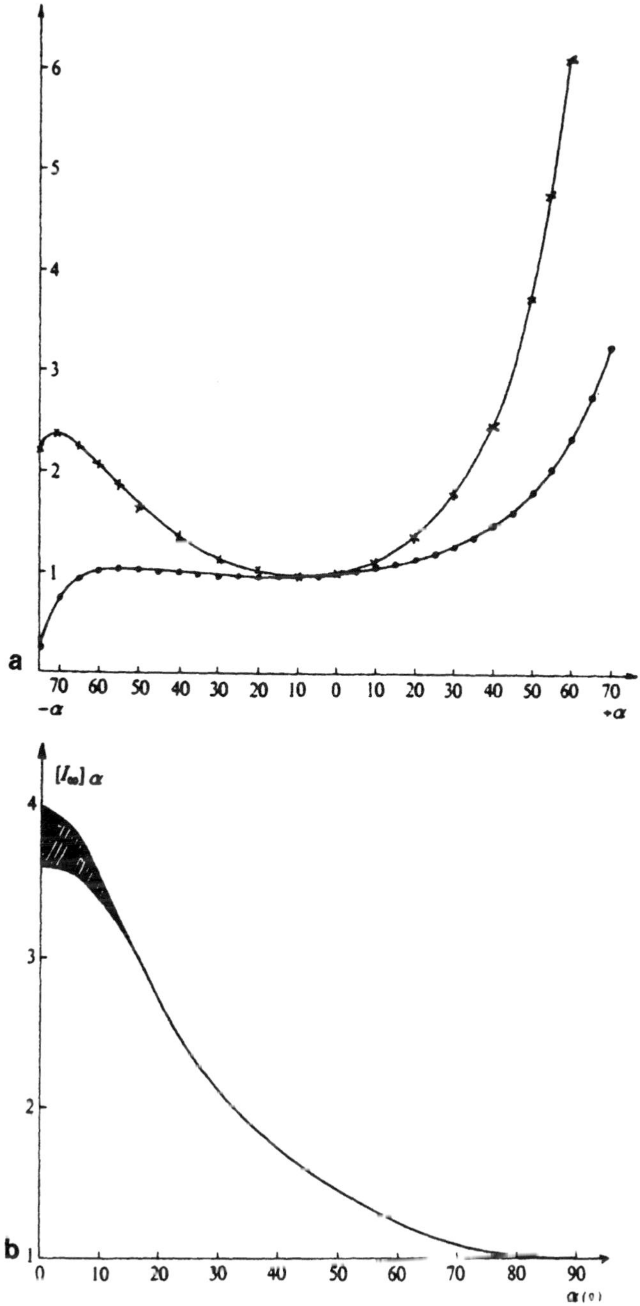

Fig. 6.20. a Experimental determination of the preferential orientation function: the *crosses* show the value of the ratio $(I_{00})_\alpha/(I_{00})_{\alpha s}$ when the plane of the sample is rotated by an angle $\pm\alpha$ on either side of the symmetrical position; the *full circles* give the theoretical absorptions A_{sym}/A. **b** The lower curve is the normalized orientation curve $p(\gamma)$

6.6 Determination of the Absolute Intensity Scale

The comparison of the corrected intensity $I_c(2\theta)$ with the profiles calculated on the basis of a given model (cf. Sect. 6.4.2) can be done using experimental diagrams in which the intensity scale is relative (arbitrary). In general, this will lead to a correct determination of the proportion and nature of the structural defects existing in the sample, but ambiguities may remain if the number of parameters taken into account in the elaboration of the model is large. These doubts decrease if the scale of the diffractograms is absolute: the comparison between the experimental and calculated profiles yields much more information and it is, for instance, easier to determine which are the defects that specifically affect the structure factor.

6.6.1 Definition of the Absolute Scale

The diffracted beam is composed of a coherent diffusion which contains the information on the structural organization of the sample, and incoherent (Compton) diffusion which must be eliminated. This is best done by determining the scattering power of the object (Guinier 1964), which is the diffracted intensity – measured in electronic units – normalized to a chosen single structural unit (chemical unit or unit cell of the crystal). It is equal to the number of electrons free to scatter the X-ray beam independently from each other and which would produce the same effect in reciprocal space as the single structural unit chosen.

This scattering power is obtained using the method of Warren and Gingrich (1934), who showed that, for very large angles of diffraction, the intensity of the experimentally observed diffraction is practically identical with the intensity of the diffusion that would be observed *if the atoms in the unit had no mutual interaction and existed in a fictitious gaseous state*, yielding a so-called gas curve. The more disordered the object, the smaller the diffraction angle at which coincidence between the experimental and the gas curve occurs.

The scattering power of the fictitious gas is

$$I_g(s) = \sum_j n_j f_j^2(s) + \sum_j n_j [t_j(s)]_{inc} \tag{6.9}$$

in which s is the modulus of the scattering vector ($s = 2 \sin \theta/\lambda$), n_j is the number of j-type atoms in the chosen structural unit, and $[(t_j(s)]_{inc}$ is the incoherent diffusion corrected for the Bray-Dirac factor for the atom j in its bonded state. The summation $\sum$ must be extended to all the atoms in the structural unit.

6.6.2 Determination of the Absolute Scale

Using Eq. (6.9), the gas diffusion curve $I_g(s)$ is first calculated for the atoms of the chosen structural unit. The following operations are then performed:

1. a total experimental diffractogram is obtained, which includes the sum of the contributions from the sample, the holder, the air molecules, etc., up to the angular region where the curve is practically monotonic; this generally requires a fairly short wave length (such as $\mathrm{Mo}K\alpha_1$);
2. a second diffractogram is obtained in *rigorously identical conditions* (diffraction geometry, accumulation time, etc.) except for the presence of the sample; this diagram represents all the parasitic contributions;
3. each of the above diagrams is corrected for absorption as in Section 6.4.2;
4. after point-by-point substraction of the two spectra, the specific contribution of the sample is obtained, which is then corrected for polarization effects (Sect. 6.4.1); the result represents the distribution of intensity $I(s)$ for the sample alone;
5. scaling to absolute values is obtained by multiplying the $I(s)$ curve by a normalization coefficient k such that, for large values of s, a best fit is obtained between the normalized intensity $kI(s)$ and the mean value of the gas curve $I_g(s)$ calculated using Eq. (6.9).

The $kI(s)$ curve is the sum, on an absolute scale, of the normalized so-called "scattering power" of the object and of the incoherent diffusion. Substraction of the latter finally yields the true scattering power to be used in the determination of the structural characteristics of the sample.

6.6.3 Examples and Applications

Figure 6.21 shows the absolute scale experimental diagram for kaolinite, obtained and corrected as described in the preceding paragraphs.

Similarly, when clay particles are suspended in a liquid, the structure and the mutual arrangement of the layers, and even the organization of any molecules of the liquid which may be fixed on them, can be determined from absolute scale diffractograms. For instance, suspensions in water yield diffractograms composed of two contributions: the diffusion from liquid water, and the diffraction from the clay particles which, in some cases, may have a number of absorbed water molecules. Since only the diffraction from the particles must be taken into account in the determination of the structural features of the solid, the diffusion from the molecules of *liquid* water must be *quantitatively* eliminated. The amount of diffracting matter taken into account in the computation of the gas curve must then be the unit cell of the mineral plus N molecules of water, in the ratio of the volumes proper to the concentration of the suspension. Corresponding to the second spec-

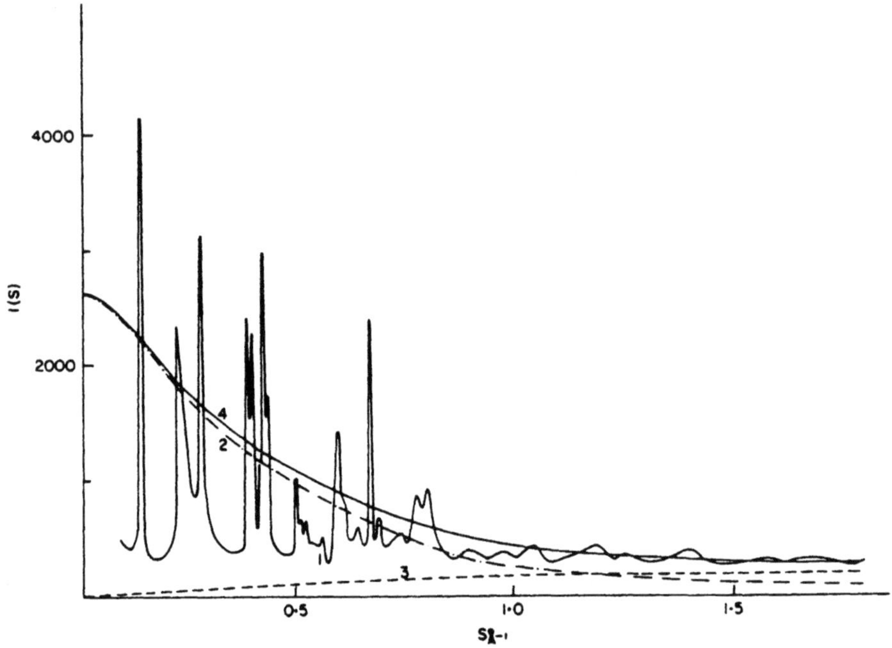

Fig. 6.21. Absolute scale diagram of a partly disordered kaolinite from Charente (France); *1* experimental diffractogram; *2* coherent diffusion; *3* incoherent diffusion; *4* gas curve

trum needed and described above, a diffractogram of liquid water is recorded and scaled to the gas curve calculated for the number N of molecules. The quantitative comparison of the two diffractograms by trial-and-error adjustment yields the number of molecules of water in the liquid state and, by difference, the specific contribution of the solvated clay (Pons et al. 1980). The organization of the water molecules interacting with the layers of clay is then obtained by comparing this fully corrected recording with the calculated band profiles. Figure 6.22 gives an example of the application of the method to a suspension of clay; the gas curve is shown in Fig. 6.22a, while Fig. 6.22b is a recording of the diffusion from liquid water scaled to the gas curve for an equal number N of molecules. The specific contribution of the clay, obtained by substracting the non-clay contributions from the corrected total diffraction, is given in Fig. 6.22c (Pons et al. 1980).

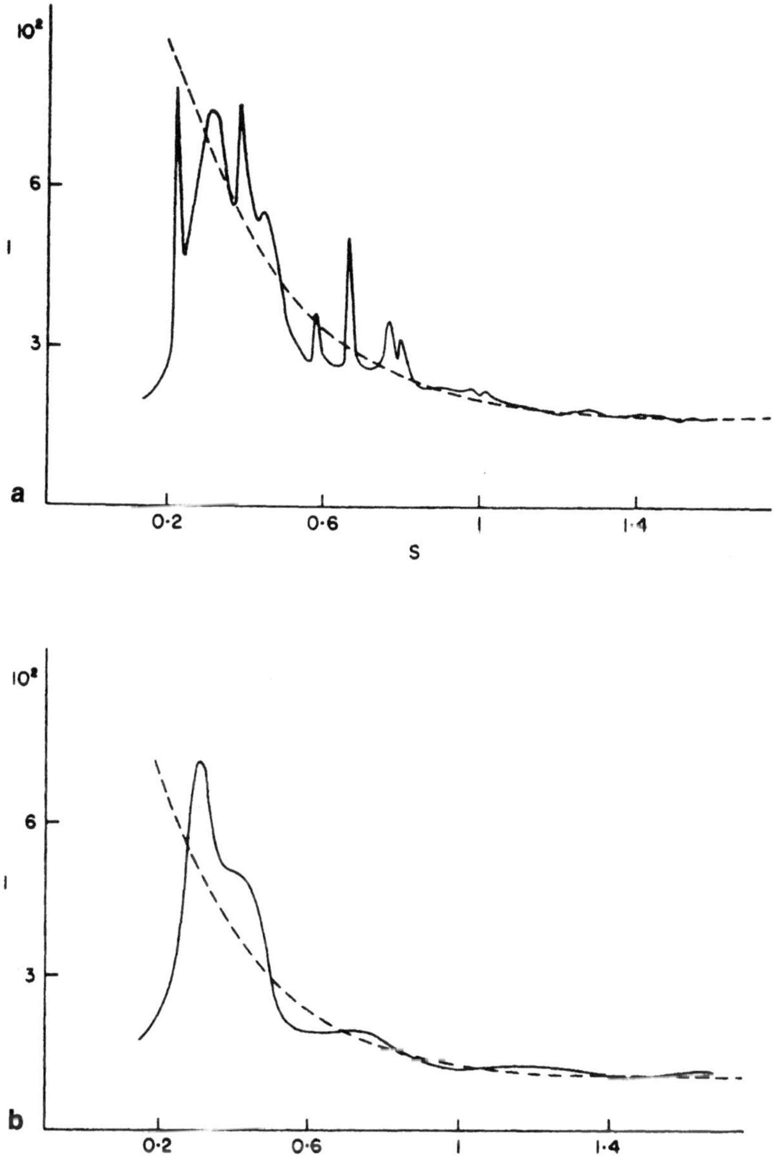

Fig. 6.22. **a** Absolute-scale raw diagram of Na^+-montmorillonite (20 g of clay in 100 g of water); **b** absolute-scale diagram of pure water; **c** absolute-scale corrected diagram of the hydrated clay

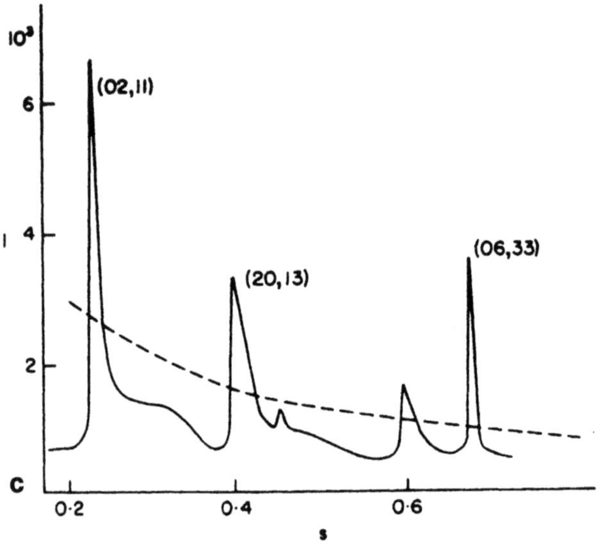

Fig. 6.22c

References

Ben Brahim J, Armagan N, Besson G, Tchoubar C (1986) Méthode diffractométrique de caractérisation des états d'hydrations. Stabilité relative des couches d'eau insérées. Clay Mineral 21:111–124

Cauchois Y (1933) Spectrographie des rayons X par transmission d'un faisceau non canalisé à travers un cristal courbé. J Phys Radium 4:6172

Debye P, Scherrer P (1916) Interference in irregularly oriented particles in Röntgen light. Phys Z 17:277–283

de Courville J, Tchoubar D, Tchoubar C (1979) Détermination expérimentale de la fonction d'orientation, son application dans le calcul des bandes de diffraction. J Appl Crystallogr 12:332–338

Guinier A (1964) Théorie et technique de la radiocristallographie. Dunod, Paris, 740 pp

Guntert DJ, Cvikevich S (1964) Preferred orientation and its effects on the (*hk*) reflections in X-ray patterns of pyrolytic graphites. Carbon 1:309–313

Hull AW (1917) A new method of X-ray crystal analysis. Phys Rev 10:661–696

Johann HH (1931) Die Erzeugung lichtstarker Röntgenspektren mit Hilfe von Konvakristallen. Phys Z 69:185–206

Klug HP, Alexander LE (1974) X-ray diffraction procedures. John Wiley & Sons, New York, 966 pp

Lemonier M, Fourme R, Rousseaux F, Khan R (1978) X-ray curved-crystal monochromator system at the storage ring DCI. New Instrum Methods 152:173–177

Plançon A (1980) The calculation of intensities diffracted by a partially oriented powder with a layer structure. J Appl Crystallogr 13:524–528

Pons CH, Ben Brahim J, Yucel A, Tchoubar D, Tchoubar C (1980) Détermination à l'échelle absolue de l'intensité diffractée aux rayons X par les argiles. Clay Mineral 15:111–120

Reynolds RC Jr (1976) The Lorentz factor for basal reflections from micaceous minerals in oriented powder aggregates. Am Mineral 61:484–491

Taylor RM, Norrish K (1966) The measurement of orientation distribution and its application to quantitative X-ray diffraction analysis. Clay Mineral 6:127–141

Warren BE, Gingrich NS (1934) Fourier integral analysis of X-ray powder patterns. Phys Rev 46:368–399

CHAPTER 7

Structural Characteristics of Carbons

7.1 General Characteristics of Carbon Materials

7.1.1 General Description

When pyrolyzed in the absence of oxidizing agents, organic substances generally form carbonaceous residues. Various types of precursors and/or methods of pyrolysis can be used, leading to materials with widely different macroscopic structures. As a first approximation, carbonaceous materials – or "carbons" – can be classified into three categories (Bouraoui and Méring 1964):

1. those which are obtained by pyrolysis of organic precursors (such as polyaromatic molecules, petroleum, or coal tars, etc.) in the condensed state, which form the basis of the polycrystalline artificial graphites;
2. those, likely glassy carbon or carbon fibers, obtained by the pyrolysis of polymers or of natural or synthetic fibers (Donnet and Bansal 1984);
3. those, like carbon blacks, resulting from gas-phase pyrolysis (Bokros 1969; Lahaye and Prado 1978) or those, like pyrocarbons, resulting from the condensation of cracked vapors (note, in this respect, that HOPG, or Highly Oriented Pyrolytic Graphite, is obtained by subsequent treatments of certain types of pyrocarbons; Moore 1973).

In all cases, the solid phases obtained at about 1000 °C are practically pure carbon and are composed of metastable lamellar structures, since the thermodynamically stable allotropic form at low pressure is hexagonal graphite. These metastable states can be modified by appropriate thermal treatments between 1000 and 3000 °C, but not all of them will evolve to graphite. The possibility of transforming a given type of carbon into graphite by mere thermal treatment seems to be determined both by the nature of the precursor and by the kinetics of the mechanisms of aromatization and polycondensation involved in the pyrolysis (Fitzer et al. 1971; Marsh and Walker 1979).

These mechanisms are highly complex and not yet fully understood, but it is nevertheless possible to state that a *graphitizable carbon* (soft carbon) is generally given by substances or compounds with low oxygen and high hydrogen content. When this type of substance is pyrolyzed in the condensed state, say between 250 and 300 °C, a liquid phase is formed, with an optical anisotropy comparable to that of discotic nematic liquid crystals (Brooks and Taylor 1965; Gasparoux 1980; Lewis and Singer 1981). The transient formation of this "mesophase" is highly beneficial to the possibility of maintaining a sufficient mobility for the polyaromatic radical intermediates created by the pyrolysis, thus ensuring their subsequent condensation into larger entities. On the other hand, *nongraphitizable carbons* find their origin in the pyrolysis of substances rich in "heteroatoms" (other than H). The pyrolysis then very often occurs in the solid state and rigid sp^3 bonds formed between carbon atoms prevent the formation of large polynuclear plane aromatic domains. The resulting structure is highly distorted and cannot reach the triperiodic regularity required for graphite. This generally occurs with carbon blacks or carbons described above as type 2. Nevertheless, all varieties of carbon which are not purely aliphatic (like diamond) include graphite-like lamellar portions and the fundamental element in their structure is always a biperiodic layer with hexagonal symmetry similar to the elementary layer in graphite. The stacking in each particle is, however, random and the spatial extension of these graphite-like elements may be very small, down to a few nanometers in diameter (Diamond 1957; Millward and Jefferson 1978).

7.1.2 Basic Features of the Graphitization Process

Graphitizable carbons will gradually evolve into graphite when heated to sufficiently high temperatures, i.e., higher than 1600 °C. The transformation may be brought to completion by letting the temperature reach about 3000 °C, at which point the carbon is found to have been graphitized. In fact, the characteristics of a given material depend on its HTT (Highest Treatment Temperature) and, to a lesser extent, on the time during which it was kept at this HTT (Fischbach 1971; Pacault 1971).

The mechanisms of this graphitization process have been extensively studied by X-ray diffraction (XRD) and the quantitative analysis of the results have led to a good grasp of the three-dimensional order associated with each step of the thermal transformation. It has been found that three-dimensional ordering of the layers can only start if the carbon layers have previously reached the structure of graphite-like layers, implying that the graphitization process depends on transformations occurring within the layers themselves. The aim of the crystallographic studies must therefore be (1) the detection and identification of the defects whose progressive disappearance constitutes the transformation of a layer of carbon into a layer of graphite, and (2) the study of stacking order in these elementary layers.

It should be noted that the examples examined in the present chapter concern carbons with HTT $>$ 1000 °C, with elementary planar domains already formed. Lower values of HTT yield substances which are more akin to oriented liquids in which the planar areas are still extremely small, with diameters in the range of 2–3 nanometers, corresponding to molecular rather than to macromolecular sizes. At HTT $\approx$ 600 °C, both intraparticle polymerization and intermolecular aromatiztion begin, resulting in the appearance of the elementary carbon layers (Guet and Tchoubar 1986). The description of such substances involves the determination of the positional correlation between small first-neighbor planar particles, full account being taken of short-range order anisotropy. This correlation function depends on the probability distribution p_3 of finding the center of first-neighbor particles in parallel stacking and on the probability $p_{1,2}$ of finding the center of first-neighbor particles in side by side packing (Ruland 1965). After heat-treatment at about 1000 °C, it is possible to consider the substances formed as solids with a high density of defects, and the crystallographic methods developed in Chapters 2 and 3 become applicable to the analysis of the structure after adaptation to the specific structural defects found in these carbon materials.

The evolution of the electronic properties of carbons with $400 < \mathrm{HTT} < 1600\,°\mathrm{C}$ is a good indication of the successive transformations in this temperature region. Up to 700 °C, there is a steady increase in the number of paramagnetic centers (Simon et al. 1977). At about 800 °C, transitions in the electrical conductivity and the value of the Hall constant indicate the onset of conducting states (Delhaes and Carmona 1981). This evolution continues up to HTT ≈ 1600 °C, when a pronounced diamagnetism becomes evident, as well as a high affinity for bromine (Maire and Méring 1967). This temperature is therefore quite decisive in the attainment of graphitization: it represents the point at which the carbon layer discontinuously "snaps" into a layer of graphite; a high mobility of the defects is noted which helps in the elimination of dislocations and the expulsion of the last heteroatoms – such as sulfur, in petroleum cokes, in which it is part of various ring systems. At the same temperature, the C-C in-plane distance becomes practically equal to 1.42 Å, its value in graphite.

This process is particularly evident in carbons such as pitch-cokes, pyrolyzed polyvinyl polymers, etc. which contain a significant proportion of interstitial carbon atoms (Bouraoui and Méring 1964). At the critical temperature T_c, these atoms relax towards potential wells of the graphitic array or are swept towards the layer edges. It is these structural changes which reveal the abrupt switch into graphite-type layers.

In the rest of this chapter, a graphite-type defect-free layer will be labeled a **β-layer**, whereas a layer which still contains intralayer defects will be called an ***α*-layer**. There is no unique description of the α structural organization since there are many possible kinds of intralayer defects, which depend essentially on the nature of the precursor. Section 7.2 is devoted to the description of various kinds of α-layers.

7.2 Structural Characteristics of the Graphitization Process

The graphitization process is the transformation of α-type carbon layers into strictly 2D periodic arrays of carbon atoms (β-type layer), and the mutual ordering of these layers into the tridimensional graphite structure.

7.2.1 Structural Study of the Carbon Layers

The fact that the elementary layer can be described by a simple 2D hexagonal lattice of carbon atoms allows a number of simplifications in the mathematical treatment and easy access to the absolute value of the interference function.

Let us thus suppose that even for imperfect layers with distortions or defects, the 2D-lattice is statistically very close to that of single layers in graphite, insofar as its symmetry and periodicity are concerned. In the latter, the 2D-cell shown in Fig. 7.1 is described by two fundamental vectors $|\vec{a}_1| = |\vec{a}_2| = 2.46$ Å, with $(\vec{a}_1, \vec{a}_2) = 120°$. The structural motif is composed of two carbon atoms at (1/3, 2/3) and (2/3, 1/3). The structure factor is given by:

$$F_{hk}(s) = f_c(s)\,A_{hk}$$

in which $f_c(s)$ **is the atomic scattering factor of carbon,** A_{hk} **is an amplitude factor** given by $A_{hk} = 2$ when $|h-k| = 3n$, (with n integer) or $A_{hk} = -1$ when $|h-k| \neq 3n$.

It should be recalled that the total intensity of the X-ray powder pattern is composed of two contributions: the first, which includes all the crystallographic information, is a coherent radiation while the second, which must be removed before the quantitative study of the intensity can be carried out, is incoherent radiation due to Compton diffusion. The elimination of the unwanted radiation requires that measurements of the intensity be referred to an absolute scale, i.e., that the scattering power P(s) of the specimen be normalized with respect to the formula of the structural motif and that it be expressed in terms of electron units. The method given by Warren and Gingrich (1934) and detailed in Section 6.6 is based on the assumption that for sufficiently large values of s, the mean value of the experimental diffractogram tends towards the "gaseous diffusion curve" of the

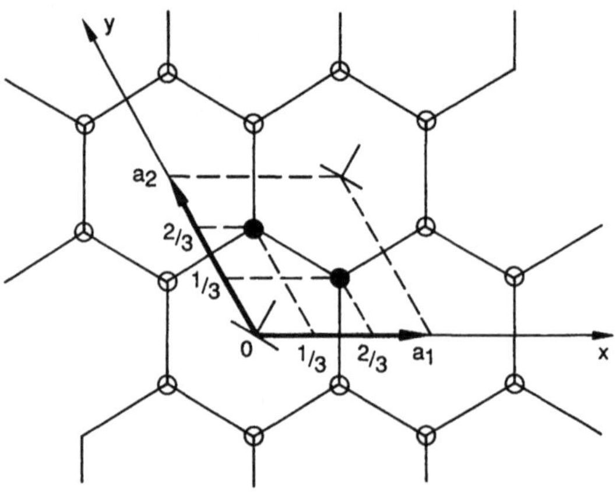

Fig. 7.1. Structure of a perfect graphite layer. Details of the unit cell ($|\vec{a}_1| = |\vec{a}_2| = 2.46$ Å)

specimen, corresponding to a fictitious state in which all the atoms in the structural motif would diffuse independently from each other. The intensity for this gaseous diffusion curve I(s) is here given by:

$$I(s) = 2\, I_{inc}(s) + 2\, f_c^2(s) \qquad (7.1)$$

in which $I_{inc}(s)$ is the contribution of a carbon atom in a covalent state to the intensity of the Compton diffusion and $f_c(s)$ is the corresponding factor for coherent diffusion. Figure 7.2 shows the fit between the experimental diagram and the calculated gaseous diffusion curve for a glassy carbon.

The interference function as defined in Chapter 2 is now given by:

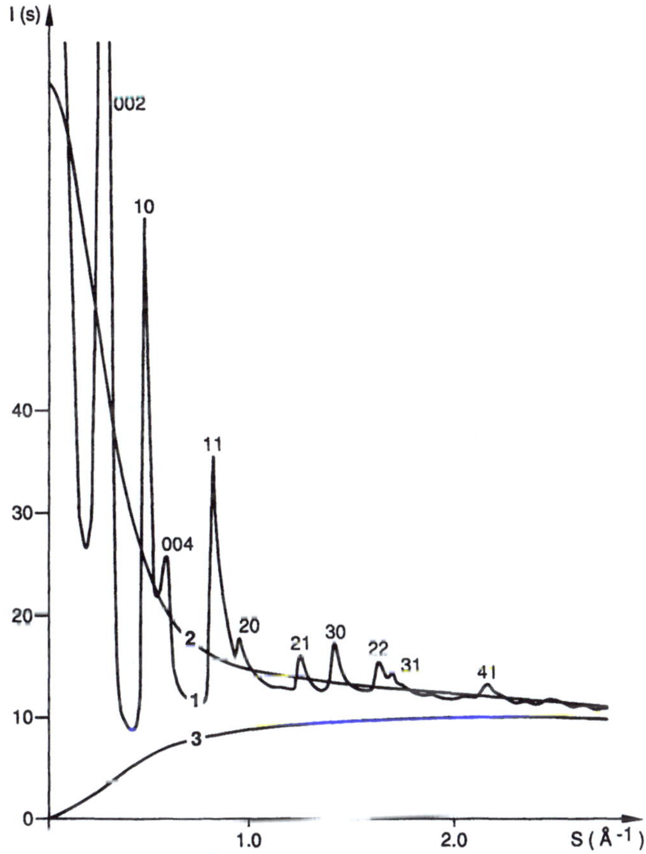

Fig. 7.2. Glassy carbon heat-treated to 2700 °C. *1* Intensity I(s) on an absolute scale (normalized to the electronic content of the 2D unit cell); *2* total intensity diffused by two independent carbon atoms; *3* Compton effect for the two carbon atoms. The *numbers* above each band give its indexation

$$G(s) = [P(s) - 2I_{inc}(s)]/2f_c^2(s) \ ; \tag{7.2}$$

the mean value of this function tends towards 1 as s increases.

Fitting the data extracted from the experimental diffractograms to the computed curves uses the theoretical results developed in Chapters 2 and 3, combined with a number of simplifications due to the specific structure of graphite, as shown hereafter.

– **Scattering intensity for a perfect graphitic layer with limited extension.**
When expressed in electron units, the intensity distribution along an (hk) rod, $I_{hk}(s)$, given in Eq. (2.4), becomes:

$$I_{hk}(s) = \frac{1}{\Omega\sigma} f_c^2(s)\,|A_{hk}|^2 G(\varepsilon_1, \varepsilon_2, \infty) \ , \tag{7.3}$$

where $f_c^2(s)|A_{hk}|^2 = F_{hk} \cdot F_{hk}^*$, $G(\varepsilon_1, \varepsilon_2, \infty)$, the layer shape factor contribution, is given by the square of the shape function Fourier transform, the in-plane area Ω of the unit cell is given by $\Omega = |\vec{a}_1 \wedge \vec{a}_2|$, and σ is the area of the layer.

– **Powder diagram for randomly oriented stacks of layers.**
As given in Eq. (2.15), the intensity of the diffraction bands is

$$I_{hk}(s) = \frac{f_c^2(s)|A_{hk}|^2}{4\pi s \Omega\sigma} \int_{\varphi} G_{hk}(Z) \cdot T(U) \cdot d\varphi \ . \tag{7.4}$$

– **Intensities of the diffraction bands given by defective carbon layers**
Because of the presence of defects in the carbon layers, the biperiodic network of atoms is deformed, the number of C atoms per unit surface may vary, and the structure factor will also vary as a function of the portion of the network explored. The evaluation of the effect on the intensity $I_{hk}(s)$ will use a derivation applied by Guinier (1964) to a powder of small imperfect crystals, after adapting it to layered crystallites.

Consider a two-dimensional crystal of limited extension in which the defects of the 2D-network are sufficiently small that the structure can still be given in terms of a mean lattice not very different from the one which would describe the defect-free lattice, and let O be an arbitrary origin. The position of the various individual lattice unit cells can be described in terms of a vector $\vec{r}$ connecting O to the crystallographically equivalent point in each cell. Let now F_n and $F_{n'}$ be the structure factors of two unit cells respectively corresponding to the vectors $\vec{r}_n$ and $\vec{r}_{n'}$. The scattering power of the crystal is given by:

$$I(s) = \sum_n \sum_{n'} F_n F_{n'}^* \exp[-2\pi i \vec{s} \cdot (\vec{r}_n - \vec{r}_{n'})] \ .$$

Writing now $r_m = r_{n'} - r_n$ and $m = n' - n$, the last expression can be transformed into

$$I(s) = \sum_m \left[\sum_n F_n F^*_{n+m} \right] \exp(2\pi i \vec{s} \cdot \vec{r}_m) \ . \tag{7.5}$$

If the structure of the crystal is statistically homogeneous, i.e., if the product $F_n F^*_{n+m}$ and the position of the vector $\vec{r}_m$ are not correlated, the term in the square brackets can be expressed as a function of the mean value $\overline{F_n F^*_{n+m}}$ of pairs separated by the distance r_m, weighted by the probability $P(r_m)$ of finding two cells with this separation in the crystal. (Note that this corresponds to defects of the first type). The vector $\vec{r}_m$ can itself be expressed as:

$$\vec{r}_m = p\vec{a}_1 + q\vec{a}_2 \ .$$

The probability $P(r_m)$ therefore corresponds to the particular value of the function $p(u_0, v_0)$ defined in Section 2.3.3 and in which $u_0 = p|\vec{a}_1|$ and $v_0 = q|\vec{a}_2|$. Writing now:

$$y(r_m) = \overline{F_n F^*_{n+m}} \ ,$$

the intensity $I(s)$ becomes:

$$I(\vec{s}) = \sum_m P(r_m) y(r_m) \exp(2\pi i \vec{s} \cdot \vec{r}_m) \ . \tag{7.6}$$

The set of vectors $\vec{r}_m$ corresponding to all the values of m is identical to the set of translation vectors (in real space) which can be represented by the set of Dirac distributions: $\delta(\vec{r} - p\vec{a}_1 - q\vec{a}_2)$, and the product $P(r_m) y(r_m)$ becomes

$$P(r_m) y(r_m) = [\delta(\vec{r} - p\vec{a}_1 - q\vec{a}_2)] * [P(r) y(r)] \ , \tag{7.7}$$

in which $P(r)$ and $y(r)$ are distributions of the continuous variable r. With this notation, Eq. (7.6) becomes

$$I(\vec{s}) = \int_\sigma \delta(\vec{r} - p\vec{a}_1 - q\vec{a}_2) [P(r) y(r)] \exp(2\pi i \vec{s} \cdot \vec{r}) d\sigma \ ,$$

where σ is here the 2D-crystal area.

Let now $T_1(s)$ and $Y_1(s)$ be the Fourier transforms of $P(r)$ and $y(r)$, respectively. The last expression immediately transforms into

$$I(\vec{s}) = [\delta(\vec{s} - h\vec{A}_1 - k\vec{A}_2)] * [T_1(s) * Y_1(s)] \ .$$

To each of the products $P(r_m)y(r_m)$ there now corresponds a function $\varphi_1(s) = T_1(s) * Y_1(s)$ in reciprocal space repeated on every node of the reciprocal lattice. However, this ideal case is never met: for Eq. (7.7) to be strictly applicable, the distribution $P(r)$ and $y(r)$ must be independent of s. Although this is true for $P(r)$, which only depends on the shape of the crystal, it is not so for $y(r)$, which is related to the structure factor so that $Y_1(s)$ varies as a function of hk and $I(\vec{s})$ must be given as

$$I(\vec{s}) \approx [\delta(\vec{s} - h\vec{A}_1 - k\vec{A}_2)] * [T_1(s) * Y_{1hk}(s)] \ .$$

It turns out that $Y_{1hk}(s)$ becomes larger as hk increases: if the atoms are only slightly displaced with respect to their position in a perfect lattice, the definition of interatomic distances is fairly accurate for close neighbors, but becomes progressively less applicable as distances increase. The lattice defects therefore show up as a broadening of the bands of diffraction which increases as h and k increase. The distribution of intensity at an hk node must then be written as:

$$I_{hk}(U, V) = \varphi_{1hk}(U, V)$$

in which the variables U and V are referred to the center of the node.

Similarly, in a powder diagram, the term $y(r_m)$ specific to an hk band corresponds to the mean value $\overline{F_n^* F_{n+m}}$ of all vectors $\vec{r}_m$ parallel to the vector $\vec{s}_{hk}$ which determines the position of the given hk node. We therefore define the function

$$\varphi_{hk}(U) = T(U) * Y_{hk}(U)$$

with

$$T(U) * Y_{hk}(U) = Q_{hk} f_c^2(s) \psi_{hk}(U)$$

so that

$$I_{hk}(s) = \frac{1}{\Omega\sigma} f_c^2(s) Q_{hk} \psi_{hk}(U) \ . \tag{7.8}$$

For the reflection considered, Q_{hk} defines the intensity factor which depends on the actual position of the atoms in the unit cell, and ψ_{hk} defines the apparent form function; it includes the function $T(U)$ and the displacement of the atoms with respect to their positions in the defect-free lattice.

The study of the lattice defects is performed by comparing the profiles of the experimental (hk) bands with the calculated diagrams, while trying to find the best possible form for the function $\psi_{hk}(U)$ and determining its variation as one goes from one reflection to the next. The analysis of this variation brings supplementary information concerning the size of the diffracting domains and the strains to which they are submitted.

The powder intensity distribution for an hk band from a defective lattice is now given by the expression:

$$I_{hk}(s) = \frac{Q_{hk} f_c^2(s)}{4\pi s \Omega \sigma} \int_{\varphi} G_{hk}(Z) \psi_{hk}(U) d\varphi \ . \tag{7.9}$$

The corresponding interference function is:

$$G_{hk}(s) = I_{hk}(s)/2 f_c^2(s)$$

Nature of the defects in the layers

The defects may be bonds between carbon atoms and residual heteroatoms, or cross-links, or Schottky defects (vacancies), or Frenkel defects (interstitial atoms) which all result in constraints on the hexagonal cycles of the carbon lattice and modifications of the structure factor modulus. Modifications of the number of C atoms per unit cell affect the modulus of the structure factor, but distortions in the lattice due to slight displacements of the atoms affect the phase factor A_{hk}. Hence $|A_{hk}|^2$ must be replaced by Q_{hk}, which can be determined experimentally. Furthermore, these distortions also slightly modify the positions of the reflections in the diagram since the experimental vectors $\vec{s}_{hk}$ are not identical to the ones found with a perfect lattice. Thus, if the 10 node of graphite is at $s_{10} = 0.4694$ Å^{-1} from the origin in reciprocal space, anthracene carbons with HTT = 1000 °C yield $s_{10} = 0.4720$ Å^{-1}, while glassy carbons yield $s_{10} = 0.4672$ Å^{-1}.

7.2.2 Examples of Structural Evolution in the Carbon Layer as a Function of the Thermal Treatment

In this section we shall consider only the two-dimensional order of α-type layers. The label α covers a wide variety of defects depending on the origin of the carbon, nevertheless, for all carbons, most of the structural evolution of prime importance for the process of graphitization occurs between 1000 and 2000 °C. Monitoring of the physical processes can be performed through a number of parameters: the diameter L_{hk} of the in-plane coherent domains, the distortion $(\overline{\delta_a^2})$ of the layers along $\vec{a}$ and/or $\vec{b}$, or the intensity factor Q_{hk}. An appraisal of the simultaneous evolution of these parameters as a function of HTT brings to light fundamental differences of behavior in the different families of graphitizable carbons.

The parameters L_{hk}, $\overline{\delta_a^2}$ and Q_{hk} used to gauge the structure of the elementary layer are obtained by the indirect method of modelization described in Chapter 3 whereby a synthetic interference function, calculated from the theoretical profiles of hk bands, is compared to the experimental distribution of intensity after correcting and scaling the latter down to absolute units. It is only when the best fit between the theoretical and experimental curves has been obtained for all the profiles that an adequate precision can be claimed for the characteristic parameters L_{hk}, $\overline{\delta_a^2}$ and Q_{hk}. To illustrate the efficiency of this indirect method, two examples are displayed on Figs. 7.3 and 7.4.

Figure 7.3a shows the fit obtained with a graphitizable anthracene coke heat-treated to 1600 °C, while Fig. 7.4 shows the fit in the case of a nongraphitizable glassy carbon heat-treated to 2700 °C. The functions consist of a number of $00l$ reflections resulting from interference between the layers, and hk bands originating from in-plane interferences. As a result of the highly asymmetrical shape of the latter, there is a partial overlap of their respective reflections.

The search for the best fit for the sample of anthracene coke led de Courville et al. (1981) to postulate the presence in the solid of two distinct phases which are the result of two competing processes in anthracene free-radical polymerization (Guet and Tchoubar 1986). The theoretical interference function shown in Fig. 7.3a was drawn up using the two models presented in Fig. 7.3b: this structural heterogeneity is connected with the presence of two possible modes of association of the aromatic building blocks formed by two sets of elementary mechanisms. The diagrams giving the thermal evolution of the coke show that it is conditioned by the structure of the mesophase up to 1600 °C (Brenasin 1979).

1. Diameter L_{hk} of coherence domains

A number of papers (Bouraoui and Méring 1965; Schiller et al. 1967; Rousseaux and Tchoubar 1977; Rouchy and Gatineau 1976; de Courville et al. 1981) have shown that the apparent size L_{hk} of the coherence domains, as deduced from $\psi_{hk}(U)$, is an inverse linear function of $|\vec{s}_{hk}|$:

$$\frac{1}{L_{hk}} = \frac{1}{L_0} + \lambda\,|\vec{s}_{hk}| \ . \tag{7.10}$$

This is illustrated in Figs. 7.5 and 7.6. The term $1/L_0$, connected with the width of the function $\psi_{00}(U) = T(U) * Y_{00}(U)$, determines the peak shape at the origin of the reciprocal lattice (cf. Sect. 7.2.1). The transform $Y_{00}(U)$, which is a function of $[y(r)]_{s=0}$, is independent of s. When the deformation of the layer is small, the values of $F_n F^*_{n+m}$ are not much different from one pair of unit cells to the other and the function y(r) is only weakly dependent on r. Consequently, $Y_{00}(U)$ is practically equivalent to a Dirac function and the shape of the 00 node is solely determined by T(U), so that *the effect of the distortion is eliminated and the determination of the true size of the layer becomes possible.* However, the larger the distortion of the layer, the more important the variation of y(r) as a function of r. The transform $Y_{00}(X)$ may have a non-negligible width and the value of L_0 obtained by linear extrapolation of Eq. (7.10) corresponds to a value lower than the actual size of the layer. Thus L_0 characterizes the mean size of a 'defect-free domain' much smaller than its actual size.

Figure 7.7 shows a typical variation of L_0 with HTT for some graphitizable and nongraphitizable carbons:

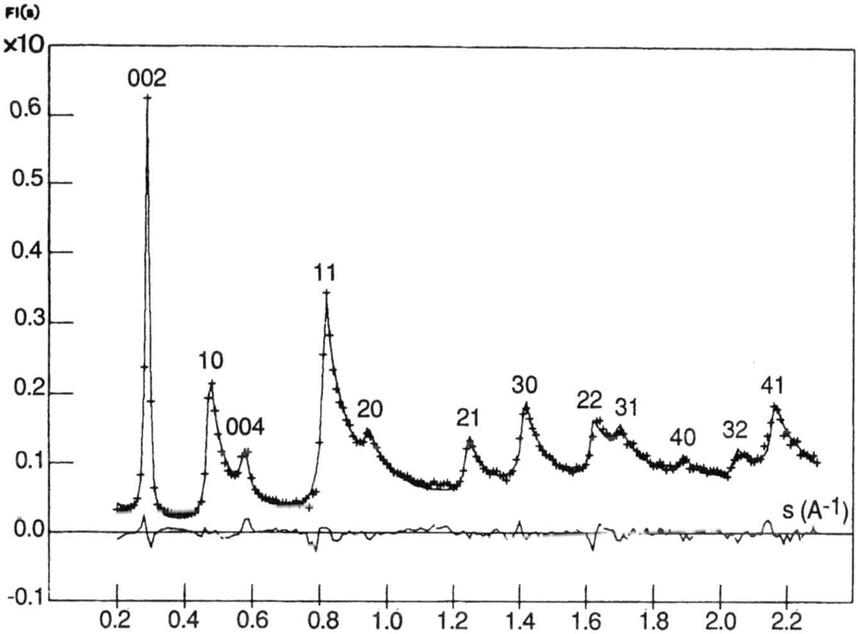

a

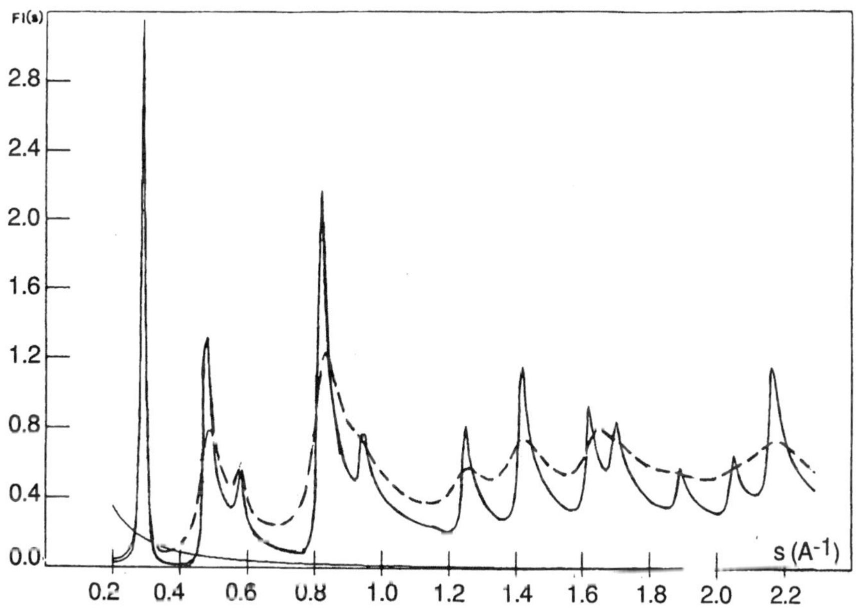

b

Fig. 7.3. Fit between the experimental and calculated interference functions of an anthracene coke heat-treated to 1600 °C: **a** resulting fit (the *crosses* represent the data); **b** calculated curves for each phase: the *dashed line* refers to a phase containing very small polymeric domains as in a mesophase heat-treated to 600 °C; the *full line* corresponds to limited size carbon layers

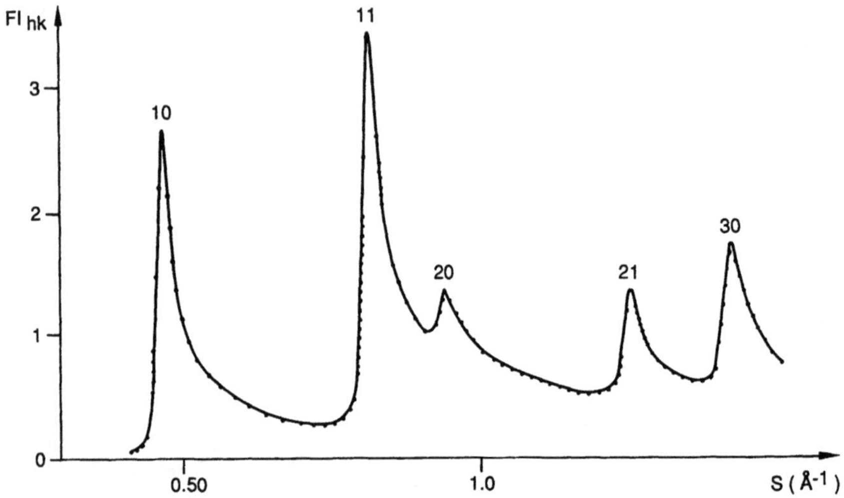

Fig. 7.4. Fit obained with a sample of glassy carbon heat-treated to 2700 °C (the *dots* represent the experimental values)

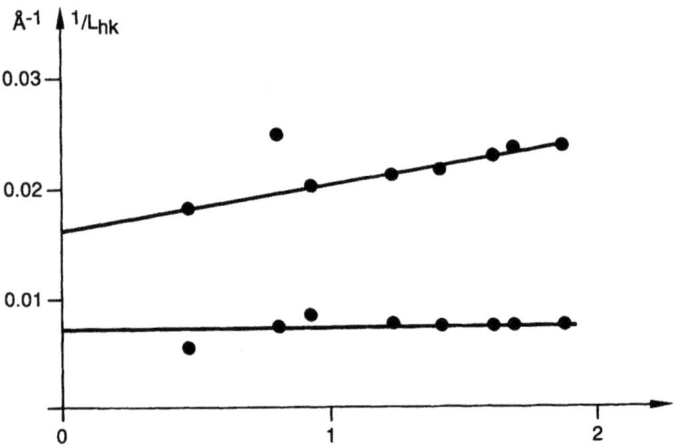

Fig. 7.5. $1/L_{hk}$ *versus* s_{hk} for the two phases of an anthracene coke heat-treated to 2000 °C

1. At 1000 °C and for all the carbons examined, the plane elementary domains are approximately of equal size.
2. Above 1600 °C, the evolution of the two families of carbons is different: in graphitizable carbons, there is a sudden increase in the size of the defect-free domains throughout the range 1600–2000 °C, probably caused by the transformation of the type α plane carbon domains to type β domains of graphite

Fig. 7.6. $1/L_{hk}$ and $1/l_{hk}$ *versus* s_{hk} for a glassy carbon CV 2700° (rectangular model)

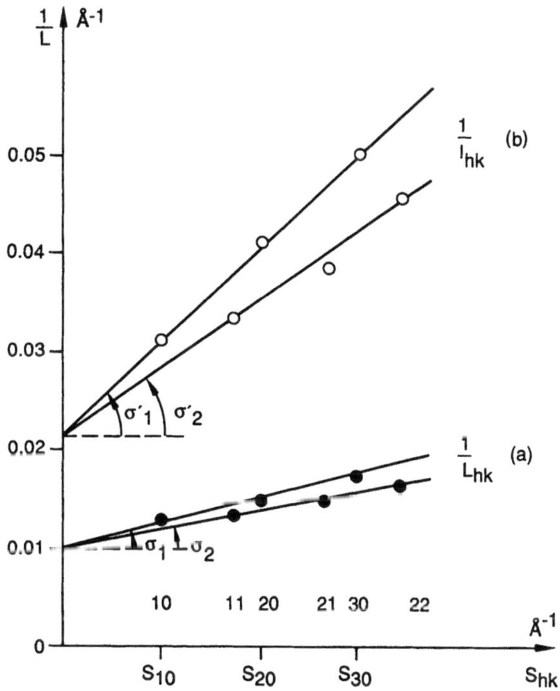

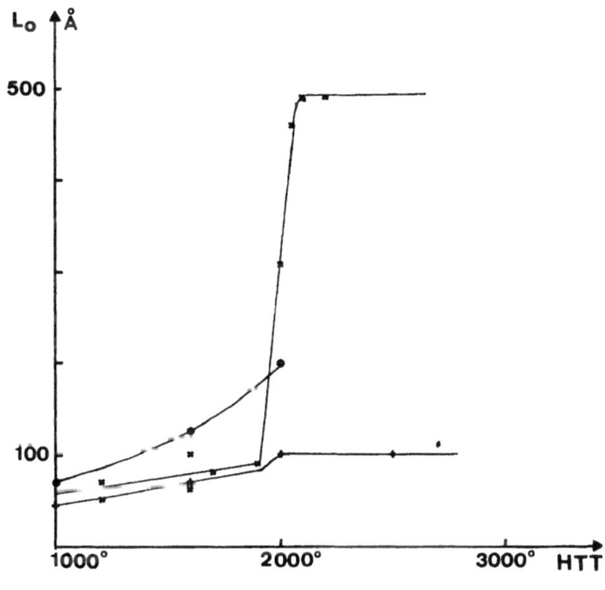

Fig. 7.7. Relationship between L_0 and HTT for some graphitizable and nongraphitizable carbons (× = pitch cokes, ● = anthracene cokes, + = glassy carbon) (*solid lines* are guides for the eye)

(Schiller et al. 1967b; Brenasin 1979). This transition is easily perceived when the parameter $\overline{\delta_a^2}$, characteristic of the distortions in the plane lattice, is studied (cf. the next section and Fig. 7.10). The critical temperature for this transition seems to be 1600–1800 °C if the carbon is free from interstitial C atoms, but it is raised to 2100 °C when the latter need to be eliminated. In nongraphitizable carbons, growth of the domains is blocked and, even at HTT = 2700 °C, their size is not greater than 100 Å.

3. The search for functions $\psi_{hk}(U)$ which will give the best fit leads to the conclusion that the ordered plane domains in graphitizable carbons are statistically isotropic, as can be expected in view of the structure of the mesophase formed initially. In contrast, a good approximation for the shape of the domains in glassy carbons would be a rectangle of width l_0 and length L_0 (Rousseaux and Tchoubar 1977). The analysis of the functions ψ_{hk} has led to relations between hk indices and domain orientation. The reasons for this anisotropy is best understood with reference to the carbonization mechanism of polyfurfuryl alcohol (the precursor of glassy carbon) detailed by Fitzer et al. (1971). As sketched in Fig. 7.8, the final state of the polyfurfuryl alcohol chain after pyrolysis is a bare chain of C atoms (step IV), and hexagonal cycles can only be formed by the creation of bonds between two of these chains placed side by side (step V). The orientation of the rectangular domains therefore places the length L_0 perpendicularly to the axis of the linear carbon chains, in the same direction as that of the increase in the size of the coherent diffraction domains (Fig. 7.8). The width l_0 corresponds to the size of the diffraction domains along the chain axis and reflects the intrinsic folding of the initial chains.

Besides XRD, high resolution electron microscopy (HREM) can also provide information of l_0 and L_0. The results from a thin section (400 to 500 Å) of a 2700 °C glassy carbon sample are given in Fig. 7.9. They were obtained by forming interferences between the 002 or 00$\bar{2}$ reflections and the 000 reflection and are therefore the projection of stacks of carbon layers on the image plane. In this image the fringes are continuous over a wide range but they present high curvatures, implying that the lamellae are highly folded. It turns out that a single layer must be considered as a mosaic of plane domains misoriented with respect to each other. For the diffraction process, these domains are not phase coherent and cannot give constructive interference. This explains why the mean dimension of a diffracting domain, as pictured by XRD, is certainly smaller than the true diameter of a whole layer. This conclusion is supported by the comparison of the length of the straight segments, measured on the HREM picture, with the size of the diameter deduced from XRD results. Even then, it is important to remember that there cannot be perfect coincidence between the results obtained by the two techniques since the HREM image shows the projection of rectangular domains in all possible directions. The fact that l_0 is practically unchanged when HTT is greater than 1000 °C (the pyrolysis temperature) implies that the topological conformation of the carbon layers is a direct result of the conformation of the chains in the polymeric precursor.

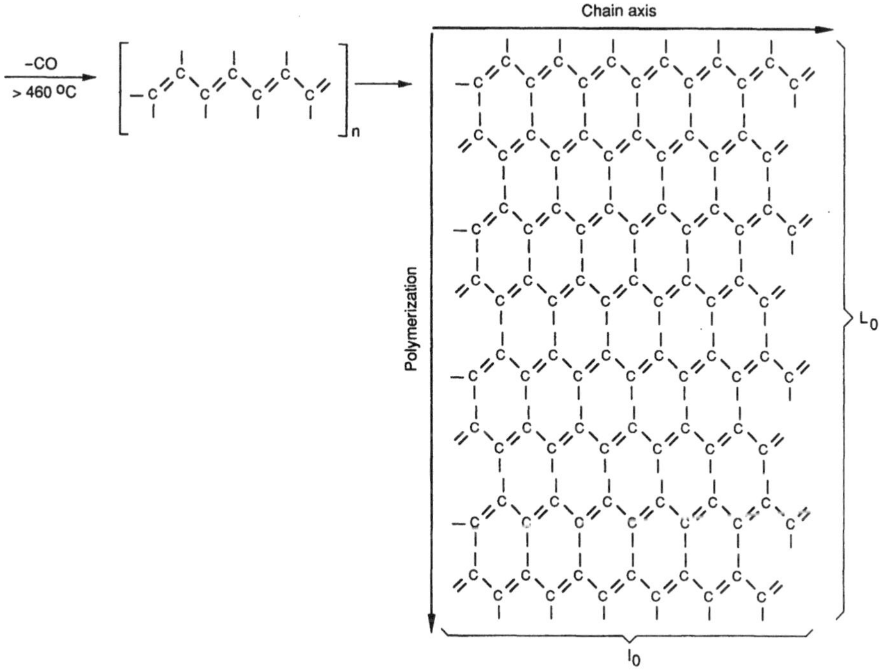

Fig. 7.8. Polymerization mechanism leading to the formation of glassy carbon with an anisotropic development of the carbon layers. (After Fitzer et al. 1971)

2. Distortions in the layer

The function $Y_{hk}(r)$ depends on the projection on $\vec{r}$ of the displacement of the atoms from their theoretical (defect-free) position. As shown by Warren and Averbach (1952), these displacements can be considered as fluctuations in the distances between the atomic rows in a direction perpendicular to $\vec{s}_{hk}$. Suppose now that the layer contains n such atomic rows separated by d_{hk}. The dimension of the domain in the direction $[hk]$ can be expressed as $L_0 = n d_{hk}$. If some of the hexagonal cycles are distorted, the atoms in the layer are not coplanar and, along a given crystallographic direction, the distance L_{hk} is the projection of L_0 on the mean plane of the layer. The distance d between two rows is $d = d_{hk} + \delta_{hk}$ and $\delta L_{hk}/L_{hk}$ varies as $n \Delta d_{hk}/d_{hk}$. The relative deviation or apparent expansion can be deduced from the elementary distance d_{hk} between two rows averaged over the whole layer. If $\overline{\delta^2_{hk}}$ is the mean square deviation of the distance, then

Fig. 7.9. Layer stacking in glassy carbon (2700 °C) visualized by lattice fringes in a 500 Å-thick cross-section. Plate: C. Clinard)

$$\frac{\Delta L_{hk}}{L_{hk}} = \frac{L_0 - L_{hk}}{L_{hk}} \approx n \frac{\sqrt{\overline{\delta_{hk}^2}}}{d_{hk}} = \frac{L_0}{d_{hk}} \frac{\sqrt{\overline{\delta_{hk}^2}}}{d_{hk}} = \frac{\sqrt{\overline{\delta_{hk}^2}}}{d_{hk}} s_{hk} L_0$$

and

$$\frac{1}{L_{hk}} = \frac{1}{L_0} + \frac{\sqrt{\overline{\delta_{hk}^2}}}{d_{hk}} s_{hk} \,. \tag{7.11}$$

from which it can be seen that the slope λ in Eq. (7.10) is a direct measure of the fluctuations since $\lambda = \sqrt{\overline{\delta_{hk}^2}}/d_{hk}$.

The value of d_{hk} is related to the crystallographic parameter a of carbon through

$$d_{hk} = \frac{a\sqrt{3}}{2(h^2 + k^2 + hk)^{1/2}} \,. \tag{7.12}$$

If the strains are isotropic (within the pseudo-bidimensional framework of the layer), λ will have the same value in all crystallographic directions in the carbon lattice and the expansion of the parameter a can be obtained from λ by the relationships:

$$\frac{\sqrt{\overline{\delta_{hk}^2}}}{d_{hk}} = \frac{\sqrt{\overline{\delta_a^2}}}{a} = \lambda \ . \tag{7.13}$$

Hence, measures of the average dimensions of the layers and of their distortions can be obtained from the intercept and the slope of the graph of Eq. (7.11). Figure 7.10 illustrates the changes of $\overline{\delta_a^2}$ with HTT for some graphitizable carbons, and, as mentioned above, it can be seen that the thermal treatment irons out the distortions by way of a clearly marked transition, whose characteristic temperature T_c depends on the nature of the precursor: it is about 2100 °C for carbons with a high content of interstitial C atoms, and in the 1600–1800 °C range for anthracene cokes, which are free from interstitials. The complete expulsion of all distortional defects reflects the transformation to the β or graphitic state. At temperatures above T_c, the thermal treatment supplies the energy necessary for mutual ordering of the β layers and the onset of tridimensional order. In glassy carbons, the values of the mean square displacements depend on the crystallographic direction, as seen in Fig. 7.11 by the different slopes of the straigth lines obtained from the results on carbons heat-treated to 2700 °C: the slopes λ_1 and λ_2 related to the length L are much smaller than the slope λ_1' and λ_2' relative to the breadth l of the domains, i.e., the displacement of the atoms in a direction parallel to l_0 and along the axis

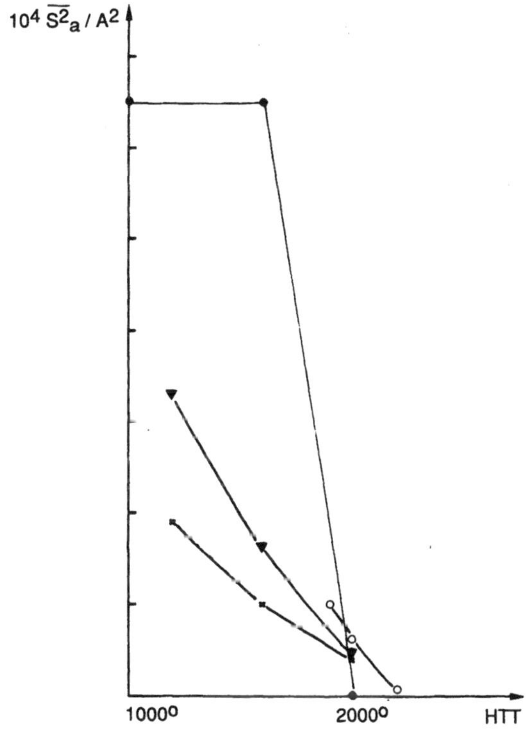

Fig. 7.10. The distortion parameter $\overline{\delta_a^2}$ versus HTT for different graphitizable carbons showing the transition of the carbon layers to graphite layers between 1600 ° and 2000 °C

of the chains is much larger than the displacement parallel to L_0 and perpendicular to the chains. Table 7.1 gives some numerical values of mean square displacements as a function of HTT for three samples of glassy carbon. Even after heat-treatment up to 2700 °C, the amplitude along l_0 is still about 50 times as important as the one observed with graphitizable carbons. The anisotropy of the distortions is itself related to the presence of curvatures and folds in the carbon layers, and associated with high mechanical stresses. Because the thermal treatment does not completely eliminate the resulting strains, the layers cannot reach the β state, and the $\alpha \rightarrow \beta$ transition does not exist, so that the graphitization process is blocked.

3. The intensity factor Q_{hk}

The evolution in the layer is also apparent in the variation of Q_{hk}, related to the mean number of atoms per unit cell and equal to 2 in graphite. Depending on the nature and origin of the carbonaceous materials, different anomalies are found which can modify the Q_{hk} values.

Presence of Interstitial Atoms. The analysis of hk bands from carbons obtained by the pyrolysis of coal tars gives abnormal absolute intensities. If Q_1 are the values of this factor measured on bands with $|h-k| \neq 3$ n and Q_3 the values mea-

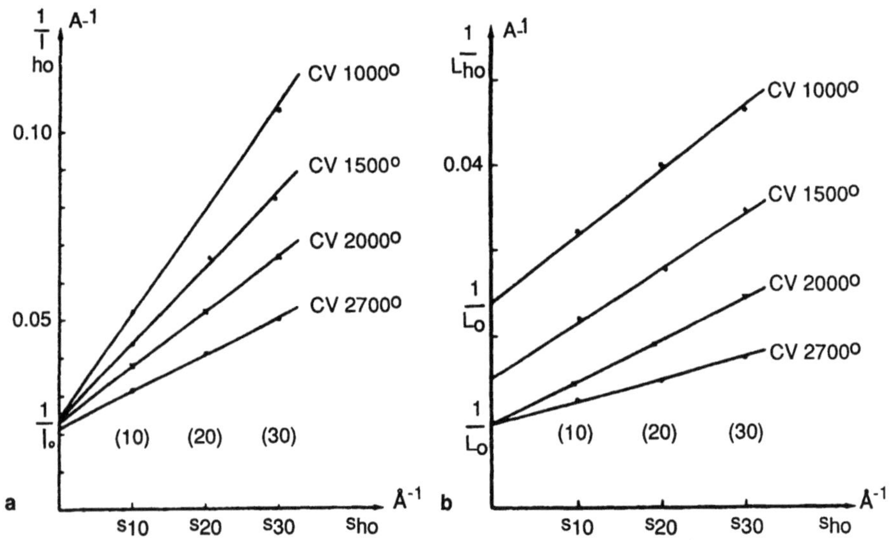

Fig. 7.11. Relationships between $1/L_{h0}$ and $1/l_{h0}$ and s_{h0} for glassy carbons as a function of HTT (rectangular model): **a** evolution of the reciprocal mean width 1/l of the coherent domains; **b** evolution of the reciprocal mean length 1/L of the coherent domains

Table 7.1. Distorsion of the 2D-lattice in a glassy carbon measured along the directions L_0 and l_0

	along $L_0 \overline{\delta_2}$ (Å²)	along $l_0 \overline{\delta^2}$ (Å²)
CV 1000°	2×10^{-4}	140×10^{-4}
CV 2000°	0.6×10^{-4}	44×10^{-4}
CV 2700°	0.2×10^{-4}	18×10^{-4}

sured on bands such that $|h-k| = 3n$, we find $Q_1/Q_3 < 0.25$, whereas this value should theoretically by 1/4 for this type of bands in any carbon with HTT ≈ 2000 °C. In the expressions giving Q_1 and Q_3, there must therefore be a term x independent of the indices. The expressions for Q_1 and Q_3 must be written as

$$Q_1 = (x-1)^2 \quad \text{and} \quad Q_3 = (x+2)^2 \ .$$

The significance of the term x is easiest seen by referring to Fig. 7.12; in the computation of the diffraction, account should be taken of the probability x of finding an interstitial atom above the center of a hexagon, with coordinates (0,0), besides the two atoms normally found in the unit cell at (1/3, 2/3) and (2/3, 1/3). Table 7.2 gives the results of a series of measurements on various industrial cokes with HTT ≈ 2000 °C: the amount of interstitial C atoms depends on the origin of the carbons. In contrast, in the case of anthracene cokes, the evolution of Q_{hk} points to the presence of vacancies in the lattice since Q_1/Q_3 is always greater than 0.25, with values 0.311 and 0.291 for HTT = 1600 and 2000 °C respectively. These values indicate an amount of vacancies lying between 0.09 and 0.06.

Nongraphitizable Carbons, Glassy Carbons. The values of Q_{hk} in Table 7.3 show wide discrepancies with respect to the theoretical values, undoubtedly owing to the disorder in the carbon network which persists up to 2700 °C. The anomalous values of Q_{hk} can be understood in terms of the mean displacement of the atoms resulting from layer distortions. Let us consider for example glassy carbon heat treated to 2700 °C. Even if we assume that the projection of the elementary unit cell is still hexagonal, we will no longer find the projections of the positions of the two atoms in the cell at the coordinates (1/3, 2/3) and (2/3, 1/3), as in graphite. The analysis of the function T(U) (cf. Chap. 2.3.4) gives a possibility of evaluating the mean displacement $\pm\Delta_1$ of the C atoms in this unit cell. The direction of the displacement is shown in Fig. 7.13, in which the new reduced coordinates (X, Y) of the atoms can be deduced from triangle O′AB, with the displacement Δ_1 taken to be about 0.04 Å:

$$\Delta_1^2 = a^2[2x^2 - 2x^2 \cos 120°] \quad \text{with} \quad a = 2.461 \text{ Å} \ ,$$

$$x = \Delta_1/a\sqrt{3} \approx 0.0094 \ .$$

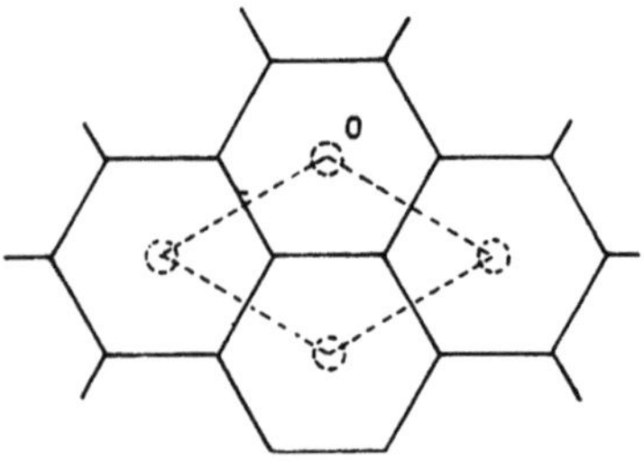

Fig. 7.12. Possible positions (circles) for interstitial C atoms above potential wells of the carbon lattice

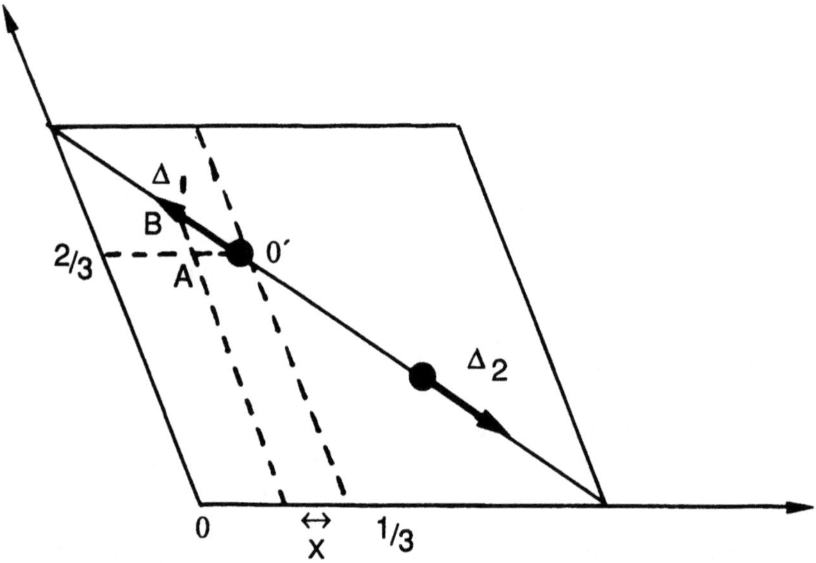

Fig. 7.13. Scale diagram of the anisotropic displacement $\vec{\Delta}_1$ and $\vec{\Delta}_2$ of carbon atoms in the plane lattice of glassy carbon

Table 7.2. Proportion of interstitial atoms in various graphitizable carbons. (After Schiller et al. 1967b)

Sample	Q_1/Q_3	Proportions of interstitial atoms; (HTT ≈ 2000 °C)
I	0.25	<0.02
III	0.235	0.02
II	0.217	0.04
V	0.208	0.05
IV	0.192	0.07
0	0.151	0.11

Table 7.3. Amplitude factors Q_{hk} deduced from the analysis of various hk bands in a glassy carbon

Q_{hk}	Q_{10}	Q_{11}	Q_{20}	Q_{21}	Q_{30}
CV 1000°	1.36	4.07	1.07	1.26	4.36
CV 2000°	1.32	3.90	0.85	1.15	4.3
CV 2700°	1.26	3.80	0.80	1.08	3.9

The new reduced coordinates are then:

$$X = 1/3 - x = 0.324$$

$$Y = 1 - X = 0.676 \ .$$

Since the square of the structure factor for a layer of carbon is

$$F^2 = 2\{1 + \cos 2\pi\,[(1/3 - 2x)(h - k)]\}$$

we finally obtain, for the

10 band: $F^2 = 1.21$ instead of 1

11 band: $F^2 = 4$ instead of 4

20 band: $F^2 = 0.62$ instead of 1

21 band: $F^2 = 1.21$ instead of 1

30 band: $F^2 = 3.87$ instead of 4

22 band: $F^2 = 4$ instead of 4

These calculated values are fairly close to the ones found experimentally for Q_{hk} (in Table 7.3), especially the abnormally large values of Q_{10} and Q_{21}. We also find that glassy carbon has approximately the same mean lattice as the one envisaged by Ergun (1976) for a pyrolytic carbon since this author used the reduced coordinates $X = 0.318$ and $Y = 0.682$ to justify the structure factor which he had found.

7.3 Organization of the Stacks

7.3.1 Structure of the Stacks in the two Graphite Polytypes

The triperiodic structure of graphite is shown in Fig. 7.14. The fundamental element is the monoatomic layer of carbon with hexagonal symmetry to which is associated a translation vector $\vec{\Delta}$ of modulus equal to the length of a C-C bond in the layer (1.42 Å) and direction parallel to one of the three C-C bonds. When referred to the hexagonal 2D-unit cell, this translation can be expressed as [1/3, 2/3] or [2/3, 1/3]. Depending on the sequence of translation vectors, two polytypes can be formed:

2H *polytype, hexagonal graphite*: the layers are alternately translated in their plane by $+\Delta$ or $-\Delta$; this leads to a stacking of noncompact layers noted ABABAB.... Perpendicularly to the layers, the period is therefore equal to twice the interlayer distance. The space group is $C6mmc$ and the unit cell contains four C atoms at (0, 0, 1/4); (0, 0, 3/4); (1/3, 2/3, 1/4); (2/3, 1/3, 3/4), with $a = 2.461$ Å and $c = 6.708$ Å.

3R *polytype, rhombohedral graphite*: the translation between all pairs of adjacent layers is the same and equal to $+(\text{or}-)\Delta$. The periodicity is now described by ABCABC... and is three times the interlayer distance. Rhombohedral graphite, which seems to be stabilized by dislocations in the layers, is formed from hexagonal graphite under the influence of mechanical shearing stresses; it has never been obtained as a pure form but always associated with the 2H polytype.

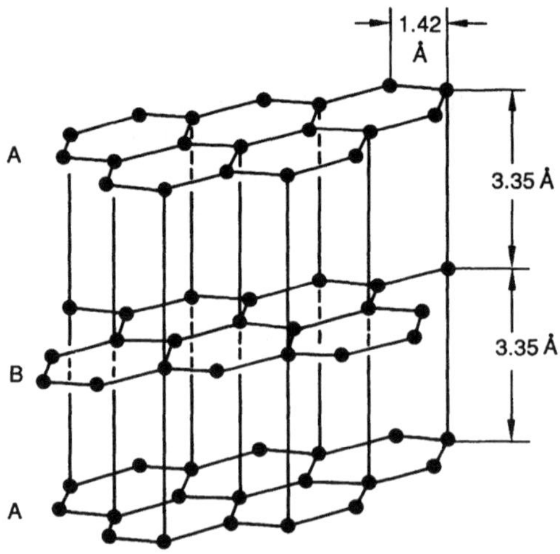

Fig. 7.14. Three-dimensional structure of 2H graphite. The letters *A* and *B* refer to the two different positions of the layers with reference to the origin of the unit cell

7.3.2 Structure of the Stacks in a Carbon Undergoing Graphitization

During graphitization, the stacks can be assimilated to an interstratification of α and β layers forming a statistical distribution of *ordered* and *disordered* intervals. As discussed above, the transformation of a defective α layer to a perfect graphite β layer is a discontinuous process starting at or just above 1600 °C (cf. Figs. 7.7 and 7.10). It has been shown that the distribution of α and β layers is random and that several types of stacking faults must be taken into account: these may include stacking faults due to the existence of the two polytypes, and misorientations between layers. A first type of fault is constituted by well-defined translation defects affecting the sequence of signs of $\vec{\Delta}$ between two ββ layers. A second type of fault is better described as a random nondefined defect (cf. Chap. 3.2.1 and 3.2.5) affecting both the αα and the ββ intervals. The degree of order therefore depends both on the number of β layers and on their distribution in the stack as well as on the nature of the α layers. The degree of order will be apprehended through the study of the modulation functions $G_{hk}(Z)$ introduced in Eq. (7.9) and discussed in Chapter 2 [cf. Eq. (2.9)] which describes the interferences between the layers.

The general expression for the modulation functions valid for a powder containing a *very large number* of stacks of finite size composed of identical layers is

$$G_{hk}(Z) = 1+2\,\mathrm{Re}\left[\sum_{M} \alpha(M) \sum_{n=1}^{n=M-1} \frac{M-n}{M} \overline{\exp[2\pi i(h\alpha_n+k\beta_n+Zz_n)]}\right] \quad (7.14)$$

in which α (M) is a normalized function describing the distribution of the thickness of the stacks, and $\vec{r}_n$ is the vector between the general origin of the coordinates and the origin of the n-th layer. The components of $\vec{r}_n$ are α_n, β_n, z_n. The discussion of the analytic form of $G_{hk}(Z)$ is, however, best done after separating the case $h = k = 0$ from the other (h, k) couples.

1. **The function $G_{hk}(Z)$ with $h = k = 0$**
When $h = k = 0$, the expression in Eq. (7.14) can be simplified to

$$G_{00}(Z) = 1+2\,\mathrm{Re}\left[\sum_{M} \alpha(M) \sum_{n=1}^{n=M-1} \frac{M-n}{M} \overline{\exp(2\pi i Z z_n)}\right] .$$

In graphitizable carbons, the distortions and imperfections in the layers bring about small fluctuations δ_c in the position of the layer along the $\vec{c}$ axis. The interlayer distances between first-nearest neighbours are therefore not very different from the mean value d_m and the difference with the mean value can be taken to constitute a disorder of the second type (Chap. 1.5.6). Furthermore, these distances are assumed to be randomly distributed (Maire and Méring 1967) so that S = 0. The expression above can then be modified to:

$$G_{00}(Z) = 1 + 2\,\mathrm{Re}\left[\sum_{M} \alpha(M) \sum_{n=1}^{n=M-1} \frac{M-n}{M} \left(1 - 2\pi^2 Z^2 \overline{\delta_c^2}\right)^n \exp\left(2\pi i n Z d_m\right)\right] \qquad (7.15)$$

in which $\overline{\delta_c^2}$ is the mean square deviation of the fluctuations. We can see that $G_{00}(Z)$ is not a periodic function: although the maxima are indeed equidistant in reciprocal space, their shape becomes progressively more diffuse. The equidistance between successive maxima is associated with a reciprocal period smaller than that found in graphite; in real or direct space, this corresponds to an interlayer spacing larger than that of graphite. It is convenient (Franklin 1951; Maire and Méring 1970) to introduce the parameter g and the relationship

$$d_m = 3.44(1-g) + 3.354\,g \ , \qquad (7.16)$$

which relates g to the interlayer distance d_m since both change with HTT. In Eq. (7.16), 3.44 Å is the interlayer distance for an $\alpha\alpha$ interval obtained at HTT = 1600 °C while 3.354 Å is the interlayer distance for pure graphite, i.e., for a $\beta\beta$ interval. The parameter g therefore varies between 0 and 1 as HTT changes from 1600 to over 3000 °C. Since it is equal to the proportion of β layers, g is often called the degree of graphitization. It is an order parameter defined in one-dimensional space. Other parameters, such as $\overline{\delta_c^2}$ and $\overline{M} = \sum M\alpha(M)$ can also be obtained from the profiles of the $00l$ reflections. This can be done either by the direct Fourier transform method of Houska and Warren (1954) (cf. Sect. 3.3.3) or using the modelization method of Méring (1949). Once d_m is known, g can be obtained by reversing Eq. (7.16).

Figure 7.15 illustrates the variation of g with HTT for five series of carbons studied by Schiller et al. (1967a). Although their shapes are similar, the curves do not coincide, and HTT is thus inadequate for the full characterization of a carbon species. On the contrary, the parameter g is a more pertinent measure of the structural characteristics; it will therefore be used instead of HTT in the rest of the present chapter. The transition $\alpha \rightarrow \beta$ is emphasized when one plots the evolution of L_c and $\overline{\delta_c^2}$ versus g. All the curves in Fig. 7.15 show a point of inflexion in the temperature range 1800–2000 °C, corresponding to a critical temperature T_c at which the formation of β layers in the carbon material begins.

Figure 7.16 shows the transition in the thickness L_c (given by $L_c = \overline{M} d_m$) of layer stackings measured along $\vec{c}$ as a function of g for various graphitizable carbons (Schiller et al. 1967a).

Finally, the description of the structure along $\vec{c}$ can also benefit from the study of the distortion parameter. The results depicted in Fig. 7.17 are typical of those observed with two kinds of carbon, respectively with and without interstitial C atoms. In both cases, $\overline{\delta_c^2}$ tends to 0 as graphitization proceeds. The disappearance of the distortions along $\vec{c}$ is paralleled by the corresponding increase in size of the interferential coherence domains which participate in the formation of the $00l$ reflections (cf. Fig. 7.16). Note that the elimination of the distortions occurs at a lower temperature in carbons which do not contain interstitials.

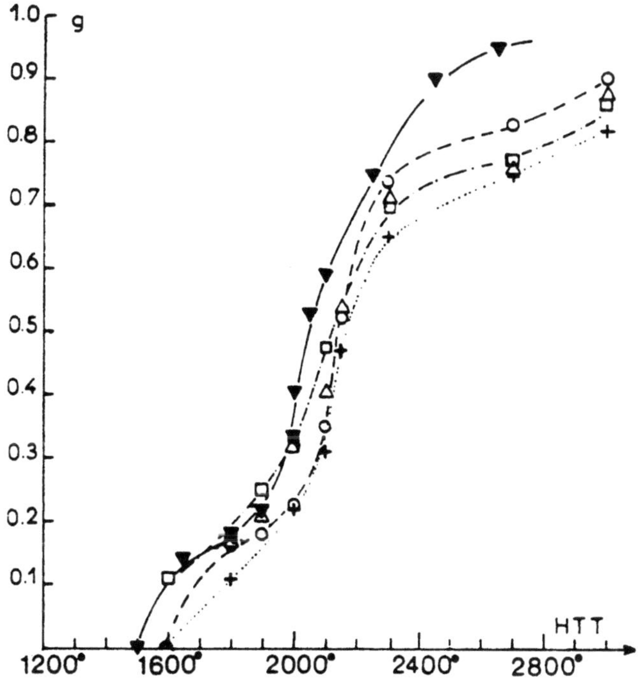

Fig. 7.15. Relationship between the parameter g and HTT: the evolution of g (degree of graphitization) reflects the 2D to 3D transition for different graphitizable carbons, identified by the different symbols

Since the fluctuations δ_c are linked with distortions of the elementary hexagons and hence with δ_a, there should be – as indeed there is – a correlation between these two types of defects, as seen in Fig. 7.18. As HTT nears 2000 °C, both $\overline{\delta_a^2}$ and $\overline{\delta_c^2}$ reach their minimum constant residual value. Beyond this temperature, the last process of graphitization starts.

2. The function $G_{hk}(Z)$ with $h \neq 0, k \neq 0$

The study of the interference functions $G_{hk}(Z)$ given in Eq. (7.14) provides a good description of the final 3D ordering process starting above the critical temperature T_c and leading to graphite by successive elimination of stacking faults which are either misorientations between layers (hereafter called T faults) or well-defined translation defects connected with the possibility of existence of the two polytypes 2H and 3R. The presence of these T faults (Maire and Méring 1970) results in the disappearance of coherent scattering in the $[hk]$ direction (in which h or k are not simultaneously equal to zero). In a substack bounded by two T defects, $z_n = n\,d_{002}$, with $d_{002} = 3.354$ Å. It has been proved that, in carbon materials, the distribution of T defects in the stack is random (R = 0); if P_1 be the probability of not finding a T defect between two adjacent layers, Eq. (7.14) becomes:

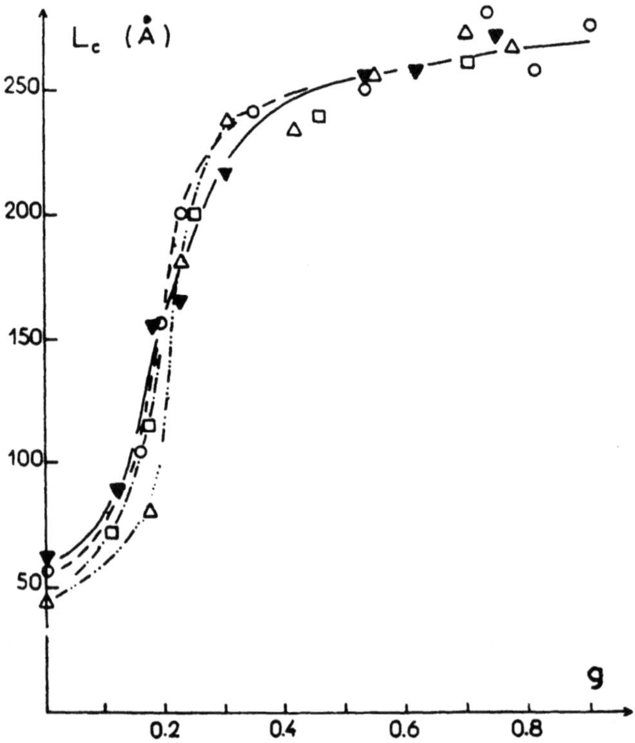

Fig. 7.16. Relationship between L_c, the size of the coherent domains along the $\vec{c}$ axis, and the parameter g, for different graphitizable carbons, identified by the different symbols

$$G_{hk}(Z) = 1+2\,\mathrm{Re}\left\{\sum_{n=1}^{n=M-1}\left[P_1\left(1-\frac{1}{M}\right)\right]^n \times (\exp 2\pi i Z d_{002})^n\,\overline{\exp 2\pi i(h\alpha_n+k\beta_n)}\right\} \tag{7.17}$$

with $P_1 = 1-P_A$, where P_A is as defined in Chapters 2 and 3. This expression implies that only ordered substacks contribute to the intensity of any given hk band. The substacks form packets of $\beta\beta$ graphite-like layers along $\vec{c}$, but the number n of layers in any substack is small compared to $\overline{M}$. The period of the function G_{hk} is $1/3.354$ Å^{-1} and it is independent of the degree of graphitization. This periodicity is, however, marred by anisotropic thermal agitation which weakens the successive maxima; apart from this restriction, the maxima of $G_{hk}(Z)$ are identical both in shape and in periodicity.

When $P_1 = 0$, i.e., when all the layers are randomly misoriented, the stack is (ideally) turbostratic and there are no Bragg reflections other than those belonging to the $00l$ series. Graphitizable carbons with HTT < 1600 °C and nongraphitizable carbons (whatever their HTT) both correspond to the case $P_1 = 0$.

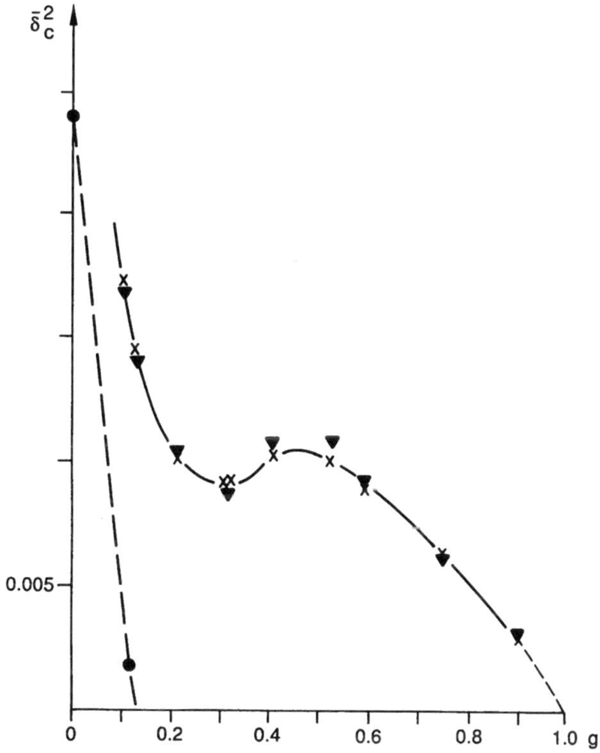

Fig. 7.17. Relationship between the fluctuation $\overline{\delta_c^2}$ of the interlayer distance and the parameter g, showing the difference in behaviour of carbons with (×), without (●) interstitial C atoms (lines are guides for the eye)

When $P_1 \neq 0$, the components α_n and β_n correspond to specific translations, well-defined with respect to the periodic 2D lattice:

$$\alpha_n = \pm 1/3 \quad \text{and} \quad \beta_n = \pm 2/3 \ .$$

The hexagonal symmetry of the layer implies the separation of the functions $G_{hk}(Z)$ into two groups: those with h, k values such that $|h-k| = 3\,n$, and those with $|h-k| \neq 3\,n$ (Maire and Méring 1970).

Bands with $|h-k| = 3\,n$. Eq. (7.17) becomes

$$G_{hk}(Z) = 1 + 2\,\mathrm{Re} \sum_{n=1}^{n=M-1} \left\{ \left[P_1 \left(1 - \frac{1}{M} \right) \right] \right\}^n \exp\left(2\pi i Z d_{002}\right)^n \ . \qquad (7.18)$$

These functions only depend on P_1 and are therefore only sensitive to the elimination of T defects.

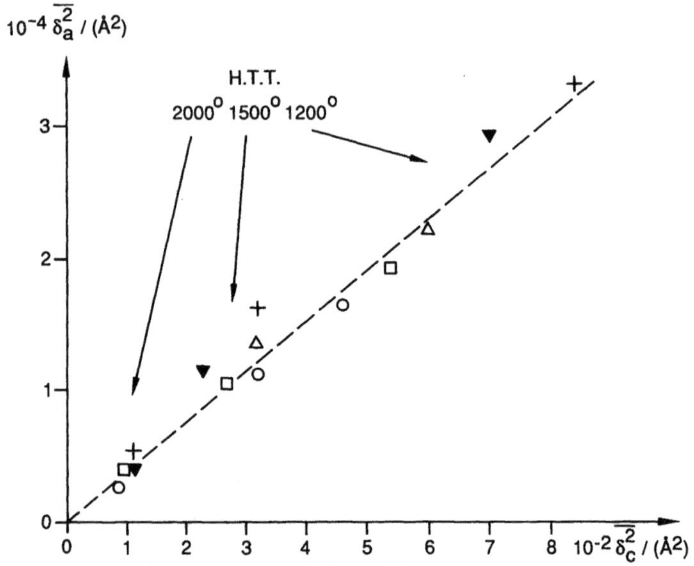

Fig. 7.18. Relationship between $\overline{\delta_\alpha^2}$ and $\overline{\delta_c^2}$ showing the correlation between the distortion in the elementary layers and the stacking order, in different carbons identified by different symbols

As soon as HTT > T_c, the modulations which appear on the profiles of the bands can be exploited to yield P_1. It should however be remembered that Eqs. (7.17) and (7.18) are only valid for a random distribution of T defects, which implies that P_1 be independent of the position of the next-nearest neighbor layer and that:

$$P_n = (P_1)^n \;, \tag{7.19}$$

where P_n is the probability of finding a T defect between two nth-nearest neighbor layers.

In Eq. (7.18), P_1 is free of h and k indices. This hypothesis is strictly verified for carbons heat-treated at temperatures above 2300 °C. A detailed analysis of the hk bands for high values of h and k showed (Tchoubar and Méring 1973) that:

- if $P_1 < 0.5$, its value depends on the indices hk of the band and, as can be seen in Fig. 7.19, P_1 is a linearly decreasing function of the square of the modulus s_{hk}^2. The exact value of P_1 is obtained by extrapolation of the line $P_1 = f(s_{hk}^2)$ to $s_{hk} = 0$. Tchoubar and Méring (1973) have shown that in this same range of P_1, there are weak fluctuations of the α_n and β_n components around the ideal values 1/3 and 2/3. In other words, the β layers in the 3D graphitic structure being formed are weakly shifted with respect to each other, probably because of residual strains due to the neighboring disordered regions.

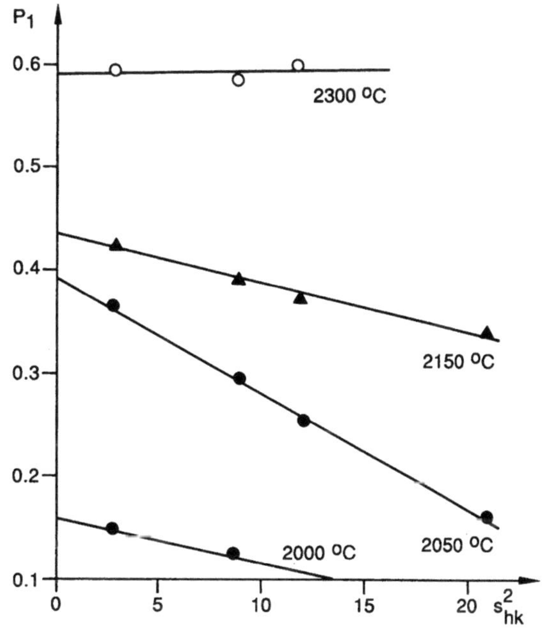

Fig. 7.19. Relationship between P_1 and s^2_{hk}: influence of the residual strains in ordered domains when $P_1 < 0.5$

– if $P_1 > 0.5$, its value is independent of the indices *hk*. Beyond this point, the ideal graphite 3D structures is set.

Comparison and Physical Significance of the Parameters P_1 and g. Both these parameters are used to describe the degree of graphitization but, strictly speaking, P_1 characterizes the 3D organization whereas g has the characteristics of a 1D order parameter. Both vary between 0 and 1 when HTT varies between 1600 and 3000 °C. The relation between P_1 and g and the true physical meaning of these parameters have been discussed by a number of authors (Maire and Méring 1970; Schiller et al. 1967); from the plot of P_1 versus g^2 (Fig. 7.20), one can deduce that:

$$P_1 = g^2 \tag{7.20}$$

The relationships in Eqs. (7.19) and (7.20), which were experimentally verified, give an unambiguous interpretation of the physical meaning of P_1 and g: any stack may be described as a sequence of both α and β layers; if g is the relative proportion of β layers, P_1 becomes the probability of finding a (ββ) pair and, by the same token, Eq. (7.16), which relates the mean period d_m along $\vec{c}$ of a partially ordered stack to the graphitization of the layers, is justified. However, in a stack with a random distribution of α and β layers (R = 0), the probability P_n of forming a substack of (n+1) β layers is

$$P_n = g^{n+1} \tag{7.21}$$

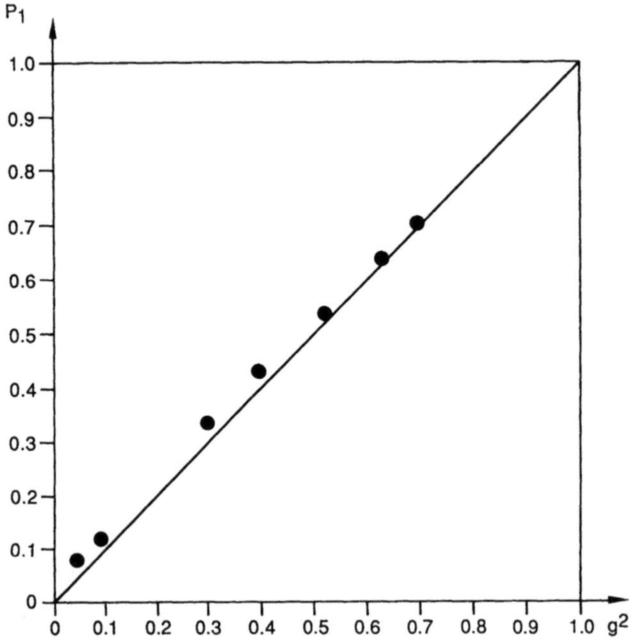

Fig. 7.20. Experimental verification of the relation $P_1 = g^2$ for a graphitizable carbon

in spite of the fact that most experiments have shown that $P_n = g^{2n}$. The justification of this experimental result requires that g be defined as *the proportion of β-type layer-faces* rather than the proportion of *layers*, and P_1, which is always given by $P_1 = g^2$, now becomes the probability characteristic of the proportion of *interlayer spaces* rather than the probability characteristic of the proportion of (ββ) pairs in the stack. Similarly, the probability P_n of finding n successive ββ interlayer spaces is given by

$$P_n = g^{2n} \; .$$

These statistical relationships observed on various carbons in the range of temperatures which trigger the 2D→3D transformation (i.e., between 1600° and 2000 °C) were interpreted by Maire and Méring (1970) as occurring via the elimination (or "sweeping away"), of interstitial atoms grafted onto the layers. This sweeping away of interstitial atoms occurs one face at a time. If α and β are now taken to refer to the **state of the layer faces**, the elementary transformation of the layers correspond to the sequences

$$\alpha\alpha \rightarrow \alpha\beta \rightarrow \beta\beta \; .$$

Furthermore, it was shown (Maire and Méring 1970; Schiller et al. 1967; Rouchy and Gatineau 1976) that the α→β transformation occurs in two steps. At first, with the faces in the α state, the interstitial atoms are randomly distributed over the layer. Subsequent heat treatments decrease the distortion and improve the periodicity of the surface potential of the layer, thus giving interstitial atoms a possibility of localization over potential wells; at this point the layer face is said to be in a state α' from which it can pass into state β. A better description of the transformation process is therefore

$$\alpha\alpha \rightarrow \alpha\alpha' \rightarrow \alpha'\beta \rightarrow \beta\beta \ .$$

The use of Simon's reagent (a solution of silver bichromate in sulfuric acid) to oxidize different carbons enabled Oberlin and Méring (1964) to point out different kinetics of oxidation associated with the four different types of interlayer spaces, namely $\alpha\alpha$, $\alpha\alpha'$, $\alpha'\beta$ and $\beta\beta$. The relative proportion of each type can then be expressed as:

$$|\alpha\alpha+\alpha\alpha'| = (1-g)^2 \ ,$$
$$|\alpha'\beta| = 2g(1-g) \ ,$$
$$|\beta\beta| = g^2 \ ,$$

where $|\ |$ denotes the relative proportion. From these proportions it is easy to compute the mean interlayer distance d_m as:

$$d_m = 3.354\, g^2 + [(3.354+3.44)/2]\, 2g(1-g) + 3.44(1-g)^2$$

from the values easily available through experimental measurements.

It would, of course, have been most interesting to study the 2D → 3D transformation using the same method of analysis on anthracene carbons, which are free from interstitial carbons and in which the transition temperature T_c is better localized in the 1600–1800 °C range (cf. Fig. 7.10). The essential difference between the two types of carbons is to be found in a lack of applicability of Eq. (7.17). Consequently, in the simulation of band profiles, Eq. (7.19) should be replaced by Eq. (7.21).

All the results presented above show that the only parameter which really represents the degree of graphitization is P_1, defined as the probability of finding a graphite-like $\beta\beta$ configuration; it is therefore the relevant order parameter.

Bands with $|h-k| \neq 3n$. Equation (7.17) cannot now be simplified and the modulations depend both on P_1 and on the signs of the translation components α_n and β_n. The functions can therefore be used for the detection of well-defined translation faults connected with changes in parity in the sequence $+\vec{\Delta}-\vec{\Delta}+\vec{\Delta}\ldots$ characteristic of the 2H polytype. A translation defect will now correspond to a sequence $+\vec{\Delta}+\vec{\Delta}$ or $-\vec{\Delta}-\vec{\Delta}$ Let η be the conditional probability

Table 7.4. Temperature dependence of the probability η (see text) for a polyvinyl coke (Maire and Méring 1970)

HTT	1800° to 2000°	2200°	2400°	2600°	2800°	2950°
η	0.5	0.39	0.28	0.29	0.28	0.27

of finding such a defect, characteristic of a 3R polytype sequence. In the case of polyvinyl cokes, the detailed analysis of the $G_{hk}(Z)$ functions (Maire and Méring 1970) differentiates between two characteristic ranges of temperature:

- for 1800 < HTT < 2200 °C, the *hk* bands are not modulated and the alternation of $+\vec{\Delta}$ or $-\vec{\Delta}$ translations is strictly random in the substacks, i.e., the sequences $+\vec{\Delta}+\vec{\Delta}$, $-\vec{\Delta}+\vec{\Delta}$, $+\vec{\Delta}-\vec{\Delta}$ and $-\vec{\Delta}-\vec{\Delta}$ have the same probability:

$$W_{++} = W_{-+} = W_{+-} = W_{--} = 0.25 \; ;$$

  and from its definition given above, we therefore have $\eta = 0.5$;
- for HTT > 2200 °C, modulations appear on the *hk* bands. These modulations increase with HTT. The parameter η can be obtained from the band profile analysis and it is found that η decreases as graphitization is improved without ever reaching 0, even at the end of the thermal treatment. As verified on a large number of different types of carbon, the limiting value of η is 0.27 (cf. Table 7.4), which means that thermal annealing does not eliminate all the structural defects in the substacks, and that it is very difficult to obtain a synthetic graphite free from any defect by mere heat-treatment. As an example, polyvinyl coke heat treated to 2950 °C yields the value $P_1 = 0.715$. This artificial graphite still contains a number of T defects at the border of the substacks which themselves contain a number of 3R defects in the 2H polytype major phase.

References

Bokros JC (1969) Deposition, structure and properties of pyrolytic carbon. In: Walker PL, Jr (ed) Chemistry and physics of carbon, vol 5. Dekker, New York Basel, pp 1–110

Bouraoui A, Méring J (1964) Diamètre des couches atomiques élémentaires dans les carbones tendres traités entre 1000 °C et 2000 °C. Carbon 1:465–469

Brenasin J (1979) Méthode permettant d'affiner l'analyse des diagrammes de poudre – exploitation d'un modèle à deux phases dans l'étude structurale des carbones d'anthracène. Thèse de Doctorat, Univ Orleans

Brooks JD, Taylor GH (1965) The formation of graphitizing carbons from the liquid phase. Carbon 3:185–193

De Courville-Brenasin J, Joyez P, Tchoubar D (1981) Méthode d'ajustement automatique entre courbes expérimentales et calculées dans les diagrammes de diffraction de poudre. Cas des solides à structure lamellaire. I: Developpement de la méthode. J Appl Crystallogr 14:17–23

Delhaes P, Carmona F (1981) Physical properties of non-crystalline carbons. In: Walker PL Jr, Thrower PA (eds) Chemistry and physics of carbon, vol 17. Dekker, New York Basel, pp 89–167

Diamond (1957) X-Ray diffraction data for large aromatic molecules. Acta Crystallogr 10:359–364

Donnet JP, Bansal RC (1984) Carbon fibers. In: Lewin M (ed) International science and technology. Dekker, New York Basel, 290 pp

Ergun S (1970) X-Ray scattering by very defective lattices. Phys Rev B 1/8:3371–3380

Ergun S (1976) Analysis of coherence, strain, thermal vibration and preferred orientation in carbons by X-Ray diffraction. Carbon 14:139–150

Fischbach DB (1971) The kinetics and mechanism of graphitization. In: Walker PL, Jr (ed) Chemistry and physics of carbon, vol 7. Dekker, New York Basel, pp 1–97

Fitzer E, Mueller K, Schaeffer W (1971) The chemistry of pyrolytic conversion of organic compounds. In: Walker PL, Jr (ed) Chemistry and physics of carbon, vol 7. Dekker, New York Basel, pp 237–397

Franklin RE (1951) The structure of graphitic carbons. Acta Crystallogr 4:253–261

Gasparoux H (1980) Carbonaceous mesophase and disc-like liquid crystals. Mol Cryst Liq Cryst 59:109–116

Guet JM, Tchoubar D (1986) Characterization of bulk mesophase by X-Ray diffraction. Fuel 65:107–112

Guinier A (1964) Théorie et technique de la radiocristallographie. Dunod, Paris, 740 pp

Houska CR, Warren BE (1954) X-Ray study of graphitization of carbon black. J Appl Phys 25:1503–1509

Lahaye J, Prado G (1978) Mechanisms of carbon black formation. In: Walker PL, Jr, Thrower PA (eds) Chemistry and physics of carbon, vol 14. Dekker, New York Basel, pp 167–281

Lewis IC, Singer LS (1981) Electron spin resonance and the mechanism of carbonization. In: Walker PL, Jr, Thrower PA (eds) Chemistry and physics of carbon, vol 17. Dekker, New York Basel, pp 1–88

Maire J, Méring J (1970) Graphitization of soft carbons. In: Walker PL, Jr (ed) Chemistry and physics of carbon, vol 6. Dekker, New York Basel, pp 125–189

Marsh H, Walker PL, Jr (1979) The formation of graphitizable carbons via mesophase: chemical and kinetic considerations. In: Walker PL, Jr, Thrower PA (eds) Chemistry and physics of carbons, vol 15. Dekker, New York Basel, pp 229–295

Méring J (1949) L'interférence des rayons X dans les systèmes à stratification désordonnée. Acta Crystallogr 2:371–377

Millward GR, Jefferson DA (1978) Lattice resolution of carbons by electron microscopy. In: Walker PL, Jr, Thrower PA (eds) Chemistry and physics of carbon, vol 14. Dekker, New York Basel, pp 1–83

Moore AW (1973) Highly oriented pyrolytic graphite. In: Walker PL, Jr, Thrower PA (eds) Chemistry and physics of carbon, vol 11. Dekker, New York Basel, pp 70–176

Oberlin M, Méring J (1964) Etude cinétique de l'oxydation en milieu liquide des carbones graphitables. Carbon 1:471–480

Pacault A (1971) The kinetics of graphitization. In: Walker PL, Jr (ed) Chemistry and physics of carbon, vol 7. Dekker, New York Basel, pp 107–153

Rouchy JP, Gatineau L (1976) Evolution de l'arrangement et des dimensions des couches d'un carbone graphitable entre 1700 °C et 2500 °C. Carbon 14:97–104

Rousseaux F, Tchoubar D (1977) Structural evolution of a glassy carbon as a result of thermal treatment between 1000 ° and 2700 °C. Pt 1: Evolution of the layers. Carbon 15:55–61. Pt 2: Tridimensional configuration of a glassy carbon. Carbon 15:63–68

Ruland W (1965) X-Ray studies on the carbonization and graphitization of acenaphthylene and bifluophenyl. Carbon 2:365–378

Schiller C, Méring J, Cornuault P, Duchaffault F (1967a) Défauts structuraux dans les carbones graphitables. Effects des traitements thermiques. Pt 1: Carbon 5:385–397
Schiller C, Méring J, Cornuault P, Duchaffault F (1967b) Défaux structuraux dans les carbons graphitables. Effects des traitements thermiques. Pt 2: Carbon 5:507–516
Simon C, Estrade H, Tchoubar D, Conard J (1977) Etude des semi-cokes d'anthracène: relation entre leurs propriétés structurales et électroniques. Carbon 15:211–218
Tchoubar D, Méring J (1973) Arrangement mutuel des couches atomiques dans les carbones tendres partiellement graphités. Carbon 11:103–111
Warren BE (1941) X-Ray diffraction in random layer lattices. Phys Rev 59:693–699
Warren BE, Averbach BL (1952) The separation of cold-work distorsion and particle size broadening in X-Ray patterns. J Appl Phys 23:467–497
Warren BE, Gingrich NS (1934) Fourier integral analysis of X-Ray powder patterns. Phys Rev 46:368–399

CHAPTER 8

The Modelization Method in the Determination of the Structural Characteristics of Some Layer Silicates: Internal Structure of the Layers, Nature and Distribution of the Stacking Faults

We shall now discuss and present the application of the methods described in Chapters 2, 3, and 4 to the study of structures of layer silicates with stacking faults. The examples have been chosen among widely distributed phyllosilicates such as kaolinites, microcrystallized micas and smectites since, as is well known, these minerals contain a number of structural defects which have been the subject of numerous studies and discussions over a number of years. Until the advent of the method of diffraction pattern modelization, the perception of these defects was based on rather intuitive structural notions which did not make a truly precise use of the diffractometric characteristics.

It is not the purpose of this chapter to give an exhaustive description of the structural characteristics of all the naturally occurring kaolinites, smectites, or micas, but rather to discuss some specific examples to illustrate the simultaneous use of diffractometric and crystallochemical data to work back to the true structural characteristics of a given mineral by the method of modelization of the XRD pattern. In each concrete example, we shall first systematically describe all the structural models which differ by the nature and concentration of stacking faults, and follow by the calculation of the XRD patterns corresponding to each of these models. It is then possible to discard the models giving diffraction patterns not in agreement with experimental data, even if these models are based on intuitive concepts which seem perfectly realistic from a crystallochemical point of view, and to retain the model which gives the XRD spectrum closest to the experimental diagram. The final step is the adjustment of its characteristic parameters in order to obtain practically identical calculated and experimental diagrams. This analytic approach gives an unambiguous evaluation of the nature of the structural defects and of their distribution in the crystals, permitting a correct determination of the actual structure of the mineral under study.

8.1 Structural Defects in Kaolinite

Natural kaolin minerals are found as three well crystallized polytypes, namely kaolinite, dickite, and nacrite. All three are made from the same type of 1 : 1 layers formed from a tetrahedral and an octahedral sheet linked together. Each SiO_4 tetrahedron in the former is linked by three of its apices to neighboring tetrahedra, thus creating a surface plane of oxygen atoms forming a more or less deformed hexagonal ring (cf. Fig. 1.2). The octahedral sheet is composed of two (O, OH) planes, one consisting of apical tetrahedral oxygens and inner hydroxyls (see Fig. 8.1) and the other of inner surface hydroxyls with two-thirds of the octahedral sites occupied by Al. (Further details can be found in the review by Bailey 1980.)

When studied by various physico-chemical methods, the kaolin minerals – and especially kaolinite – all display extreme diversity. Thus, the X-ray diffraction patterns from individual samples differ in the amount of smearing shown by various reflections which then look like diffraction bands corresponding to 2D structure types. These peculiarities are due to the presence of stacking faults, and a number of attempts have already been made at trying to understand their nature (Brindley and Robinson 1946; Murray 1954; Mitra and Bhattacherjee 1970; Noble 1971; Plançon and Tchoubar 1977a, b; Tchoubar et al. 1982). However, all the solutions proposed fall short, from the crystallochemical point of view, of a proper description of the mineral.

In the present section, we shall present the results of the latest work on this subject, showing how the structural models of kaolinite have progressively evolved to include the crystallochemical parameters necessary to account for all the features of the diffractograms. However, before describing the different models and the XRD patterns to which they give rise, it will be necessary to have a good grasp of the general structural characteristics of the three polytypes found in kaolin minerals and to compare precisely the structure of kaolinite and dickite if we are to understand the reasons behind the choices of structural models that were made.

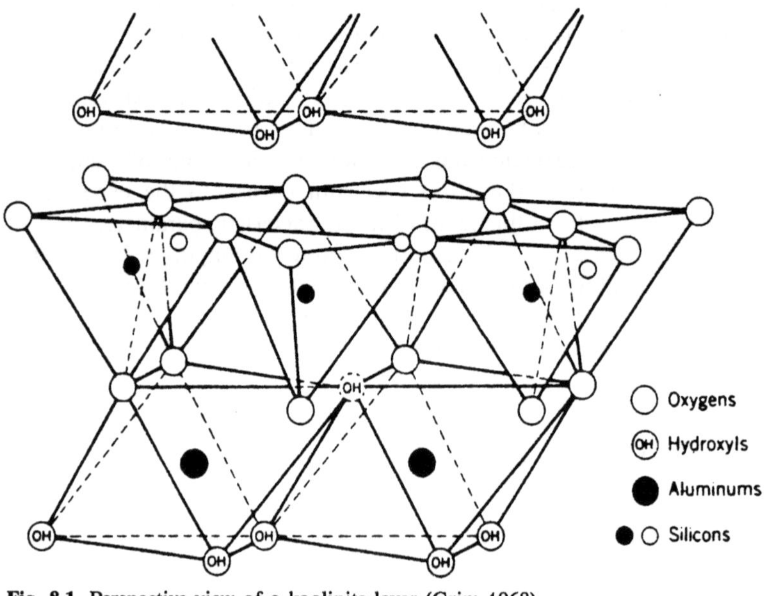

Fig. 8.1. Perspective view of a kaolinite layer (Grim 1968)

8.1.1 Common Features of the Layers in Kaolin Minerals

The ideal structure of an individual 1 : 1 layer can be described in terms of a centered unit cell with an angle $\gamma = 90°$ between $\vec{a}$ and $\vec{b}$, themselves in the ratio $b/a = \sqrt{3}$. The specific feature of the kaolin minerals is the presence of Al^{3+} ions in two out of three octahedral sites, so that one-third of the cavities are vacant. This creates three types of nonequivalent octahedral sites (Fig. 8.2) labeled A, B, and C by Bailey (1980), and the 1 : 1 layers will hereafter be identified and designated by the letter corresponding to their type of vacant site. Figure 8.3 shows a fragment of a B layer. Adjacent layers are always stacked so as to ensure pairing of the inner-surface hydroxyls and the basal oxygens in the interlayer space, a feature which favours the formation of long hydrogen bonds.

In ideal kaolinite, successive layers are shifted by $-\vec{a}/3$, forming a one-layer monoclinic cell. In ideal dickite, successive layers are not only shifted by $-\vec{a}/3$ but also rotated by $\pm 120°$, so that, for fixed $\vec{a}$ and $\vec{b}$, the vacancies alternate between B and C sites and a two-layer unit cell is formed. Figure 8.4 shows the pairing of the basal oxygens and the hydroxyl groups in the kaolinite and the dickite interlayer spaces. The two-layer periodicity in nacrite results from the rotation of adjacent layers by $\pm 60°$.

8.1.2 Common Features of the 1 : 1 Layers in Dickite and Nacrite

Refinements of dickite and nacrite crystal structures have shown that there are definite differences between the actual and ideal structures described in the preceding section. In both minerals, the counter-rotation of the bases of the octahedra leads to an increase in the dimensions of the vacant octahedron. Rotation of the tetrahedra transforms the hexagonal ring in the tetrahedral sheet to ditrigonal (cf. Figs. 1.2 and 8.4). Single, individual layers have no elements of sym-

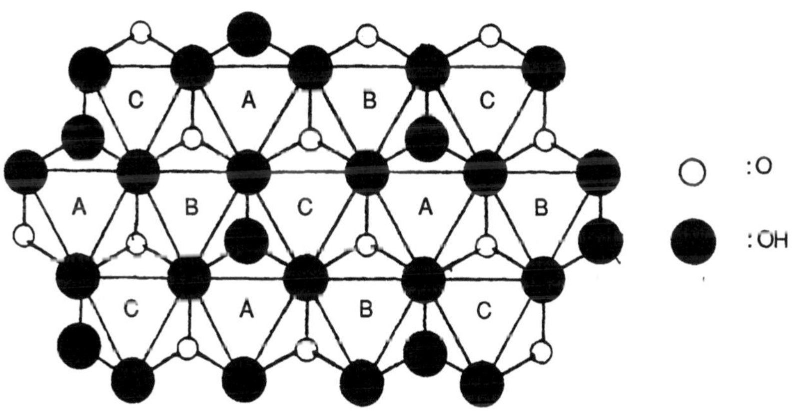

Fig. 8.2. Idealized octahedral 1 : 1 layer of kaolinite showing the different types of octahedral sites

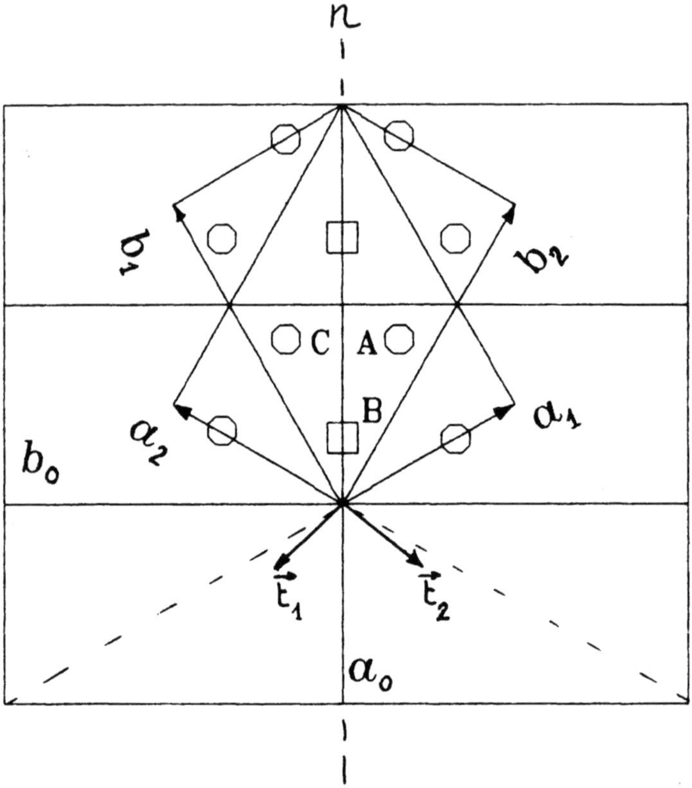

Fig. 8.3. Possible unit cells and a_i, b_i parameters for a dioctahedral 1:1 layer. (After Bookin et al. 1989a)

metry except translations. Although cations in a layer seem to be related by reflection in a mirror plane *m* (Fig. 8.4), this is not a true symmetry element in the space group *Cc*.

The basal tetrahedral surface is corrugated in both structures owing to the displacement of the basal oxygens O_1 (cf. Fig. 8.4) lying in the *m* plane inside the layer. In both dickite and nacrite, deviation of the hydroxyl surfaces from planarity results from the displacement of the independent OH groups (labeled O_5 in dickite in Fig. 8.4) beyond the octahedral sheet. The displaced basal oxygens and hydroxyl groups form one of the hydrogen-bond pairs. This suggests that the OH displacement is not an inherent property of the 1:1 layers but a reaction to the stretching of the O-OH bond resulting from the depression of a basal oxygen.

The dickite structure has another peculiarity described by Bookin et al. (1989a), who showed that the atomic positions in its 1:1 layer are related, to a high degree of accuracy, by reflection in a mirror plane passing through the center of the hexagonal ring formed by surface oxygens and parallel to *m*. Thus, the plane *m* of pseudosymmetry mentioned above indeed reflects the layer symmetry although the space group does not require it.

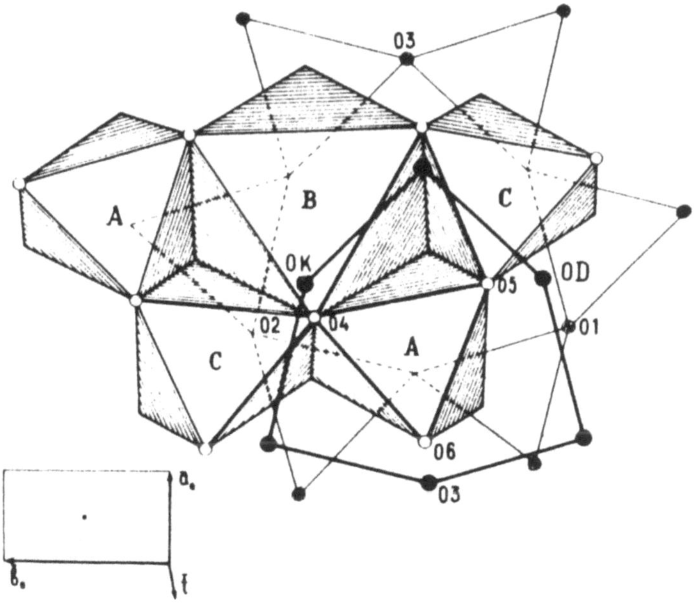

Fig. 8.4. Dioctahedral 1:1 layer with a vacant B-site. (After Bookin et al. 1989a). *Open circles* inner surface hydroxyls; *full circles* basal oxygens; the *thick line* shows the hexagonal ring of neighboring tetrahedral sheets; $O_K = O_1$ in kaolinite; $O_D = O_1$ in dickite; A, B, C are the octahedral sites. Plane m passes through O_1 and O_5.(After Bailey 1980)

8.1.3 Characteristics of the 1:1 Layer in Kaolinite

Early structural refinements of the kaolinite structure were made 30 years ago using oblique-texture electron diffraction (Zvyagin 1960) and X-ray single crystal diffraction (Drits and Kashaev 1960), but the accuracy attained was insufficient. Both X-ray and neutron powder diffraction have lately been used for refinement of 1:1 phyllosilicate structures (Adams and Hewat 1981; Suitch and Young 1983). The poor agreement between the results of structural refinement powdered dickite by the Rietveld method (Suitch and Young 1983) and of a single crystal (see, for example, Joswig and Drits 1986) shows that the data presented by Suitch and Young (1983) only allow discussion of the structural motif but reveal no fine structural details, perhaps because preferential orientation in the sample, which is an important source of errors, was not taken into account.

Let us now consider the vacant sites in the octahedral sheet of a kaolinite layer. Bailey (1980) supposed that both B and C sites can be vacant and that two kinds of crystals built from B and C layers would be diffractionally indistinguishable. However, all the kaolinite structure refinements mentioned above, when referred to a conventional unit cell with $\alpha > 90°$ and $\gamma < 90°$, indicate that it is always the B site which is vacant and that Bailey's assumption is therefore only correct for an ideal kaolinite structure having $\alpha = \gamma = 90°$. Moreover, all the computed diffractograms calculated on the basis of structures with $\alpha > 90°$ and $\gamma < 90°$ and either all the B or all the C sites vacant show substantial differences in the intensity distribution (Bookin et al. 1989a). The comparison between these curves and the experimental results leads to the conclusion that regular kaolinite always consists of B and never of C layers, in terms of the conventional unit cell.

Another important point concerns the structure of the interlayer space in kaolin minerals: this depends essentially on the amount of rotation of the tetrahedral and octahedral bases which modifies the O-OH distances: they increase when the octahedral OH base rotates, and decrease when it is the tetrahedral base which rotates (cf. Fig. 8.4). The angles of rotation reported in various studies are given in Table 8.1, which shows that the structure proposed by Zvyagin (1960) and by Drits and Kashaev (1960) for kaolinite implies a shorter length for the hydrogen bonds than that in dickite, other parameters being equal.

Another characteristic of the interlayer space is related to the existence of vacant octahedral sites in the octahedral sheets. Their presence leads to corrugation of the tetrahedral basal plane owing to displacements into the layer of basal oxygens lying on the *m* planes. In order to keep a short hydrogen bond, the corresponding OH groups on the surface of the adjacent layer are shifted outside the layer. This feature of the structure of the interlayer space of dickite and nacrite was described respectively by Newnham (1961) and by Blount et al. (1969), and Bookin et al. (1989b) concluded that the model of matching corrugated surfaces is also valid for kaolinite.

8.1.4 Comparison of the Kaolinite and Dickite Unit Cells

Bookin et al. (1989a) determined the unit cell parameters in several samples of dickite and compared them with those of kaolinite from Keokuk. All the dickite samples yielded the same values for the unit cell parameters (to within ±0.001 Å), namely $a_D = 5.151$ Å; $b_D = 8.943$ Å; $c_D = 14.430$ Å; $\alpha = 90°$; $\beta = 96.75 \pm 0.01°$; $\gamma = 90°$ and $c_D \sin \beta/2 = 7.165$ Å. The parameters of the Keokuk kaolinite were the same as those given by Suitch and Young (1983): $a_K = 5.153$ Å; $b_K = 8.941$ Å; $c_K = 7.403$ Å; $\alpha = 91.69°$; $\beta = 104.86°$; $\gamma = 89.82°$; $c_K \sin \beta = 7.152$ Å.

Table 8.1. Rotation angles of polyhedra in kaolin minerals

Minerals	Rotation of tetrahedra (°)	Octahedral (O, OH) bases rotations (°)	Octahedral O bases rotations (°)	Reference
Dickite	6.7	8.3	8.0	Joswig and Drits (1986)
Nacrite	7.3	7.1	5.4	Blount et al. (1969)
Kaolinite	11.5	5	3	Zvyagin (1960)
Kaolinite	10.5	6.5	4	Drits and Kashaev (1960)
Kaolinite (Keokuk)	7	6	8	Suitch and Young (1983)

It is now possible to compare the translation between adjacent layers in dickite and kaolinite. In the former, the interlayer shift component $\vec{t}_a$ parallel to $\vec{a}$ (with $\vec{t}_a = c \cdot \cos\beta/2 = 0.335\,\vec{a}$) is close to $\vec{a}/3$. If the origin of the unit cell is placed at a point invariant with respect to rotation of the layer and vacancy displacements, e.g., at the center of the hexagonal ring, the interlayer shift $\vec{t}_b$ along $\vec{b}$ can also be found. This component – which is zero in the idealized structure – is actually equal to $0.024\,\vec{b}$. Since the shifts are of opposite sign in two adjacent pairs, the two-layer unit cell of dickite is monoclinic.

The in-plane components $\vec{t}_a$ and $\vec{t}_b$ of the translation vector obtained from the unit cell parameters of the Keokuk kaolinite are $(-0.369\,\vec{a},\ -0.024\,\vec{b})$. If the structures of dickite and kaolinite are analyzed in terms of the same coordinate system, i.e., the initial B-type layer in dickite, the $\vec{t}_b$ components for both structures will then not only have the same value but also the same sign. Hence, $\vec{t}_b$ does not depend on the position of the vacancy in the layer but is affected only by the particulars of the asymmetric hydroxyl network of the preceding layer. In kaolinite, $\vec{t}_a$ differs substantially from its homologue in dickite. It follows (cf. Fig. 8.4) that a minor change in $\vec{t}_a$ does not affect the length of the O_5-O_D bond, stretches O_6-O_3 and shortens O_K-O_4, so that the mean length of the hydrogen bond is preserved. However, both the increase in $\vec{t}_a$ and the sign of $\vec{t}_b$ cause a shortening of the O_4-O_K bond which is the longest owing to the tilt in the tetrahedra.

In the same way, the comparison of $c_D \sin\beta/2$ of dickite and $c_K \sin\beta$ of kaolinite leads to the conclusion that if the layer thickness is assumed to be constant in both minerals, the change in repeat distance along $\vec{c}^*$ must result from a decrease in interlayer separation which favors shortening of hydrogen bonds in kaolinite.

Lastly, we can compare the unit cell in-plane parameters of individual layers of kaolinite and dickite. As given above, $\vec{a}$ and $\vec{b}$ are practically identical, with $b/a > \sqrt{3}$, but the angles γ are slightly different. There is, however, another way of comparing these parameters which leads to important supplementary conclusions, as will now be seen.

In order to elucidate the effect of the lack of hexagonal symmetry of the lattice in the actual 1:1 layers on the values of the cell parameters, consider the dioctahedral 1:1 layer as in Fig. 8.3, with its three possible sets of axes, with the corresponding parameters (a_i, b_i, γ_i) (i = 0, 1, 2). In the ideal layer the three sets would of course be totally equivalent. If the parameters are known for one of the cells, say (a_1, b_1, γ_1), those of the other two cells are readily obtained by the relation in Eq. (1.2):

$$
\begin{aligned}
4a_i^2 &= a_1^2 + b_1^2 \pm 2a_1 b_1 \cos\gamma_1 \\
4b_i^2 &= 9a_1^2 + b_1^2 \mp 6a_1 b_1 \cos\gamma_1 \\
\cos\gamma_i &= (a_i^2 + b_i^2 - 4a_1^2)/2a_i b_i
\end{aligned}
\tag{8.1}
$$

(i = 0, 2), in which the upper sign corresponds to the cell (a_0, b_0, γ_0). The experimental (i = 1) and calculated (i = 0, 2) parameters for Keokuk kaolinite are given in Table 8.2. It can be seen that the sets (a_1, b_1, γ_1) and (a_2, b_2, γ_2) are practically identical, and hence that the corresponding unit cells are related by mirror symmetry rather than by a rotation axis, i.e., that they are enantiomorphic. As expected, the (a_0, b_0, γ_0) cell is practically orthogonal. Thus, although the lattice of a layer is not strictly hexagonal, even for regular kaolinite, it is not totally asymmetric. To emphasize the presence of a mirror plane in the 2D-lattice of a layer, the unit cell of kaolinite can be expressed in terms of (a_0, b_0, γ_0), giving $a_0 = 5.167$ Å, $b_0 = 8.917$ Å, $c_0 = 7.403$ Å, $\alpha_0 = 102.0°$, $\beta_0 = 98.8°$, $\gamma_0 = 90°$; a_0 of kaolinite is now greater than a_D of dickite, whereas $b_0 < b_D$, so that $b_0/a_0 < \sqrt{3}$.

In order to differentiate between the occupancy of octahedral sites by cations in the 1:1 layers of dickite and kaolinite, let us refer once more to Fig. 8.3. If the a_1 and b_1 axes in this figure correspond to the orthogonal a, b axes of dickite, two successive layers of this structure will then alternately have vacant B and C octahedral sites, but if the a_1 and b_1 axes are those of a kaolinite unit cell, the vacant sites will all be type B in this mineral.

The comparison between dickite and kaolinite therefore leads to the following conclusions:

1. The dickite layer unit cell is much closer to the kaolinite (a_1, b_1) layer unit cell with the acute angle than to the (a_0, b_0) rectangular unit cell.
2. In order to have a common unit cell, the B and C layers alternating in the dickite structure must be slightly distorted, so that the γ angle becomes equal to 90°.

8.1.5 Models for the Stacking Faults in Kaolinite

Diffraction patterns from kaolinites vary in resolution, intensity, and position of the peaks [see, for example, the review by Brindley (1980)]. As the basal $00l$ reflections indicate a fairly large number of coherently scattering parallel layers, the periodicity along $\vec{c}^*$ must therefore be unaffected by the stacking faults. Furthermore, different types of reflections do not have identical sensitivity to the presence of stacking faults: reflections having the index $k \neq 3$n are shifted, broadened or disappear more rapidly than those having $k = 3$n. Thus, even a qualitative analysis imposes limitations on possible models for the stacking faults.

Table 8.2. Characteristics of an individual 1:1 layer of regular Keokuk kaolinite, as a function of the possible values of i (see text)

Possible i values	1	2	0
a_i	5.153	5.153	5.167
b_i	8.941	8.941	8.917
γ_i	89.82	89.82	90.00
$(b_i/a_i)^2$	3.010	3.010	2.978

1. Model in which the stacking faults are displacements by $\pm\vec{b}/3$

The model in which defects are random shifts by $\pm\vec{b}/3$ was proposed by Brindley and Robinson (1947), who assumed that the sheet of hydroxyl groups in an idealized layer would coincide with itself after such additional displacements. The verification of the validity of this model by Plançon and Tchoubar (1975, 1977b) involved a systematic study of the "theoretical" diffractograms corresponding to varying proportions of additional random translations equal to $\vec{b}/3$ and $-\vec{b}/3$ added to the vector $\vec{t}_0$ characterizing the normal displacement between two adjacent layers. The comparison of these diagrams with the experimental distribution of intensity given by samples of kaolinite containing different degrees of structural disorder showed that it was not possible to obtain acceptable coincidence of the theoretical and experimental diffractograms for all the samples of kaolinite when only $\pm\vec{b}/3$ translation defects were considered. This is shown in Fig. 8.5 for the (02, 11) band for two different samples. The theoretical curve (full line) in Fig. 8.5a corresponds to a probability $P_T = 0.46$ for the occurrence of translation faults equal to either $\vec{b}/3$ or $-\vec{b}/3$; agreement with the experimental points (circles) in the diffractogram of a sample of kaolinite with a high degree of disorder is fairly

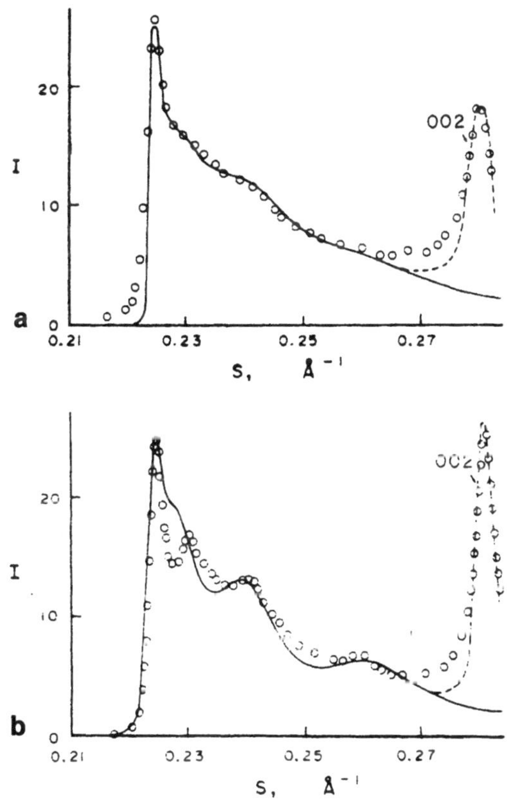

Fig. 8.5. Calculated (*full lines*) and experimental (*circles*) profiles of the (02, 11) diffraction bands of a kaolinite. The model used in the computation assumes that the defects are $\pm b/3$ translation faults. Curve **a** corresponds to a kaolinite from the district of Charente (France), to a kaolinite curve **b** corresponds is from Georgia. (After Plançon and Tchoubar 1975)

satisfactory. On the other hand, the value of P which gives the best agreement with the experimental profile in Fig. 8.5b is $P_T = 0.39$. The kaolinite is better ordered, but lack of agreement between the theoretical and experimental profiles is evident.

The inadequacy of the model involving only $\pm\vec{b}/3$ defects is understandable both from a crystallographic and from a crystallochemical point of view: although the relative positions of the octahedral cations in one layer and the tetrahedral cations in the adjacent layer are the same when the translation vector is $\vec{t}_0$ or $(\vec{t}_0+\vec{b}/3)$, a displacement by $(\vec{t}_0-\vec{b}/3)$ places all the octahedral cations of one layer exactly above Si atoms of the nearest layer. As noted by Newnham (1961) and confirmed by Giese (1982), this arrangement – which is electrostatically unfavorable – corresponds to a so-called "monoclinic kaolinite" which has not been found to occur naturally. An improvement of the model therefore requires that the probabilities characteristic of shifts leading to an unfavorable stacking be reduced or even excluded.

2. Model in which the stacking faults are rotations by ±120°

The model proposed by Murray (1954) is based on the assumption that, after ±120° rotations, a layer coincides with itself except for the position of the vacant octahedral site. Figure 8.3 proves that this model is totally unrealistic (Bookin et al. 1989). After rotation by ±120° around the center of the hexagonal ring, the basal oxygens in the ring will indeed occupy nearly the same positions, but the centers of other rings will no longer coincide: since the moduli of $\vec{a}_1$ and $\vec{a}_2$ are the same, $\vec{a}_1$ can always be made to coincide with $\vec{a}_2$ by clockwise rotation, but the angle of rotation is not equal to 120° and the directions of $\vec{b}_1$ and $\vec{b}_2$ differ by 0.4°, so that the center of the hexagonal ring, which is separated by about 18 Å (twice the parameter *b*) from the axis of rotation, shifts by 0.1 Å in addition. It is only by violation of the periodicity of the layer that an important accumulation of this error can be avoided, but this would result in small coherent domains, in contradiction with the experimental data that these domains must be hundreds of Å units long (Plançon and Tchoubar 1977b, Tchoubar et al. 1982). Counter-clockwise rotation will make $\vec{a}_1$ coincide in direction with $\vec{a}_0$ but not in length, while $\vec{b}_1$ and $\vec{b}_0$ will then not even be parallel. In other words, if the layers were not distorted, the initial and rotated 1:1 layers could not be described by a common lattice.

The crystallographic aspects of this model were also studied by Plançon and Tchoubar (1976), who compared the experimental and calculated profiles of the intensity distribution and concluded that a model with stacking defects constituted solely by ±120° rotations is incapable of reproducing the experimental profiles of more or less disordered kaolinites.

3. Octahedral vacancy displacement model

Plançon and Tchoubar (1977b) proposed a model which has some of the features of the models examined above and is based on the assumption that a crystal consists of identically oriented layers in which vacancies can be located at any single A, B, or C position for a given layer, thus overcoming the difficulty discussed in the preceding paragraph. Each layer is associated with a specific interlayer translation coinciding with or differing by $\pm\vec{b}/3$ from the defect-free translation in kaolinite. For a block of layers of the same type, the translation is chosen so as to ensure a kaolinite-like stacking sequence.

The parameters in this model are the proportion of each type of layer and the possibility of transition from one layer type to another. Plançon and Tchoubar (1977b) used a set of parameters leading to segregation of layer types to form fragments consisting of layers of the same type. Decrease in the probability of stacking faults should lead to increasing thicknesses of defect-free fragments and, at the limit, kaolinite would consist of physical mixtures of three different crystals. The conventional unit cell parameters calculated using the translations chosen by Plançon and Tchoubar (1977b) are presented in Table 8.3. The model predicts equal abundance for the three types of kaolinite, whereas only the first unit cell is found in natural samples. Moreover, fragments consisting of A layers with the translation vector $\vec{t}_0-\vec{b}/3$ would correspond to "monoclinic kaolinite", with an unfavourable arrangement of cations in adjacent layers.

Tchoubar et al. (1982) used another set of variables with the vacancy displacement model. Their version implies a prevalence of B-type layers randomly alternating with A- or C-types so that thick blocks of energetically unfavorable substacks cannot be formed. The proportion of different-type interlayers, however, now increases, with layer sequences B-C-B and B-A-B becoming the most frequent. Of these two, the first seems the same as that in dickite and a number of papers have examined the formation of dickite nuclei within a kaolinite crystal (Plançon and Tchoubar 1977b; Brindley et al. 1986). In fact, it is the B-C sequence which actually approximately corresponds to that in dickite and sequences C-B and A-B correspond to "monoclinic kaolinite".

While studying the polymorphism of kaolin minerals, Bailey (1963) showed that with a proper choice of common directions for the $\vec{a}$, $\vec{b}$ and $\vec{c}$ axes of kaolinite and dickite, the two mineral species only differ by the position of the vacant octahedral sites; the sequences will then be B-B-B-... in kaolinite and B-C-B-C-... in dickite. This led him to conjecture that disorder in kaolinite is connected with the random replacement of B-layers by C-layers. The verification of this model by Bookin (un-

Table 8.3. Parameters of the conventional unit cells for defect-free fragments having different vacancy positions

Vacancy position	B	C	A
Translation	$\vec{t}_0$	$\vec{t}_0+\vec{b}/3$	$\vec{t}_0-\vec{b}/3$
a (in Å)	5.155	5.166	5.170
b (in Å)	8.959	8.940	8.932
c (in Å)	7.408	7.389	7.301
α (in °)	91.68	87.89	90.41
β (in °)	104.87	104.26	101.42
γ (in °)	89.94	89.90	89.87

published results) was done be calculating the theoretical XRD diagram corresponding to structures containing varying proportions W_{BC} of (B, C)-pairs, from pure dickite ($W_{BC} = 1$, $W_{BB} = 0$) to pure kaolinite ($W_{BC} = 0$, $W_{BB} = 1$). The results are shown in Fig. 8.6, in which the intensities are given in arbitrary units. The diagrams were calculated on the basis of translations between (*i*,*j*)-pairs corresponding to (B, C)-pairs in dickite and (B, B)-pairs in kaolinite, namely $\vec{t}_a = -0.335\,\vec{a}$, $\vec{t}_b = -0.024\,\vec{b}$ and $\vec{t}_a = -0.369\,\vec{a}$, $\vec{t}_b = -0.024\,\vec{b}$, respectively (cf. Sect. 8.1.4).

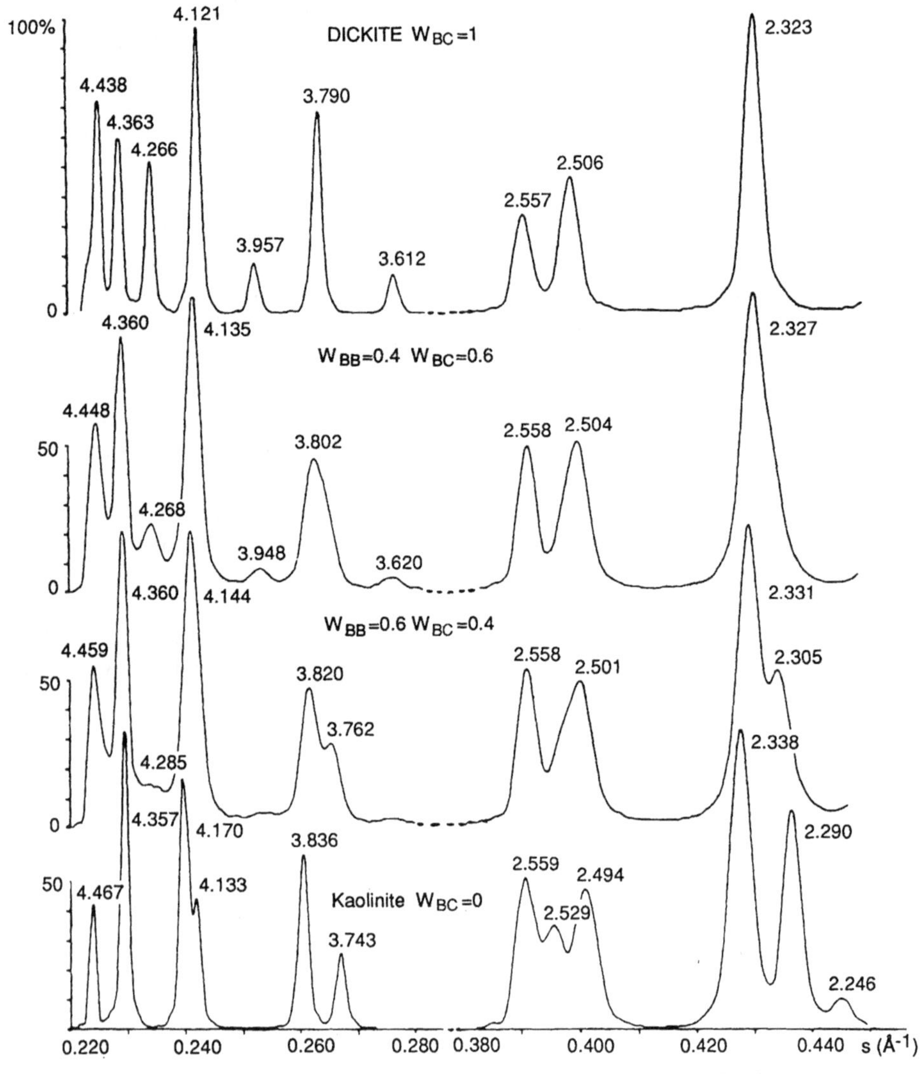

Fig. 8.6. Calculated X-ray diffraction profiles for models of stackings varying between pure kaolinite ($W_{BC} = 0$) and pure dickite ($W_{BC} = 1$). (Bookin unpublished results)

The diagrams in Fig. 8.6 show that, once again, this model does not account satisfactorily for the intensity distribution in the experimental diffractograms of kaolinite. Furthermore, there are experimental results which exclude the existence of different types of layers in kaolinite, at least in sizable proportions. Samotoin (1966) studied the growth pyramids in kaolin minerals by electron microscopy decoration method and found that the growth steps formed characteristic patterns; kaolinite shows continuous terraces (cf. Fig. 8.7a) whereas the pattern in dickite is that of intersecting steps due to unequal growth rates associated with the different adjacent layers in a given direction (Fig. 8.7b). These intersecting steps were never observed in kaolinite, implying that most of the layers are identical.

4. Model containing only enantiomorphic B layers

The symmetrical arrangement of atoms with respect to the n plane in kaolinite layers suggests a simple model for the stacking faults (Bookin et al. 1989a). Suppose that the displacement between two adjacent B layers be $\vec{t}_1$. When a regular crystal is formed, all the successive B-layers are shifted by the same vector $\vec{t}_1$. A stacking fault will appear whenever two successive layers are related by a plane n. In other words, n constitutes a glide plane for these two layers. The periodicity and cation distribution are conserved but the $\vec{t}_1$ translation vector has been changed to $\vec{t}_2$ (see Fig. 8.3). The formation of a "defective" B-layer would therefore bring about only minor changes in the electrostatic energy of layer interaction and H-bonding energy. It seems logical that $\vec{t}_2$, which is related to $\vec{t}_1$ by the same glide plane n, will be the new translation between the defective fragment and the next stacking fault. The second defect-free fragment thus formed would be an enantiomorph of the first. In terms of this model, stacking faults would result from micro intergrowths of right- and left-handed kaolinite crystals. As was done with dickite, we can assume that the pseudosymmetry in kaolinite layers is close to true symmetry, in which case a layer reflected in the n mirror-plane will be self-coincident, and the n and m planes of kaolinite then coincide (cf. Fig. 8.3). The model of enantiomorphic B-layers therefore leads to an alternation of identical layers stacked with symmetrical translation vectors $\vec{t}_1$ and $\vec{t}_2$.

In terms of the conventional (a_1, b_1, γ_1) system, the components of the projection of $\vec{t}_1$ on the (a, b) plane are $(-0.369\,\vec{a}_1,\ -0.024\,\vec{b}_1)$. The corresponding components of $\vec{t}_2$ $(-0.352\,\vec{a}_1,\ 0.304\,\vec{b}_1)$ are readily calculated from $|\vec{t}_1| = 1.901$ Å and the angle $\delta = 7°$ between $\vec{t}_1$ and $-\vec{a}_1$. Vectors $\vec{t}_1$ and $\vec{t}_2$ differ by the vector $\vec{\tau} = (0.017\,\vec{a}_1,\ 0.328\,\vec{b}_1)$ and it can easily be shown that $\vec{\tau} = (0, \vec{b}_1/3)$ if $b_1/a_1 = \sqrt{3}$ and $\vec{t}_1 = -\vec{a}_1/3$. The present model is therefore an extension of that of Brindley and Robinson (1947) to the actual crystal structure. The difference lies not only in the more realistic value of $\vec{\tau}$ (which replaces the $\pm b/3$ shifts) but also in the exclusion of the unfavorable stackings which would be the result of a $\vec{t}_1 - \vec{\tau}$ translation and, finally, in the correspondence with the features of the growth pyramids which were observed with kaolinite, since displacement of identical layers would have no effect on the growth rate in any direction.

The parameters needed for the quantitative treatment of the model are the short-range order factor (S = 1), the proportion W_i of the translation $\vec{t}_i$, and one of the

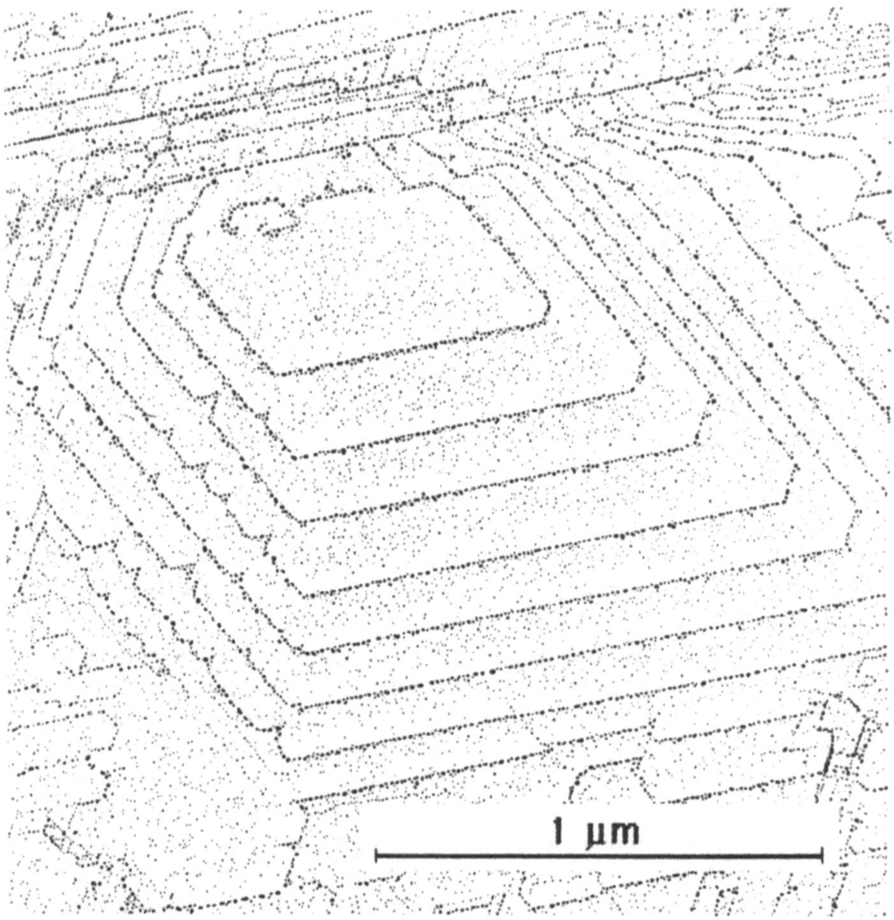

a

Fig. 8.7. Electron microscopy decoration patterns of kaolinite (**a**) and dickite (**b**) single crystals (Samotoin et al. 1980)

probabilities $P_{ij}(i, j = 1, 2)$ of finding a translation $\vec{t}_j$ after the translation $\vec{t}_i$. The equivalence of the left- and right-handed unit cells of kaolinite suggests that the two forms are equiprobable in a sufficiently large stack of crystallites, hence that $W_1 = W_2 = 0.5$. All the P_{ij} can then be calculated if one of them, say P_{11}, be specified. The variation of this parameter from 0 to 0.5 corresponds to a change from ordered to random alternation of $\vec{t}_1$ and $\vec{t}_2$, while the range from 0.5 to 1 corresponds to increasing segregation. A crystal with growth defects corresponding to $P_{11} > 0.5$ is shown schematically in Fig. 8.8, while Fig. 8.9 presents the calculated X-ray powder pattern in the (02, 11) region. For $P_{11} = 1$ we would have a sample consisting of equal proportions of regular right- and left-handed kaolinite crystals, and $P_{11} = 0.9$ would correspond to a mixture of regular and twinned crystals. A decrease of P_{11} to 0.5 leads to smoothing of modulations and

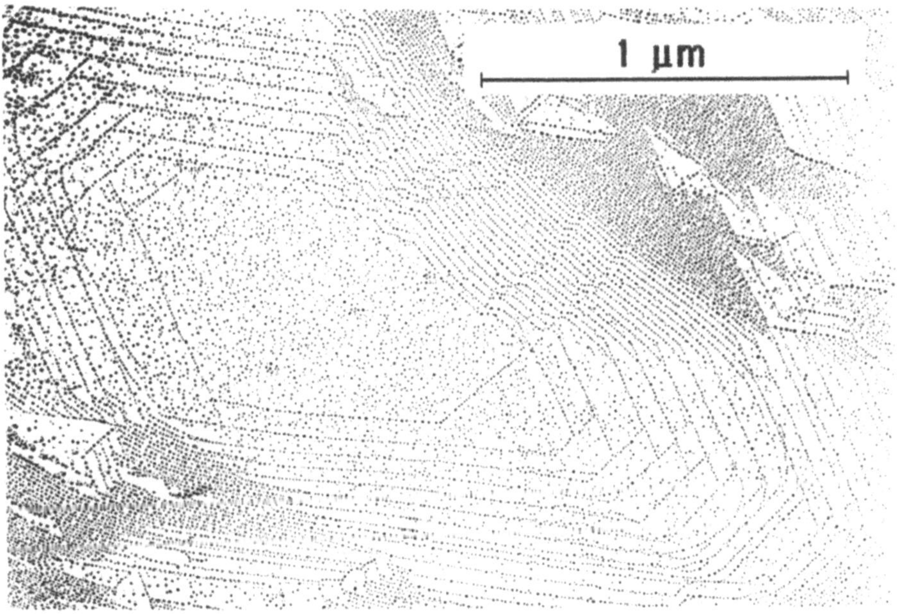

b

a redistribution of intensity. Although maximum disorder corresponds to $P_{11} = 0.5$, it is found that maximum smearing occurs at $P_{11} = 0.65$ for the given angular range. For $0 \leqslant P_{11} \leqslant 0.5$, modulations appear, corresponding to *d*-spacings unusual in kaolinite. The limit case $P_{11} = 0$ yields a powder pattern given by a two-layer polytype of kaolin. The XRD pattern in the (20, 13) reflection region changes little when P_{11} varies from 0 to 1.

The set of parameters discussed above is suitable for a model based on a symmetrical 1:1 layer. In defective crystals, distortions in the layers may be such that the (a_1, b_1, γ_1) and (a_2, b_2, γ_2) cells would not be exactly equivalent. If this occurs,

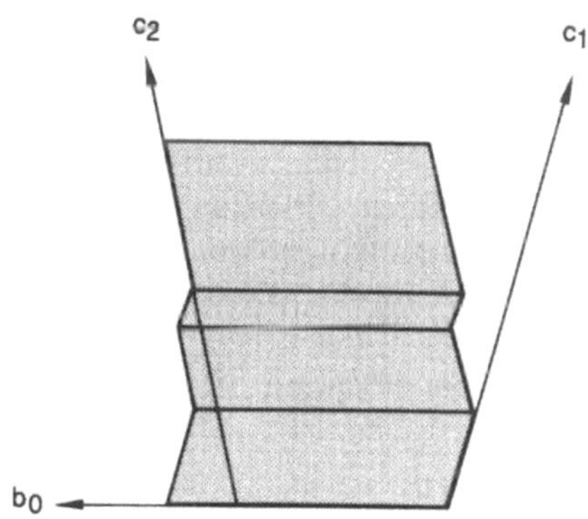

Fig. 8.8. Defective kaolinite crystal with a low defect content (growth fault). (After Bookin et al. 1989b)

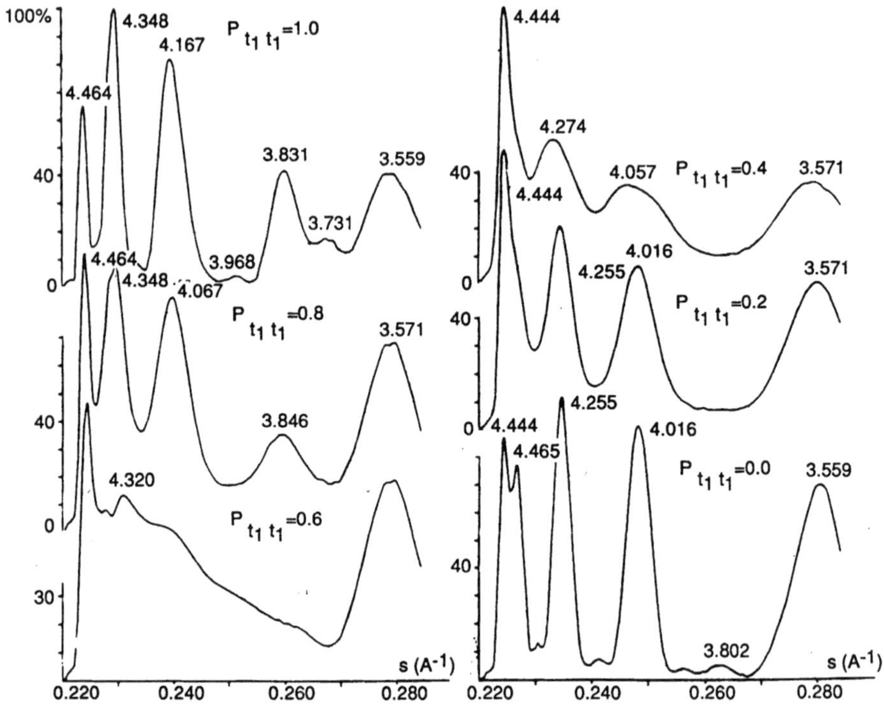

Fig. 8.9. Calculated X-ray diffraction pattern for a single-layer model of kaolinite in the (02, 11) reflection region (growth faults). (After Bookin et al. 1989a)

the proportion of one of the translation vectors may be greater than that of the other. An additional parameter W_1, also ranging from 0 to 1, is then required, but this does not affect the essence of the model.

5. Model containing enantiomorphic B and some C layers

X-ray patterns calculated for a model containing only enantiomorphic B-layers show that varying P_{11} from 0 to 1 only affects the (20, 13) reflection region to a small degree, insufficient to render it comparable to the experimental (20, 13) profiles.

In order to obtain a sufficient modification of the intensities in this region of the XRD pattern, Plançon and Tchoubar (1977b) used a "monoclinicity parameter" C_M amounting to small modifications of the angles α and β of the mean unit cell. An identical effect can, however, be obtained by admitting a small proportion of C-layers in the structure, since the comparison of kaolinite with dickite has shown that deviation of γ_1 from 90° leads to inequivalence of B- and C-sites

(Bookin et al. 1989a). In regular kaolinite, the vacancy in the unit cell is always next to the acute angle γ_1 (Fig. 8.3). Owing to the minor deviation of γ_1 from 90°, one may assume that a vacancy, as if by mistake, could occupy a C-site. As stated in Section 8.1.4, the interlayer shift is unambiguously defined by the position of the vacant site in the preceding layer, for a choice of z direction going from a tetrahedral to an octahedral sheet, and the translation vector for a B-C sequence is t_1. Since kaolinite containing only C-layers has never been found, defective fragments containing a considerable number of C-layers should be absent in the model and the probability of finding a B-layer after a C-layer must be greater than that of finding C-C pairs. Note, in this connection, that the y-component of the translation vector for the layer pairs C-C and C-B, has a positive sign, i.e., $t_1' = -0.369\,\vec{a}_1,\ 0.024\,\vec{b}_1$.

The proportion of C-layers and the P_{ij} values are the parameters needed in the quantitative treatment of this model. If a random distribution and a small concentration are assumed for the C-layer (the simplest case), B-C-B sequences will be found to be the most probable. For $W_2 \leqslant 0.1$, changes in XRD patterns of the (02, 11) reflections are similar to those described for the model containing only enantiomorphic B-layers. A fundamentally new difference in the (20, 13) regions is found in the hkl and $h\bar{k}l$ reflections which are drawn nearer to each other, as observed with natural kaolinites and attributed to α getting closer to 90°. The concept of the "accidental" occupation of a wrong site by a vacancy is common to the present model and to the one proposed by Plançon and Tchoubar (1977b). However, the set of translations used changes both the crystallochemical nature of the stacking faults and the consequences of their presence on the diffraction pattern.

8.1.6 Comparison Between Calculated and Experimental XRD Patterns

We shall now give an account of the results obtained in this field by Plançon et al. (1989a, b) with the last model described in the preceding section. These authors analyzed a number of kaolin samples with crystallographic characteristics similar to those described by Brindley et al. (1986), to which was added one more sample from the Charente district (France) already described by Plançon and Tchoubar (1977b). In the (02, 11) and (20, 13) regions, this set of samples yields a fairly wide cross-section of diffractograms in which the reflections vary from broad 2D bands to well-individualized hkl lines. The X-ray diffractograms were obtained with Cu$K\alpha$ radiation and a graphite monochromator on a Siemens B500 spectrometer. Intensities were measured every 0.01° and expressed in terms of counts per 5 s. The diffraction pattern around the (02, 11) and (20, 13) regions, i.e., with $9° < \theta < 13°$ and $17° < \theta < 21°$, respectively, were sampled under these conditions. The $00l$ bands were used, among other purposes, to estimate the average number of layers in the coherent domains, and all the diffractograms were obtained under the same experimental conditions. To avoid experimental deformation of the intensity profiles due to preferential orientation of particles in the samples, the latter were prepared as a powder of small spheres, using the procedure described by Snyder (cf. Plançon et al. 1989a). Figure 8.10 shows some of the experimental profiles thus obtained.

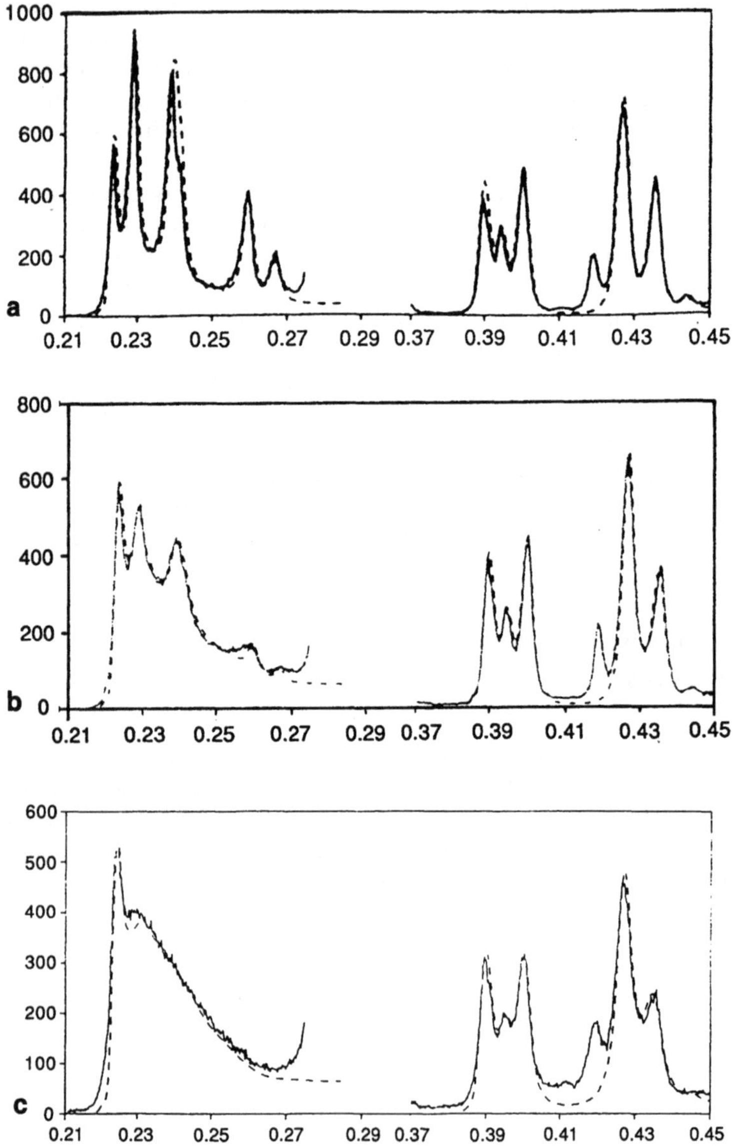

Fig. 8.10. Experimental and calculated X-ray diffraction patterns for kaolinites having different degrees of structural disorder: (a) weak; (b) intermediate; (c) strong. (After Plançon et al. 1989a)

The computation of the intensities given by the model for kaolinite with enantiomorphic B and some C-layers was done using the relationship in Eq. (3.17) which, in the absence of preferential orientation and with a thickness sufficient to eliminate the need for the factor α(M), could be simplified to:

$$I_{hk}(s) = \frac{1}{\Omega \sigma s} \int \mathrm{Tr} \cdot \mathrm{Re}\{[F][W][R]\} T(U)\, d\varphi \tag{8.2}$$

in which

1. Each element (i, j) of matrix [F] is equal to $F_i^* F_j$ and depends on the structure of the B and C layers (i, j = 1, 2 = B, C). The atomic coordinates for layer B are as given by Drits and Kashaev (1960) and those of the C-layers are obtained from those of the B-layer by inversion of the y-coordinates. In all cases, the parameters of the unit cell in the (a, b) plane are those of Goodyear and Duffin (1961), i.e., $a = 5.155$ Å, $b = 8.959$ Å and $\gamma = 89.94$ Å.
2. [W] is a diagonal matrix in which $W_{11} = W_B$ and $W_{22} = W_C$.
3. Matrix [R] is expressed as a function of the matrix [Q] as in Eq. (3.15). [Q] depends only on the displacements between the layers, and its elements are functions of the amplitudes of the translations between layers and of their probability of occurrence. The vector $(\vec{t}_1)_{BB}$ between the layers of a (B, B)-pair is that of the unit cell determined by Goodyear and Duffin (1961), with slight modifications:

$$(\vec{t}_1)_{BB} = -0.3689\,\vec{a} + 0.0223\,\vec{b} + 7.155\,\vec{n}$$

in which $\vec{n}$ is a unit vector perpendicular to the (a, b)-plane. The translation $(\vec{t}_2)_{BB}$ is obtained from $(\vec{t}_1)_{BB}$ by symmetry with reference to the pseudo-mirror-plane passing through the centers of the vacant octahedra. Taking into account the fact that the 2D unit cell is centered, the translation is

$$(\vec{t}_2)_{BB} = -0.3488\,\vec{a} + 0.3051\,\vec{b} + 7.155\,\vec{n}\ .$$

The translation $(\vec{t}_0)_{BB}$ lies in the (a, b) plane, coincident with the pseudo-mirror-plane and its modulus is equal to that of $\vec{t}_1$ and $\vec{t}_2$

$$(\vec{t}_0)_{BB} = -0.3151\,\vec{a} - 0.3151\,\vec{b} + 7.155\,\vec{n}\ .$$

The translations $(\vec{t}_k)_{BC}$ (with k = 0, 1, 2) are the same as the translations $(\vec{t}_k)_{BB}$, and the translations $(\vec{t}_k)_{CC} = (\vec{t}_k)_{CB}$ are obtained by inverting the y-component of $(\vec{t}_k)_{BB}$.

Because of the symmetry of the B and C layers, it is assumed that the ratio of the probabilities is independent of the type of layer (R = 0). Thus, if $(\vec{P}_k)_{ij}$ be the probability of the translation $(\vec{t}_k)_{ij}$, the ratios $(P_2)_{ij}/(P_1)_{ij}$ and $(P_0)_{ij}/(P_1)_{ij}$ do not depend on i or j. With this hypothesis, it is only necessary to specify the values of P_1, P_2 and P_0 to characterize the defects due to the translations $\vec{t}_1$, $\vec{t}_2$ and $\vec{t}_0$. For stackings containing only these translations, the element (i, j) of the matrix [Q] is given by

$$Q_{ij} = \sum_{k=0}^{2} (P_k)_{ij} \exp [2\pi i \vec{s} \cdot (\vec{t}_k)_{ij}] \quad . \tag{8.3}$$

Furthermore, and as shown by Giese (1982), on one hand the cohesion energy between kaolinite layers has a very broad minimum and, on the other hand, the mean length of the hydrogen bonds is not much affected by small deviations of the translation vectors $\vec{t}_1$ and $\vec{t}_2$ from their ideal values. It is therefore logical to suppose that there may be fluctuations in the relative positions of the layers about the projections of the vector $\vec{t}_1$ and $\vec{t}_2$ on the (a, b) plane. This structural defect (of the second type, as described in Chap. 3.3) may be taken into account by multiplying each element Q_{ij} given in Eq. (8.3) by the factor

$$\exp [-\pi^2 (h^2 \delta_x^2 + k^2 \delta_y^2)] \quad , \tag{8.4}$$

where δ_x and δ_y are the characteristic widths (in fraction of the unit cell parameters) of the distributions of the interlayer translations along a and b. For the set of samples under consideration, best agreement was obtained with $\delta_x = \delta_y$, and a single parameter δ will therefore be indicated in the description of the various samples.

The comparison of the intensity profiles calculated on the basis of the above model with the experimental values led to the following conclusions:

1. *Kaolinites with an intermediate degree of structural disorder,* which yield X-ray diffractograms with weak modulations in the region of the (02, 11) reflections (cf. Fig. 8.10b), are correctly described by the last model, and the values of the parameters characteristic of the disorder in their structure are regrouped in Table 8.4. The kaolinites described by Brindley et al. (1986) correspond to the samples with the references III-1, IV-L, and V-G in Table 8.4. These samples contain about 93% of B and 7% of C-layers. The probabilities P_1, P_2 and P_0 attached to the translations $\vec{t}_1$, $\vec{t}_2$ and $\vec{t}_0$ are respectively about 0.58, 0.3 and 0.04. The average thickness of the interferential coherence domains is that of 40 layers ($M = 40$), which justifies the neglect of the distribution function $\alpha(M)$ since its effect on the profiles is quite negligible. The value of δ was found to be $0.04a \approx 0.2$ Å. Lastly, it was noted that there are, in this category of kaolinites, species which present an important background in the (20, 13) region and in which the maxima of the $\bar{1}31$ and $1\bar{3}1$ reflections are strongly drawn towards each other: this is particularly true for sample C3 in Table 8.4 from Charente (cf. Fig. 8.10c). The interpretation of these diffractograms requires, on one hand, the presence of an important proportion (up to 15%) of C-layers randomly distributed ($P_{CC} = W_C$) in the stack and, on the other hand, an increase of the parameter $\delta \approx 0.08a = 0.4$ Å.

Table 8.4. The parameters describing the defect structure for the intermediate defect kaolinites: samples III-1, IV-L and V-G (from Brindley et al. 1986), and C3, from Charente (Plançon and Tchoubar 1977). The last two lines gives the parameters for the two phases present in the low-defect kaolinites

Sample	δ	W_c	P_1	P_2	P_0	$\Delta\alpha^0$	M	R_{02} (Å)	R_{20} (Å)
III-1	0.04	0.07	0.60	0.28	0.05	0	40	225	175
IV-L	0.04	0.06	0.58	0.32	0.04	−4	40	225	175
V-G	0.04	0.07	0.57	0.33	0.03	−2	40	225	175
C3	0.08	0.15	0.55	0.25	0.05	−4	30	200	200
Nearly defect-free phase	0.02	0.00	0.98	0.02	0.00	0	40	225	175
Defective phase	0.02	0.03	0.60	0.32	0.05	0	40	225	175

2. *Weakly disordered kaolinites*, which yield X-ray diffractograms with well separated 02*l*, 11*l* and 20*l*, 13*l* reflections (cf. Fig. 8.10a), constitute a second category of natural species. The reflections 02*l*, 11*l* are sharp with a clearly visible separation of the $\bar{1}\bar{1}1$ and $\bar{1}11$ reflections and with non-negligible intensities for the $0\bar{2}1$ and 021 reflections. Another characteristic of the powder patterns in this group is the presence of an important background between the 02*l* and 11*l* reflections. The simultaneous presence of narrow reflections and of this background is incompatible with any theoretical intensity distribution calculated on the basis of a *single model* of structural defects since the introduction of any type of defect softens the modulation of the XRD profile, with reflections becoming broader and less intense. It is for this reason that neighboring reflections such as $\bar{1}\bar{1}1$ and $\bar{1}11$, which are well separated in kaolinites with few defects, rapidly become indistinguishable as the number of defects increases, while reflections with smaller intensities and higher indices such as $0\bar{2}1$ and 021, tend to disappear into the background. Given this situation, namely sharp peaks associated with a low level of defects and an important background between the peaks associated with a high level of defects, a model consisting of a mixture of two kinds of kaolinite was used, one nearly defect-free and the other highly defective, yielding a 2D-type (02, 11) band with hardly any modulation (Plançon et al. 1989b). The parameters of the models used for the nearly defect-free and for the defective kaolinite phases are given respectively in the penultimate and in the last line of Table 8.4. The agreement between the experimental diffractogram obtained with a weakly disordered kaolinite sample and the intensity calculated for a mixture of 60% of well-crystallized and 40% of highly disorganized phase can be judged from Fig. 8.10a. The need to bring in a mixture of two phases is not just a convenient working hypothesis; it has been experimentally confirmed by Tchoubar et al. (1982) who fractionated by centrifugation natural kaolinite (sample E3) into two phases with quite different proportions of structural defects. Similarly, the study by Deluca and Slaughter (1985) of the diffraction profiles of 00*l* reflections in well-crystallized natural kaolinites also led them to the conclusion that two phases were present.

8.2 Distribution of the Cations in the *cis* and *trans* Octahedral Sites of Dioctahedral Smectites

Smectites are microdivided lamellar silicates composed of 2:1 layers (cf. Chap. 1.4) between which are found molecules of water and cations which neutralize the negative charge of the layers and which can be replaced by other cations through ionic exchange. As indicated in Chapter 1 the octahedral sheet in the 2:1 layers can have either all or only 2/3 of its octahedral sites occupied by cations, and the layers are then said to be respectively *trioctahedral* or *dioctahedral.* It was shown in Section 1.4 that in 2:1 dioctahedral layers, the sites containing the cations may be *trans* or either *cis-1* or *cis-2.* For this type of layer, the problem is therefore the determination of the distribution of cations in the different categories of sites. However, one of the characteristics of these microdivided minerals is that the stacking is practically totally random along the $\vec{c}$ axis, so that the XRD powder patterns are composed of (hk) bands with marked 2D features and the determination of the distribution of cations in the octahedral sites therefore constitutes an arduous problem.

Based on studies of the hydration properties of montmorillonites from Camp Bertaux and Wyoming, Méring and Glaeser (1954) suggested that the *trans* octahedra of the 2:1 layers of these minerals were occupied by cations. They showed that, in the presence of bivalent interlayer cations, this model of the 2:1 layers is favored owing to a better local charge balance (Glaeser and Méring 1954).

Méring and Oberlin (1971) approached the problem of octahedral cation distribution by analyzing the selected area electron diffraction (SAD) patterns from monocrystals with a single 2:1 layer thickness. They concluded that the 2:1 layer of nontronite (Pfaffenreuth) was *trans*-vacant, while that of Wyoming montmorillonite was *cis*-vacant. Subsequent studies using the SAD method showed that there were beidellites with centrosymmetrical layers and with charge localized in the tetrahedral sheets (Güven et al. 1977), as well as noncentrosymmetrical layers and charge equally distributed over the tetrahedra and the octahedra (Besson and Tchoubar 1972).

This experimental procedure is, however quite delicate, and it is also difficult to interpret the point diffraction patterns. In the first place, the sample thickness is not always that of a single layer. In the second place, modifications of the intensity distribution are inevitable because of the local deformations of single layers and their structural distortions resulting from the interaction with the beam. Finally, the diffraction data obtained for separate single layers may fail to reflect the octahedral cation distribution in the 2:1 layers forming thicker particles.

Several authors (Goodman et al. 1976; Rozenson and Heller-Kallai 1977) tried to compensate for the insufficient reliability of the diffraction data by using Mössbauer spectroscopy. It became a tradition to resolve the experimental spectra of the Fe^{3+}-containing smectites as well as glauconites and celadonites into two doublets, one of which was related to the Fe^{3+} ions occupying *cis* octahedral positions and the other to the Fe^{3+} ions occupying *trans* octahedral sites. This approach was recently found to be erroneous by Dainyak (1980) and by Bookin et al. (1978, 1979), who showed that one of the necessary conditions for a reliable interpretation of Mössbauer spectra from phyllosilicates containing Fe^{3+} ions was the availability of data on cation distribution in the unit cell, averaged over the volume of the crystal.

8.2.1 Preparation Techniques for Smectite Samples Used in the Diffractometric Determination of the Distribution of Cations in Octahedral Sites

In order to obtain XRD patterns with hkl reflections of significant intensity usable in the determination of the localization of the cations in the octahedral sites of natural samples, it is essential that the stacking order of the layers be improved. A number of techniques are available for this purpose:

1. In the first, the cations usually present between the layers (such as Na^+, Ca^{2+} or Mg^{2+}) are replaced by K^+. The first advantage in using this cation is that its diameter is practically identical with the size of the hexagonal cavity on the surface of the layers (cf. Chap. 1.4), and secondly that it is only weakly susceptible to hydration. Owing to these two features, K-exchange leads to a first reorganization of some of the layers in the stack, so that the projection of the centers of the hexagonal cavities on either side of an interlamellar space are superposed in the (a, b) plane.
2. Mamy and Gaultier (1976) further improved the degree of structural order in the stack by submitting the sample to a number of wetting and drying (WD) cycles after the exchange of cations with K^+. During these cycles, energy is spent in reorienting adjacent layers, leading to an increase in mica-like interlayers. For those adjacent layers where the hexagonal cavities are not coaxial, the K^+ cations remain exchangeable and hydratable. Even small deviations of the azimuthal orientation of adjacent layers lead to significant consequences since interlayer K^+ cations will then sit in an "uncomfortable" anionic environment and the interlayer spaces will not lose their ability to swell. Each new wetting will then lead to hydration of these interlayer spaces, after which drying leads to further reorganization of the layer stacking. Note that K^+ cations in well-ordered mica-like interlayer spaces are no longer exchangeable and that these interlayer spaces have lost most of their ability to swell.
3. In the technique used by Besson et al. (1983b) to improve layer stacking order, the interlayer cations which are naturally present in the mineral are replaced by Cs^+ ions through ion exchange. There are some advantages to this:

   - the diameter of Cs^+ is large and, consequently, the increase in layer stacking order is important even in the absence of WD cycles: the degree of order of the stack in a Cs-exchanged smectite is similar to the one obtained with a K-smectite after many tens of WD cycles;

- preferential orientation in the powder decreases in Cs-exchanged smectites as compared to equivalent samples containing K^+;
- lastly, because of the greater scattering power of Cs^+, the hkl reflections from a Cs-containing smectite are more intense and sharper than those from K-treated samples.

Examples of the practical use of diffractometric methods on samples "reorganized" using the technique of K-exchange followed by WD cycles or, more simply, after Cs-exchange, will now be examined.

8.2.2 Determination of the Octahedral Cation Distribution in K-Smectites by Oblique Texture Electron Diffraction

The smectite is first K-saturated by leaving it 3 h in a 1 N K_2CO_3 solution maintained at 70 °C. It is then filtered and washed to remove any adhering salt solution. XRD patterns of the K-saturated samples show d spacings of 10.2 to 11.0 Å. The dry K-saturated sample is then mixed with distilled water, dried at 80–100 °C then rewetted, and this procedure repeated 70 to 100 times. The samples are then studied by oblique texture electron diffraction (OTED). To this purpose, a drop of suspension is placed on a thin collodium film support. After drying, a highly oriented specimen is obtained. When such a specimen, made up of a set of single crystals with perfect structures, is inclined relative to the beam, a diffraction pattern is obtained in which the reflections are distributed over a series of ellipses (Zvyagin 1967). The $02l$ and $11l$ reflections are grouped on the first ellipse, the $20l$, $13l$ on the second, etc. The sequence of reflections with different indices which may be expected on the first ellipse of the 1 M mica pattern is represented schematically in Fig. 8.11.

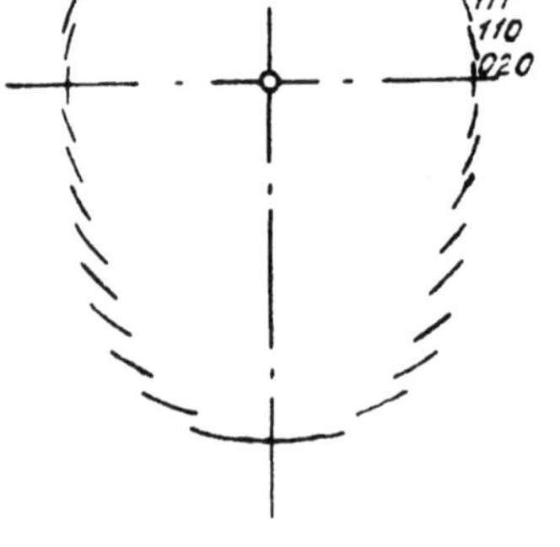

Fig. 8.11. Arrangement of reflections on the first ellipse of the oblique texture electron diffraction pattern

The most important advantage of OTED, as compared with XRD, is the possibility of obtaining a two-dimensional distribution of intensities for finely dispersed layer-silicates.

Zvyagin (1967), Zvyagin et al. (1979) and Drits (1987) demonstrated the effectiveness of this method for revealing various subtle structural features of clay minerals. According to their results, it is possible to conclude that the character of the interaction of electrons with clay particles is purely kinematic. Hence, the intensities of reflections may be used without any dynamic corrections to test the validity of the structural model.

The effectiveness of the OTED method is reduced in the case of smectites which dehydrate in vacuum, with concomitant change in the initial structure. However, as relatively high structural ordering can be achieved at the same time, the diffraction data may be used to obtain additional information on the structural features of the 2:1 layers, their azimuthal orientations, etc.

In practice, and because of the disordered structure of their stacks, all naturally occurring smectites lead to an OTED pattern similar to the one shown in Fig. 8.12a. After 70–100 WD cycles, there are distinguishable reflections on the first and second ellipses (cf. Fig. 8.12b, 12c, 12d). It can therefore be concluded that after 70–100 WD cycles, most of the 2:1 layers in the particles of the majority of K^+-smectites have identical azimuthal orientations.

Analysis of the geometry of the reflections and their indices indicate that the smectites studied have monoclinic one-layer unit cells similar to those of 1 M micas, with the parameters listed in Table 8.5. It is evident from Tables 8.5 and 8.6 that all (Al, Mg)-smectites (beidellites and montmorillonites) may be divided into three groups according to the value of $c \cos \beta$. In the first, $|c \cos \beta|$ is considerably greater than $a/3$ (about 0.37 a to 0.38 a), in the second it is close to $a/3$ (0.34 a to 0.35 a) and, in the third, it is equal to or less than $a/3$. The intensity distribution on the first ellipse is similar for (Al, Mg)-smectites from the same group, but differs noticeably for smectites from different groups. A marked difference is observed for smectites with maximum and minimum $c \cos \beta$ values: for instance, the strongest reflections on the patterns of (Al, Mg)-smectites from the first group with the maximum $|c \cos \beta|$ values (0.38 a) are $\bar{1}12$ and 112, and between them lies a weak 022 reflection. On the diffraction pattern of the (Al, Mg)-smectites with minimum $|c \cos \beta|$ values, the 022 reflection is, in contrast, the strongest, the $\bar{1}12$ and 112 reflections are weak and the intensity of the $\bar{1}13$ reflection is markedly increased. On the patterns of the smectites with $|c \cos \beta|$ in the range 0.34 a to 0.35 a, the intensities of the reflections are practically equal.

Smectites with high octahedral Fe^{3+} contents ($Fe^{3+}/Al > 0.9$) have similar intensity distributions on the first ellipse, $\bar{1}12$ and 112 being the most intense reflections. The $|c \cos \beta|$ values range from 0.35 a for Fe-smectites (sample 11) to 0.36 a–0.37 a for nontronites. In sample 16, which shows a relatively high octahedral Al content ($Fe^{3+}/Al = 0.91$), $|c \cos \beta| = 0.38\,a$. The decrease in $|c \cos \beta|$ in the case of nontronites is as expected, since the ionic radius of Fe^{3+} is larger than that of Al^{3+}, which reduces the difference in the sizes of *cis* and *trans* octahedra of 2:1 layers. As a consequence, the shift between the upper and lower

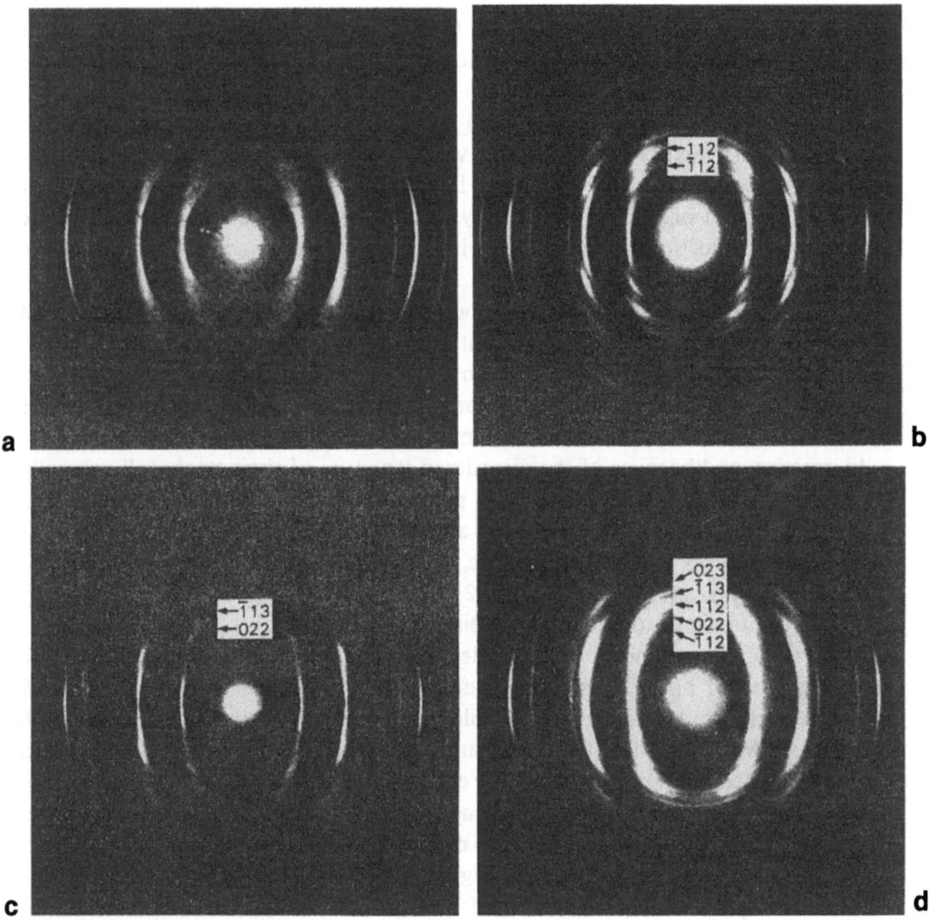

Fig. 8.12. Oblique texture electron diffraction pattern **a** of natural smectite with turbostratic structure, **b** of Fe^{3+}-montmorillonite, **c** of Wyoming montmorillonite, **d** of clay from the Panama Basin (after 70–100 WD cycles). The angle between the initial electron beam and the normal to the surface of the sample is 60°

tetrahedral sheets within the 2:1 layer is closer to $-a/3$ in nontronites than in montmorillonites and beidellites.

The patterns of the Red Sea Fe-montmorillonites (sample 11, Tables 8.5 and 8.6), Wyoming montmorillonite (sample 2) and Panama montmorillonite (sample 10) (typical representatives of the different groups) are presented in Fig. 8.12b, c, d, e. The distribution of intensities in Fig. 8.12d is intermediate between those of Fig. 8.12b and 12c. The data show that a series of $|c \cos \beta|$ values from 0.38 a to 0.32 a is possible for (Al, Mg)-smectites with continuous transfers in the intensity distribution between the extreme cases indicated.

Table 8.5. Unit cell parameters and occupancies of the *trans*-octahedral sites (P_1%) of the smectites studied. Note: The correspondence between sample number and the nature of the smectites is given in Table 8.6

Sample number	a (Å)	b (Å)	c (Å)	β (°)	$\lvert c\cos\beta/a\rvert$	P_1 (%)
1	5.18	8.97	10.07	99.6	0.32	75 – 100
2	5.18	8.98	10.10	99.5	0.32	75 – 100
3	5.18	8.97	9.95	100.4	0.35	50 – 75
4	5.18	8.98	10.08	100.4	0.35	0 – 25
5	5.18	8.98	10.10	99.5	0.32	75 – 100
6	5.19	9.00	10.10	101.3	0.38	0
7	5.19	9.00	10.10	101.3	0.38	0
8	5.18	8.97	10.20	101.3	0.38	0
9	5.18	8.98	10.13	99.8	0.33	75 – 100
10	5.18	8.98	10.08	101.0	0.37	25 – 50
11	5.23	9.06	10.20	100.3	0.35	0
12	5.18	8.98	10.05	101.4	0.38	0
13	5.18	8.97	10.10	100.5	0.35	50 – 75
14	5.18	8.98	10.10	99.6	0.32	75 – 100
15	5.18	8.98	10.08	100.2	0.34	50 – 75
16	5.20	9.01	10.20	101.3	0.38	0
17	5.26	9.12	10.10	101.0	0.37	0
18	5.29	9.17	10.10	101.0	0.36	0
19	5.26	9.12	10.14	101.0	0.37	0
20	5.28	9.14	10.14	100.7	0.35	0

The division detailed above of smectites into three groups (Tsipursky and Drits 1984) on the basis of the relative values of $\lvert c\cos\beta\rvert$ is associated with the following distribution of vacancies in the octahedral sites: values of $\lvert c\cos\beta\rvert > a/3$ correspond to smectites with vacant *trans* sites; those in which the *cis* sites are vacant have $\lvert c\cos\beta\rvert < a/3$; those in which the vacancies are statistically distributed between *cis* and *trans* sites have $\lvert c\cos\beta\rvert \approx a/3$. This variation is connected with the fact that the relative translation of the two tetrahedral sheets within the same layer is not always the same since it depends on the nature of the nonvacant octahedral sites. In the OTED diagrams, this is seen by the relative displacement of the positions of the hkl reflections on a given ellipse. Hence, mere semi-quantitative analysis of the unit cell parameters is enough to differentiate *cis*-vacant from *trans*-vacant smectites, as confirmed by calculation of the intensities of the hkl reflections (Tsipursky and Drits 1984). Calculations were made for models with the same composition as Ascan smectite (sample 12), but with variants of cation distribution over octahedral sites. The intensities of the $11l$ and $02l$ reflections appeared to be most sensitive to the changes in cation distribution (cf. Table 8.7). The gradual transfer from *trans*-vacant to a model in which the proportion of occupancy is the same for all octahedral sites leads to a decrease in intensities of the strongest 020, $\bar{1}12$, and 112 reflections and an increase in intensities of the weaker $\bar{1}13$ and 022 reflections. With equiprobable occupation of the octahedral sites, the calculated intensities for the whole group of these reflections are equalized. With

Table 8.6. Crystallochemical formulas on an $O_{10}(OH)_2$ basis for 20 different smectites defined in the notes[a]

Sample number	Nature and amount of cations									
	Interlayer				Tetrahedra		Octahedra			
	Ca	Na	K	Mg	Si	Al	Al	Fe^{3+}	Fe^{2+}	Mg
1	0.22	0.01	0.04		3.95	0.05	1.38	0.18		0.44
2	0.06	0.20			3.96	0.04	1.54	0.18		0.26
3	0.06	0.17	0.09		3.98	0.02	1.38	0.14	0.01	0.48
4	0.13	0.02	0.04		3.91	0.09	1.36	0.39	0.02	0.24
5	0.03	0.31	0.02		4.00		1.51	0.10		0.39
6	0.15	0.01	0.03		4.00		1.40	0.26		0.34
7	0.18	0.05	0.02		3.98	0.02	1.32	0.26		0.41
8	0.06	0.17	0.07		4.00		1.39	0.26	0.05	0.30
9	0.13	0.25			3.83	0.17	1.47	0.10		0.34
10	0.15	0.20	0.10		3.89	0.11	0.89	0.62	0.03	0.46
11	0.11	0.04	0.30		4.00		0.20	1.13	0.38	0.29
12	0.06	0.21	0.27		3.71	0.29	1.64	0.05	0.01	0.31
13	0.10	0.05	0.05	0.27	3.73	0.27	1.05	0.37		0.57
14		0.46(a)			3.86	0.14	1.68			0.32
15	0.07	0.01	0.11	0.15	3.41	0.59	1.57	0.37	0.01	0.05
16	0.11	0.03	0.16		3.53	0.47	0.96	0.88	0.02	0.26
17	0.12	0.05	0.01	0.17	3.45	0.55	0.33	1.59		0.08
18	0.23	0.03	0.01		3.49	0.51 (b)		1.87	0.17	
19	0.16	0.03	0.01	0.05	3.65	0.35		1.92		0.08
20		0.02	0.40	0.04	3.46	0.54(c)	0.15	1.85		

[a] Notes: (a) corresponds to the sum of interlayer cations obtained by the radioactive sodium method by Besson (1980); (b) corresponds to Fe^{3+} instead of Al^{3+} cation in tetrahedral sites; (c) corresponds to $0.40\,Al^{3+} + 0.14\,Fe^{3+}$ in tetrahedral sites.

Correspondence between sample numbers and the nature of the smectites: (1) Bentonites, Ascan (Drits and Kossovskaya 1980); (2) Montmorillonite, Wyoming (Mamy and Gaultier 1976); (3) Bentonite, Ascan (Drits and Kossovskaya 1980); (4) Hydrothermal alteration product of dacite-andesite, Kunashir Island (Drits and Kossvskaya 1980); (5) Bentonite, Oglalyn from Dr. V.I. Eirish, Geological Institute, Kazan, USSR; (6)(7) Bentonites, Gumbrin, from Dr. V.I. Eirish, Geological Institute Kazan, USSR; (8) Bentonite, Sarygiuk, from Dr. V.I. Eirish, Geological Institute, Kazan, USSR; (9) Ancient soil in sedimentary rocks, Kazakhstan (Drits and Kossovskaya 1980); (10) Clay from Panama Bassin, Pacific Ocean (Drits and Kossovskaya 1980); (11) Modern sediment, Red Sea (Boutouzova et al. 1979); (12) Bentonite Ascan (Drits and Kossovskaya 1980); (13) Clay from geothermal field, Kamchatka (Drits and Kossovskaya 1980); (14) Beidellite Rupsroth (Besson 1980); (15) Clay from geothermal field, Kamchatka (Drits and Kossovskaya 1980); (16) Modern red clay from Pacific Ocean (Tsipursky et al. 1978); (17)(19) Weathering crust of gneisses (Tsipursky et al. 1978); (20) Nontronite, Garfield (Besson et al. 1982)

the transfer from equiprobable occupation of octahedral sites to the model with *cis*-vacant sites, the $\bar{1}12$, 112, and 023 reflections are weakened and the 022 and $\bar{1}13$ reflections are strongest. The intensities of the $02l$ and $11l$ reflections for the three limit cases considered (i.e., *cis*-vacant, *trans*-vacant and statistically homogeneous distributions) are shown in Fig. 8.13.

Table 8.7. Calculated intensities of the 02l and 11l reflections for models of smectites with different octahedral cation distributions

hkl	1	2	3	4	5	6
020	840	690	240	140	160	140
110	20	50	100	140	270	400
$\bar{1}11$	350	250	60	10	60	100
021	60	30	20	20	60	90
111		10	60	90	160	260
$\bar{1}12$	580	480	470	430	430	320
022	170	230	420	530	590	730
112	630	540	530	480	470	360
$\bar{1}13$	130	170	230	370	380	480
023	210	170	180	200	210	160
113	10	20	40	100	90	120
$\bar{1}14$	10		10	10	10	10

Columns 1 – 6 correspond to the following cation occupancies of the *trans*, *cis*-1 and *cis*-2 octahedral sites: (1) 0.0; 1.0; 1.0 (model 1). (2) 0.25; 0.75; 1.0. (3) 0.5; 0.5; 1.0. (4) 0.67; 0.67; 0.67 (model 3). (5) 0.85; 0.25; 1.0. (6) 1.0; 0.0; 1.0 (model 2)

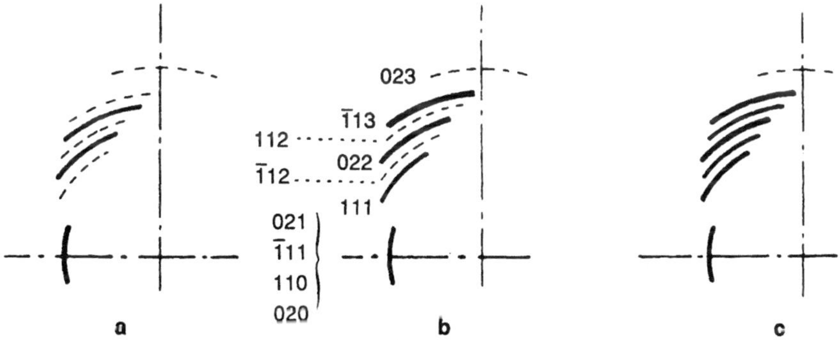

Fig. 8.13. Schematic oblique texture diffraction pattern for different models of octahedral occupancy in the dioctahedral 2 : 1 layers; (a) model with vacant *trans*-octahedra; (b) model with filled *trans*-octahedra; (c) model in which the average occupancy of each octahedral site by cations is 2/3

In order to evaluate the influence of the different structural factors on the intensities of the 02l and 11l reflections for each category of cation distribution, ideal structural models were considered, each with different chemical compositions. Comparison of the data obtained shows that the dominant factor determining the main features of the 02l and 11l reflection intensity is the distribution of cations in the octahedral *cis* and *trans* sites. The specific cation composition of the 2:1 layers and their structural distortions mainly effect the absolute values of the intensity rather than the relative distribution. Analysis of the variations of the 02*l* and 11*l* intensities taking into account the above factors shows that the distribution of cations over the available sites may be established using the data presented in Table 8.7 based on the comparison of calculated and experimental intensities. Note, in this respect, that since the experimental intensities were obtained visually, the accuracy of the determination is not better than 20%.

These results show the possibilities offered by the method of oblique texture electron diffraction. Unfortunately, neither the method not the apparatus adapted to its use are truly widespread. Furthermore, when the method is limited to the exploitation of photographic data, it remains only semi-quantitative.

8.2.3 Determination of the Distribution of Octahedral Cations in K^+-Nontronites by XRD

Nontronites are dioctahedral smectites in which the cations present in the octahedral sites are principally Fe^{3+}. The sample studied by Besson et al. (1983a) was from Garfield, Washington (USA). Its chemical analysis leads to the structural formula $(Si_{6.91}Al_{1.06}Fe^{3+}_{0.03})(Fe^{3+}_{3.96}Mg_{0.04})O_{20}(OH)_4M^+_{1.13}$, calculated on the basis of an anionic network $O_{20}(OH)_4$. In the natural state, the interlayer cations are essentially calcium.

The determination of the distribution of cations in the octahedral layer was performed by comparing the experimental powder pattern to X-ray diffractograms calculated on the basis of various structural models, using the method described in Chapters 2 and 3. The three curves in Fig. 8.14, spanning the region of the 02*l* and 11*l* reflections, were calculated assuming (a) that all the *trans* sites are vacant (model *a*, Fig. 8.14a); or (b) that one of the categories of *cis* sites (*cis*-1 or *cis*-2) is vacant (model *b*, Fig. 8.14b); or (c) that the cations are statistically and homogeneously distributed among the three categories of sites (model *c*, Fig. 8.14c). The atomic coordinates of the unit cell, which are required for the calculation of the synthetic spectra, were obtained using regression equations (Drits 1971, 1975) on the basis of the structural formula and the *b* parameters determined experimentally. In Table 8.8, these coordinates are given for a layer with unoccupied *trans*- and *cis*-octahedral sites (column 1 and 2, models *a* and *b*, respectively). For model *c*, the atomic coordinates (column 3) were obtained by proper averaging of the corresponding coordinates of models *a* and *b*. In order to take into account the disorder in the stacking of the layers, a 20% concentration of stacking faults was introduced in the calculations for all three models.

The diagrams in Fig. 8.14a, b and c are typical of powder patterns given by smectites with an important proportion of stacking faults. The most intense reflections are those which correspond to the "head" of the band, i.e., the 020 and 110 reflections, which are found to be joined and give a single peak. Qualitatively the

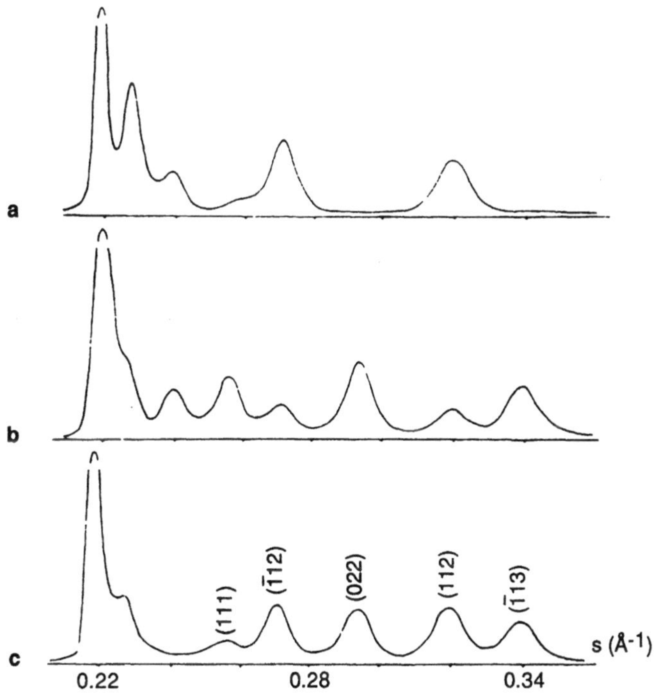

Fig. 8.14. Calculated X-ray diffraction of (02, 11) bands for the three models described in Fig. 8-13, with identical (a), (b) (c) labels

diagrams exhibit the same characteristics as those described above (Sect. 8.2.2) for the three models:

for model a: The $\bar{1}12$ and 112 are clearly more intense than the 111, 022, and $\bar{1}13$ reflections;
for model b: on the contrary, the 111, 022 and $\bar{1}13$ reflections are obviously more intense than the $\bar{1}12$ and 112;
for model c: the 111, $\bar{1}12$, 022, 112, and $\bar{1}13$ reflections are about equally intense.

It should be emphasized once again that these characteristics of X-ray diffractograms are valid for all the dioctahedral smectites, whatever the chemical composition of the 2:1 layers and the kind of distribution of cations (random or ordered) among the occupied sites. Calculations show that the ordered distribution of cations among the occupied octahedral sites slightly modifies the absolute intensities but preserves the same sequence of strong and weak reflections. Hence, the properties of the patterns described above for each model are general and can be exploited to characterize the distribution of cations in the octahedral sites of different smectites, particularly those having a high degree of order in layer stacking.

Table 8.8. The atomic coordinates as a function of the occupancy of the octahedral sites

Atoms	Vacant *trans* positions ($M_2 = M'_2 = 1$; $M_1 = 0$)		Vacant *cis* positions ($M_1 = M'_2 = 1$; $M_2 = 0$)		Average occupation ($M_1 = M_2 = M'_2 = 2/3$)		
	x	*y*	*x*	*y*	*x*	*y*	z (Å)
OH	0.375	0	0.312	0.979	0.333	0	1.11
O_1	0.312	0.688	0.375	0.667	0.333	0.667	1.08
O_2	0.312	0.312	0.312	0.354	0.333	0.333	1.08
O_3	0.329	0.5	0.361	0.511	0.350	0.5	3.16
O_4	0.054	0.742	0.086	0.752	0.075	0.742	3.37
O_5	0.054	0.258	0.086	0.269	0.075	0.258	3.37
T_1	0.314	0.672	0.351	0.674	0.333	0.667	2.72
T_2	0.314	0.328	0.355	0.346	0.333	0.333	2.72
M'_2	0.5	0.833	0.5	0.833	0.5	0.833	0
M_2	0.5	0.167	0.5	0.167	0.5	0.167	0
M_1	0	0	0	0	0	0	0
K	0.317	0	0.342	0.008	0.333	0	5.05

The qualitative comparison of the calculated (Fig. 8.14) and experimental (Fig. 8.15) patterns obtained from Garfield K-nontronite indicates that model *a*, with the empty *trans* sites, is to be preferred, especially since it is particularly sensitive to the intensity of the 111 ($s = 0.255$ Å^{-1}) and $\bar{1}13$ ($s = 0.340$ Å^{-1}) reflections which are absent in the experimental pattern.

For a more quantitative determination of the structural characteristics, different variants directly derived from model *a* were tried (Besson et al. 1983a) to obtain a fit with minimal discrepancy between the calculated and experimental values of intensities for the (02, 11) domain of the diffractogram. A first step in the calculation consisted in admitting a small proportion (5, 10 or 15%) of occupancy of *trans* sites, to provide a check on the validity of conclusions yielded by some older Mössbauer spectroscopy determinations (Goodman et al. 1976; Russel et al. 1979; Heller-Kallai and Rozenson 1980) and also to test the sensitivity of the method of modelization of X-ray diffractograms. In the second step, the possibility of having various proportions of different types of stacking faults was examined: random

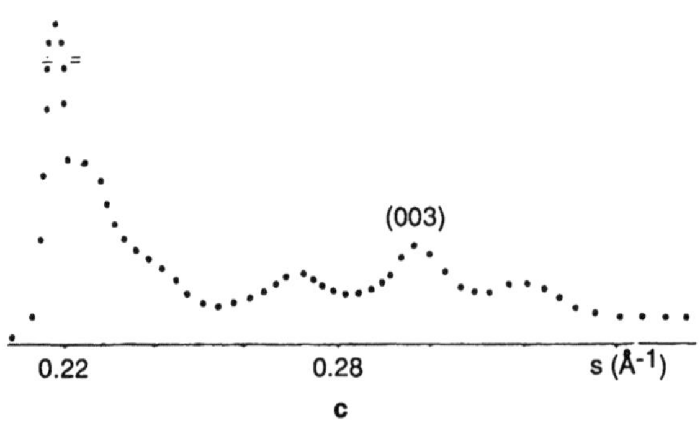

Fig. 8.15. Experimental (02, 11) diffraction band in the X-ray diffractogram obtained from Garfield nontronite

faults (cf. Chap. 1.6.5), defects corresponding to the presence of additional translations equal to $\pm\vec{b}/3$, and faults corresponding to rotations by $\pm n 2\pi/6$. The comparison of the diagrams calculated on these bases for the range of the $02l, 11l$ and $20l, 13l$ reflections indicate that the Garfield nontronite is devoid of these last two types of defects.

Figure 8.16 shows (on a relative arbitrary scale) the fit between the experimental and theoretical X-ray diffraction profiles calculated using the parameters and coordinates given in Table 8.9, in which t_{kk} denotes mica-type translations between adjacent layers, P_T and P_R are the probabilities of existence of $\pm\vec{b}/3$ translational and $\pm n 2\pi/6$ rotational faults, and P_A is the probability attached to random stacking faults. Agreement between experimental and calculated values was obtained for a ten layer-thick domain of coherence and an in-plane radius of 130 ± 10 Å determined from the $02l, 11l$ reflections, or of 80 ± 10 Å from the $20l, 13l$ reflections.

From this quantitative study, it was possible to deduce that, in nontronite, almost all the cations are in *cis* octahedral sites. Taking into account the precision of the recordings and the defects present in the K^+ nontronite after 100 WD cycles, the occupancy of the *trans*-octahedral sites cannot exceed 10% of the total number of sites.

To complete the description of this example of application of the modelization method, it should be pointed out that another problem connected with the internal structure of the layers in Garfield nontronite was solved simultaneously. From a crystallochemical point of view, it was interesting to find out if Fe^{3+} cations could

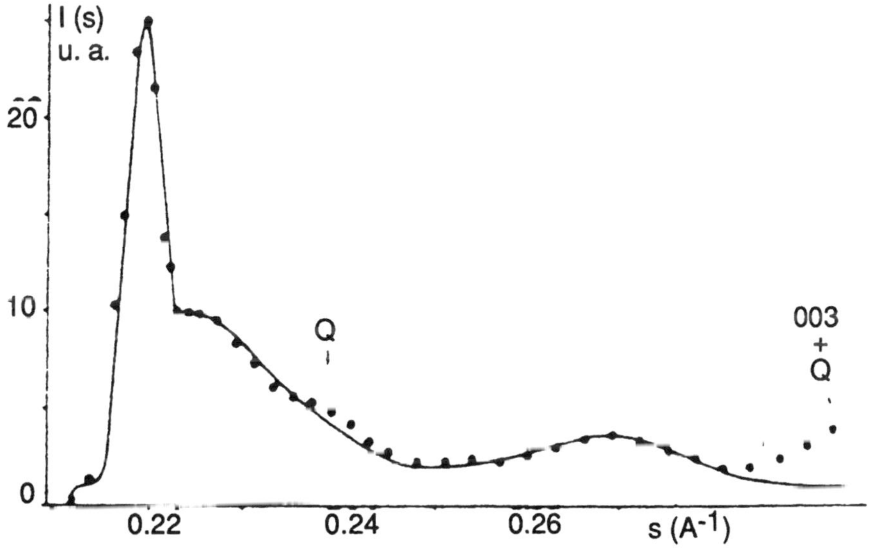

Fig. 8.16. Agreement between the experimental and calculated X-ray diffractograms of the (02, 11) band from Garfield nontronite. The excess intensities labelled Q are due to a small amount of quartz

Table 8.9. Coordinates and structural parameters obtained for the Garfield nontronite

Atoms	x	y	z (Å)
OH	0.375	0	1.11
O_1	0.312	0.688	1.08
O_2	0.312	0.312	1.08
O_3	0.339	0.5	3.16
O_4	0.049	0.737	3.37
O_5	0.049	0.263	3.37
T_1	0.315	0.672	2.72
T_2	0.315	0.328	2.72
M_2'	0.5	0.833	0
M_2	0.5	0.167	0
M_1	0	0	0
K	0.317	0	5.05

$b = 9.14 \pm 0.01$ Å; $\alpha = 6 \pm 1°$; $P_A = 65 \pm 5\%$;
$P_T = 0$; $P_R = 0$; $\vec{t}_{KK} = -0.367\,\vec{a}$;
$d_{001} = 10.1 \pm 0.05$ Å; $M = 10 \pm 1$ layers

occupy tetrahedral sites, with Al^{3+} ions present in the octahedral sheet. For the calculation of the structural formula, a rule was used according to which the Si deficiency in the tetrahedral positions was compensated at the expense of Al. The X-ray analysis presented here shows that such a distribution of the isomorphous cations over octahedral and tetrahedral positions can be incorrect: if one compares the atomic coordinates in the initial and final models, it is found that some of these coordinates are different, more particularly those of the basal oxygens. This is due to the fact that the real tetrahedral rotation angle α is larger than that estimated with the help of the regression equations (Drits 1971, 1975) using the Si-Al content of the tetrahedra. The regression equations yielded $\alpha = 4°$, whereas $\alpha = 6°$ was needed to obtain a good agreement of the experimental and calculated profiles. Several hypotheses may be proposed to explain this difference:

a) some of the tetrahedral substitutions may be $Si^{4+} \rightarrow Fe^{3+}$ instead of $Si^{4+} \rightarrow Al^{3+}$;
b) the relation which allows b to be calculated on the basis of the composition of the $Si_{8-x}Al_x$ tetrahedra may be incorrect;
c) the chemical analysis of the mineral may be in error.

Of these three hypotheses, the first was confirmed experimentally by Osthaus (1954) who, by selective chemical analysis, concluded that 20% of the total Fe^{3+} was to be assigned to the tetrahedral positions in Garfield nontronite. The regression equations of Drits (1971, 1975) will yield $\alpha = 6°$ if one starts from the structural formula

$$(Si_{6.91}Al_{0.77}Fe^{3+}_{0.32})(Fe^{3+}_{3.67}Al_{0.29}Mg_{0.04})(OH)_4O_{20}M^{+}_{1.13}$$

consistent with the Mössbauer spectral analysis and leading to the structural model yielding good agreement between the experimental and calculated XRD bands. Recently, Cardile et al. (1986) studied a series of nontronite samples by Mössbauer spectroscopy and confirmed the presence of Fe^{3+} cations in the tetrahedra of Garfield nontronite.

8.2.4 Analysis of the XRD Powder Patterns from Dioctahedral Cs-Smectites

The advantages and drawbacks associated with the method of Cs-exchange were discussed in Section 8.3.1. The present Section will therefore be limited to the description of the results obtained by Besson et al. (1983b) in the determination of the structural characteristics of two dioctahedral smectites with Si-Al substitutions in the tetrahedral sheets: beidellite from the Black Jack Mine (USA) and beidellite A from Rupsroth (Bavaria).

The ion exchange was performed by placing the sample of natural smectite in 1 N CsCl solution (sometimes heated to 50 °C to speed up the process) then washing it repeatedly to eliminate any adhering solution. After dehydration at 100 °C for 24 h, the sample was placed in a diffraction chamber specially contrived to maintain it in the anhydrous state at 1.3 Pa in the presence of P_2O_5. No WD cycles were performed.

The experimental tracings for the (02, 11) domain of reflections shown in Fig. 8.17a (beidellite from the Black Jack Mine) and Fig. 8.17b (Rupsroth beidellite) were both obtained with $CuK\alpha_1$ radiation, using a linear localization detector. As compared to the equivalent diagram in Fig. 8.18 given by K-exchanged samples of the same Black Jack Mine beidellite with 100 WD cycles, the bands in the present tracings are better resolved, with a higher signal/background ratio.

Thanks to the better quality of the XRD patterns from Cs-beidellites, an important difference beween the diffractograms of the two beidellites studied becomes evident: the intensity of the 111 and 022 reflections is greater than that of the 021 and $\bar{1}12$ reflections in Rupsroth beidellite, whereas the reverse is true for the sample from the Black Jack Mine. As explained in the preceding section, this difference is connected with the specific distribution of the cations in the octahedral sites and, even by a mere qualitative examination of the diffractograms, it is easy to deduce that most *trans* sites are respectively vacant or occupied in Black Jack Mine or Rupsroth beidellites.

However, when quantitative agreement between the experimental and calculated spectra is sought, a specific difficulty crops up connected with the two possible layer stacking sequences in smectites:

- the relative position of successive layers may be such that the hexagonal surface cavities which flank the interlayer space lie face to face; this may happen either when there are no stacking faults in a pair of successive layers, or when there

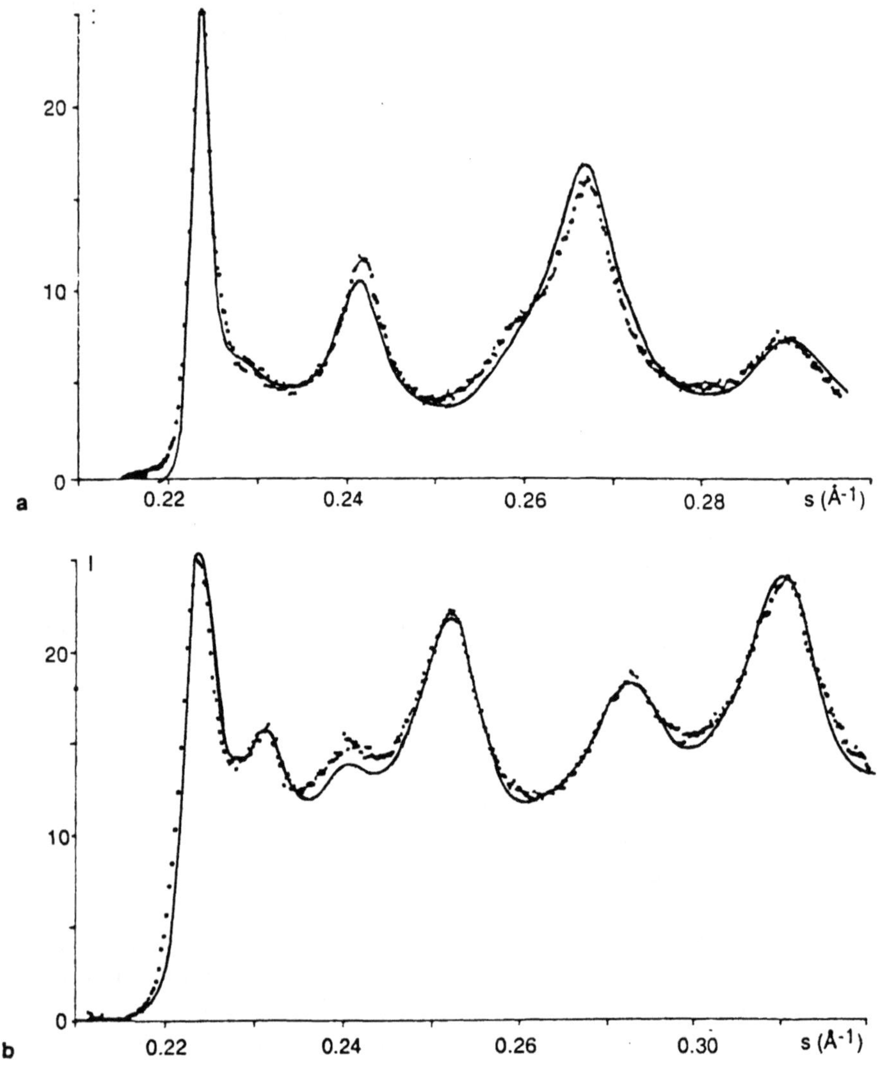

Fig. 8.17. Experimental and calculated X-ray diffraction profiles of (02, 11) bands from two anhydrous Cs-beidellite; **a** Black Jack Mine beidellite, **b** Rupsroth beidellite

is a $\pm n2\pi/6$ rotation fault. This so-called "symmetric" situation is shown in Fig. 8.19a;

– the relative position of successive layers may lead to hexagonal surface cavities which are not lying directly opposite each other, either because of the presence of translation defects or arbitrary rotations, or because of the presence of supplementary translations equal $\pm\bar{b}/3$; this "nonsymmetric" situation is shown in Fig. 8.19b.

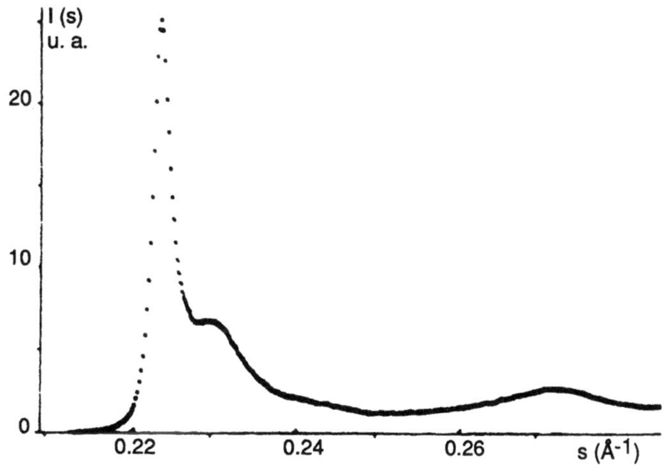

Fig. 8.18. Experiment X-ray diffraction profile of the (02, 11) band from K-beidellite from the Black Jack Mine

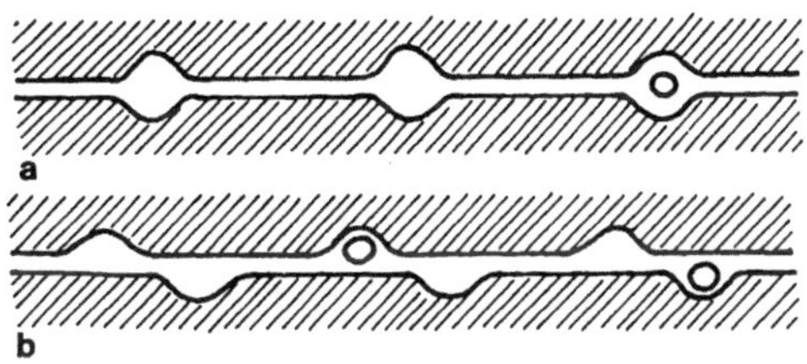

Fig. 8.19. Schematic diagram of the interlayer space between two 2:1 layers; **a** symmetrical position; **b** nonsymmetrical position

The existence of these two types of arrangements has direct consequences on the basal distances when the compensating cation has a large radius, e.g. Cs^+: with this ion, the calculated interlayer distances must be $d_1 = 10.8$ Å or $d_2 = 11.6$ Å respectively for the "symmetric" and "nonsymmetric" cases. The coexistence of two arrangements with corresponding different values of basal distances within the same stack is expected to lead to *interstratification* which would show up by the presence of intensity maxima between the two positions corresponding to $1/d_1$ and $1/d_2$. This is confirmed by the diffractograms given by Cs-beidellite A in which one finds $(d_{001})_{\text{apparent}} = 11.1$ Å. By applying the modelization method to the profile of the basal reflections, Besson (1980) determined the proportion of each component in the interstratification: the apparent distance corresponds to 70% d_1 and 30% d_2 in Cs-beidellite from Black Jack Mine, whereas Cs-beidellite A contains 65% d_1 and 35% d_2.

The correct analysis of the *hk* band profiles in Cs^+-beidellites however requires a more complete description of the arrangement of the layers corresponding to the "nonsymmetric" case which, it will be recalled, is caused by the presence either of random or of well-defined $\pm \vec{b}/3$ translations. Since statistically random translations destroy the interferential coherence for the *hk* bands, the diagram will tend towards that of a 2D structure. If, on the other hand, the "nonsymmetric" configuration is induced by supplementary $\pm n\vec{b}/3$ translations, the diffracted beams will still interfere but there will be interstratification and the two basal distances d_1 and d_2 must appear in the matrices used in the calculation of the band profiles. It is therefore necessary to take into account three possible translations:

– the one corresponding to two layers in the "symmetric" configuration:

$$\vec{t}_1 = \vec{t}_0 + \vec{u}_z d_1 \ ;$$

– those corresponding to the "nonsymmetric" cases created by the supplementary $\pm\vec{b}/3$ translations, namely:

$$\vec{t}_2 = \vec{t}_0 + \vec{b}/3 + \vec{u}_z d_2 \ ;$$

and

$$\vec{t}_3 = \vec{t}_0 - \vec{b}/3 + \vec{u}_z d_2 \ .$$

Let now W_1, W_2 and W_3 be the proportions of layers translated by $\vec{t}_1, \vec{t}_2$, and $\vec{t}_3$. If the parameter R, characteristic of a possible interaction between the defects, is zero and the probabilities of meeting any (i, j)-pair are proportional to the abundances of the types of layers, and if we further assume that the translations $+\vec{b}/3$ and $-\vec{b}/3$ are equally probable and let P_T be the possibility of finding them among the non-random translations, the three matrices become:

$$[\varphi] = \begin{bmatrix} \exp(2\pi i\vec{s}\vec{t}_1) & 0 & 0 \\ 0 & \exp(2\pi i\vec{s}\vec{t}_2) & 0 \\ 0 & 0 & \exp(2\pi i\vec{s}\vec{t}_3) \end{bmatrix}$$

$$[P] = \begin{bmatrix} 1-P_T & P_T/2 & P_T/2 \\ 1-P_T & P_T/2 & P_T/2 \\ 1-P_T & P_T/2 & P_T/2 \end{bmatrix}$$

and

$$[W] = \begin{bmatrix} 1-P_T & 0 & 0 \\ 0 & P_T/2 & 0 \\ 0 & 0 & P_T/2 \end{bmatrix}$$

The matrix $[Q] = [\varphi]\,[P]$ is then calculated; this matrix is degenerate and the only nonzero eigenvalue is given by

$$\lambda = (1-P_T)\exp(2\pi i\vec{s}\cdot\vec{t}_1) + (P_T/2)\,[\exp(2\pi i\vec{s}\vec{t}_2) + \exp(2\pi i\vec{s}\vec{t}_3)]$$

from which we obtain

$$\lambda = \exp(2\pi i \vec{s}\vec{t}_0)\,[(1-P_T)\exp(2\pi i Z d_1) + (P_T \cos 2\pi k/3)\exp(2\pi i Z d_2)]$$

or

$$\lambda = [\exp(2\pi i \vec{s}\vec{t}_0)]\,\lambda_1 \ .$$

The eigenvalue of the matrix is therefore the product of two factors, one of which corresponds to the interference term connected with the translation $\vec{t}_0$ between adjacent layers while the second, λ_1, is the interference term due to the existence of interstratification of the two basal distances d_1 and d_2. Furthermore, the presence of well-defined rotation defects equal to $\pm n 2\pi/6$ produces five possible translations i.e. $\vec{t}_1, \vec{t}_2, \ldots \vec{t}_5$, derived from $\vec{t}_0$ by a 60°, 120°, ... 300° rotation. In the concrete case of a stack simultaneously containing $\pm b/3$ translations, random stacking faults, and well-defined $2\pi n/6$ rotation defects, the general matrices [F], [W] and [P], resulting from Eqs. (3.24), (3.25) and (3.26), become

$$[\varphi] = \lambda_1 \begin{bmatrix} \exp(2\pi i \vec{s}\vec{t}_0) & 0 & \ldots & 0 \\ 0 & \exp(2\pi i \vec{s}\vec{t}_1) & \ldots & 0 \\ \vdots & \vdots & & \vdots \\ 0 & 0 & \ldots & \exp(2\pi i \vec{s}\vec{t}_5) \end{bmatrix} \tag{5.5}$$

$$[W] = \begin{bmatrix} (1-P_R) & 0 & \ldots & 0 \\ 0 & P_R/5 & \ldots & 0 \\ \vdots & \vdots & & \vdots \\ 0 & 0 & \ldots & P_R/5 \end{bmatrix} \tag{5.6}$$

$$[P] = (1-P_A) \begin{bmatrix} (1-P_R) & P_R/5 & \ldots & P_R/5 \\ (1-P_R) & P_R/5 & \ldots & P_R/5 \\ \vdots & \vdots & & \vdots \\ (1-P_R) & P_R/5 & \ldots & P_R/5 \end{bmatrix} \tag{5.7}$$

in which P_A is the probability of random stacking faults.

For both beidellites under consideration, the use of these matrices in the general relationship of Eq. (3.17) led to the best agreement between the experimental profile of the (02, 11) band and the one calculated using, for each beidellite, the parameters shown in Table 8.10. For the Rupsroth mineral, the model was based on fully occupied *trans* sites whereas these sites were all taken to be vacant in beidellite from the Black Jack Mine.

The fits are presented in Fig. 8.17, in which the full lines represents the calculated spectra and the points indicate the outline of the experimental profiles. The misfit with the 021 reflection of the Rupsroth beidellites is probably due to the presence of a small proportion of *trans* sites devoid of cations in the sample, in contradiction with the model used for the calculation of the theoretical profile which assumed that all *trans* sites were occupied. Lastly, it is gratifying to note that

Table 8.10. Specific structural parameters for Rupsroth and Black Jack Mine beidellite, obtained by fitting of the calculated and experimental XRD patterns

	Rupsroth beidellite	Black Jack Mine beidellite
b	9 Å	9 Å
P_A	0.15	0.30
P_T	0.20	0
P_R	0.20	0.50
$\vec{t}_0$	$-0.30\,\vec{a}$	$\approx -0.42\,\vec{a}$

the values obtained for $|\vec{t}_0|$ confirm the chosen distribution of cations in octahedral sites since it is found that the value of this translation is such that $|\vec{t}_0| < a/3$ for layers with *cis*-vacant sites, and $|\vec{t}_0| > a/3$ for layers with *trans*-vacant sites.

8.3 Determination of the Distribution of Cations and Water Molecules in the Interlamellar Spaces of Dioctahedral Smectites

The properties of smectites are tightly bound to the possibility shown by these minerals of taking up molecules of water by hydration of the cations in the interlamellar space; depending on the nature of these cations and on the amount of water taken in, the mutual arrangement of the layers and the structure of the interlamellar space may be modified. The problem is therefore the accurate determination of the distribution of cations and water molecules, and of the change in the arrangement of the adjacent layers caused by the degree of hydration.

We shall first give a fairly detailed account of the study of Na-exchanged Rupsroth beidellite A in the homogeneous two-water-layer state (sample E_2) corresponding to 90% relative humidity (R.H). This will be followed by a somewhat less detailed presentation of the results obtained with a "single water layer structure"(sample E_1) of the same Na-beidellite A (R.H. = 40%). All these results are due to Ben Brahim (Ben Brahim et al. 1983, 1984; Ben Brahim 1985).

8.3.1 Experimental Conditions

The composition of the Rupsroth Na^+-beidellite was

$$(Si_{7.73}Al_{0.27})(Al_{3.36}Mg_{0.64})O_{20}(OH)_4Na_{0.91}\cdot n\,H_2O\ .$$

The sample was placed in a special diffraction chamber [1] (cf. Fig. 8.20) and submitted to a stream of air issuing from a wash-bottle [2] containing sulfuric acid of definite concentration giving a known partial pressure of water vapor. The air was circulated through plastic tubing by a force pump [3]. It was possible to isolate the chamber from the gas circuit by valves and to connect it to a two-stage vacuum pump capable of maintaining a pressure of about 0.13 Pa. Once the sample was in the chamber, it was therefore possible to have the sample under different partial pressures of water vapor, to perform humidification and drying cycles, or even to put the sample under vacuum without having to handle it, with the additional advantage that the diffractograms were always given by the same clay particles. It was thus possible to measure in situ the absorption coefficient corresponding to specific degrees of hydration of the sample (Ben Brahim et al. 1983).

8.3.2 Analysis of the Profile of the 00*l* Reflections from a Two-Water-Layer Na-Beidellite (Sample E_2)

The goals of the study were the accurate determination of the quantity of water retained under 90% humidity, and the determination of the *z* ordinates of the water molecules and of the Na^+ cations. To obtain these data, the 001, 003, and 005 reflections (the only ones with a measurable intensity) were used.

The calculation of the intensity of the wave diffracted along a (00) rod was done using Eq. (2.43). In a homogeneous hydrate, the layers are equidistant and the

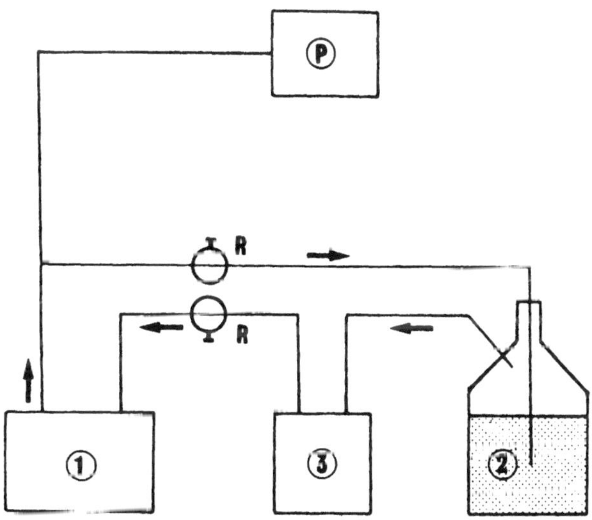

Fig. 8.20. Lay-out of the pumping and controlled humidity circuits. *R* taps; *P* vacuum pump; *1* X-ray diffraction chamber; *2* solution for relative humidity control in the diffraction chamber; *3* force pump

modulation function $G_{00}(Z)$ is limited to the shape function connected with the apparent size of the crystallites measured perpendicularly to their midplane. The function $G_{00}(Z)$ is then simply

$$G_{00}(Z) = \sum^{M} \frac{\alpha(M)}{M} \frac{\sin^2(M\pi Z d_{001})}{\sin^2(\pi Z d_{001})} \tag{8.8}$$

in which

- d_{001} is the basal interlayer distance,
- $\alpha(M)$ is the function connected with the statistical distribution of the crystallite thicknesses along $\vec{c}^*$ (cf. Chap. 3.1.2).

Table 8.11 gives the calculated values of the ratio I_{001}/I_{003} and I_{001}/I_{005} as a function of the amount of intercalated water, normalized to a single unit cell, with the molecules placed on either side of each 2:1 layer, at 6 Å of the midplane of the octahedral sheet. It is clear that the intensity ratio I_{001}/I_{005} can be correlated to the amount of intercalated water, whereas the ratio I_{001}/I_{003} seems to be quite insensitive to this parameter.

A similar calculation in which the amount of water was taken to be 8 molecules per unit cell indicates the sensitivity of these ratios to variations of the z ordinates within the interlamellar space. As seen in Table 8.12, it is the ratio I_{001}/I_{003} which is particularly affected by even slight changes in z.

The following conclusions can be drawn from these results:

- The dimension of the mean interferential coherence domain along $\vec{c}^*$ corresponds, on the average, to 16 ± 1 layers.
- An analysis based solely on the $00l$ series of the two-water-layer Na^+-beidellite with two layers of water leads to an ambiguity in the relation between the water content and the z coordinates of the cations: the profiles and the intensities of the $00l$ reflections can be interpreted:

Table 8.11. Evolution of the intensities of the basal reflections of sample E_2 as a function of the quantity of intercalated water in each unit cell (i.e., per eight tetrahedral sites)

Number of molecules of water / Relative intensities	$6H_2O$	$8H_2O$	$10H_2O$
I_{001}/I_{003}	≈47	47	47
I_{001}/I_{005}	28	≈20	≈13

Table 8.12. Evolution of the relative intensities of the basal reflections of sample E_2 as a function of the z ordinates (in Å) of the water molecules (8 molecules per unit cell)

Relative intensities \ z ordinate (Å)	±5.75	±6.0	±6.25	±6.50
I_{001}/I_{003}	≈97	≈47	≈26	≈16
I_{001}/I_{005}	≈24	≈20	≈18	≈21

a) either by assuming that the Na^+ ions are at the middle of the interlamellar space, as is generally the case with polyvalent cations in hydrated silicates (Méring and Glaeser 1954) and that each unit cell contains ten molecules of water of hydration at ±6.2 Å with respect to the midplane of the 2:1 layers;
b) or by supposing that the Na^+ ions are partially within the ditrigonal cavities at ±4.3±0.1 Å, and that there are 10.8 molecules of water at the ordinate 6.3±0.1 Å.

8.3.3 Qualitative Description of the (02,11), (20,13) and (04,22) Bands given by Two-Water-Layer Na^+-Beidellite

The next step was the determination of the *x,y* coordinates of the sites occupied by the water molecules and the proportion of occupancy of these sites. This required:

- the determination of the structural modifications in the layers due to the presence of the water-cation polyhedra in the interlamellar space;
- the determination of the nature and concentration of the stacking faults;
- the analysis of the *hk* bands to permit a choice between the two possibilities which could not be differentiated merely by the analysis of the 00*l* reflections.

Figure 8.21 a and b shows the experimental (02, 11), (20, 13) and (04, 22) diffraction bands obtained from homogeneously hydrated two-water-layer Na^+-beidellite, while Fig. 8.22a and b gives the (02, 11) and (20, 13) bands for the same beidellite in vacuum (1.3 Pa in the presence of P_2O_5). In both series of diagrams, the bands are practically nonmodulated, indicating a tendency to 2D-structure. However, even a simple qualitative comparison of the experimental curves indicates obvious modifications in the profiles and in the relative intensities of the diffraction bands. Since the recording parameters and the diffracting particles are the same in the two sets of diagrams, the changes in the spectra must be due to one of or both the following causes:

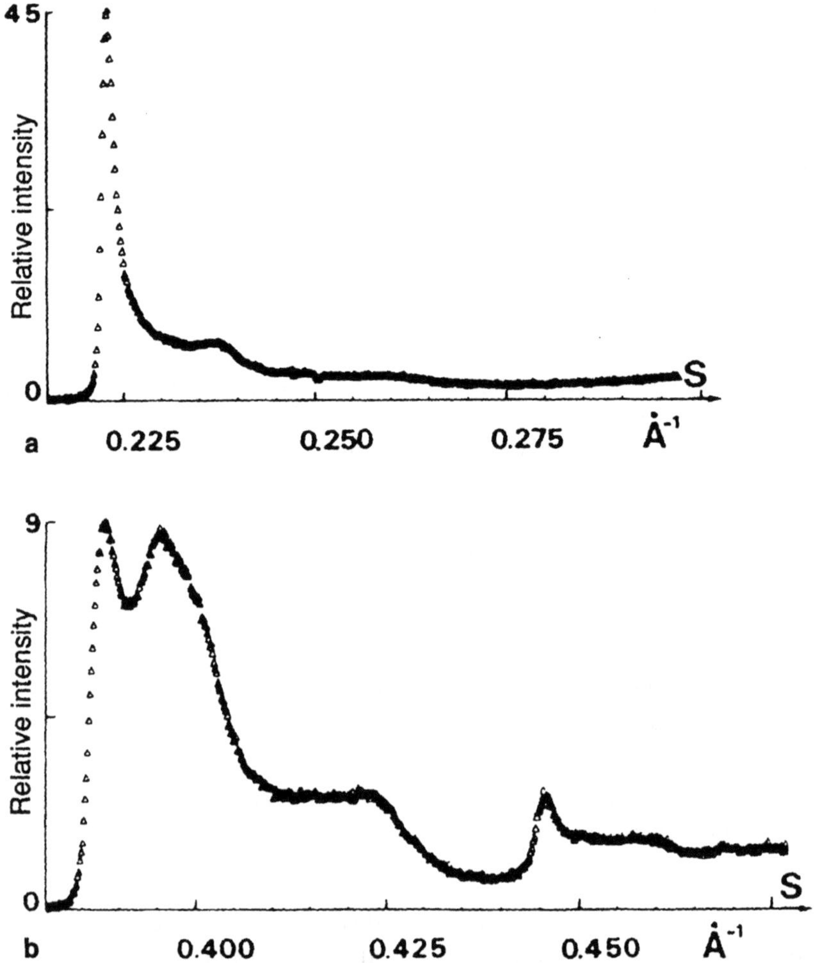

Fig. 8.21. Experimental X-ray diffraction profiles from hydrated Na-beidellite with two layers of water. Each point corresponds to an experimental value and the relative scale was normalized to 100 for the 001 reflection. Curve **a**: (02, 11) band; curve **b** (20, 13) and (04, 22) bands

- during the hydration process, a mutual rearrangement by slippage of the layers may occur, caused by the geometry and the forces of interaction of the interlamellar polyhedral hydrated cations;
- there may be a change in the structure factor caused by the contribution of the molecules of water.

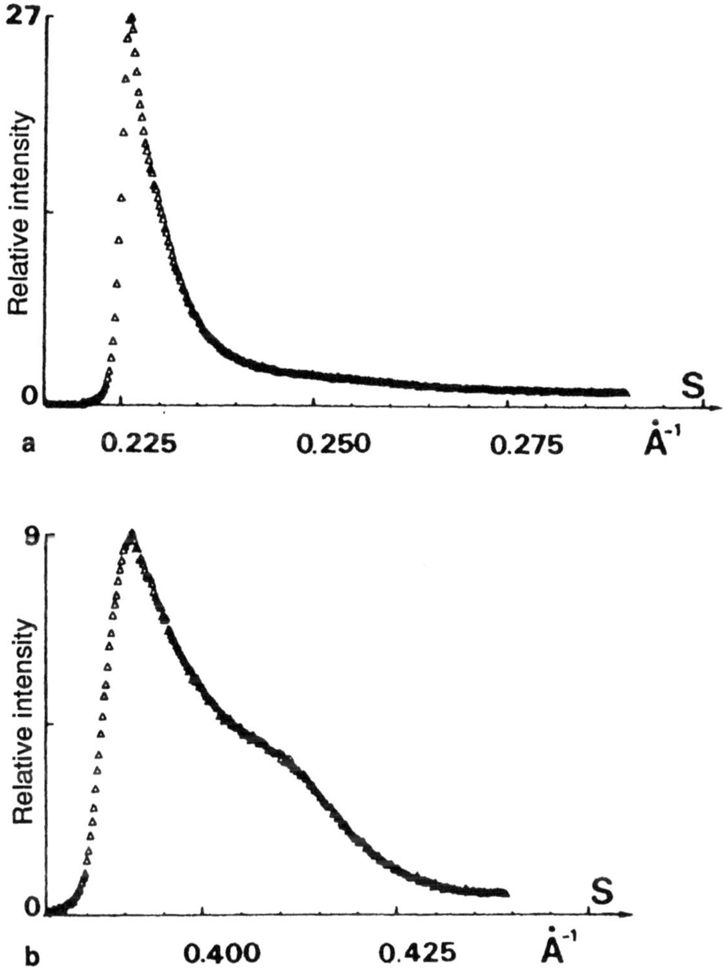

Fig. 8.22. Experimental profiles of the diffraction bands from anhydrous Na-beidellite (same remarks as in the caption of Fig. 8.21). Curve **a**: (02, 11) band; curve **b**: (20, 13) band

8.3.4 Determination of the *x*, *y* Coordinates of the Sites Occupied by Water Molecules

Studies on hydrated lamellar silicates (Barshad 1949; Mathieson and Walker 1954; Bradley and Serratosa 1960; Shirozu and Bailey 1966; de la Calle 1977; Alcover and Gatineau 1980; Pons et al. 1980) have shown that the water molecules in the (a, b) plane are found distributed in the three types of sites shown in Fig. 8.23 a (in which no account has been taken of the proportion of occupancy or possible steric hindrances):

- sites hereafter called A_1, of which there are twelve in each unit cell (six on each side) and whose projection coincides with or is very close to the surface oxygens of the layers;
- eight sites whose projection coincides with the tetrahedral cations Si or Al: these will be referred to as sites B;
- four sites, called C, whose projection coincides with the centers of ditrigonal surface cavities.

The fact that the B and C sites can be deduced from each other by changing their y coordinate to $y \pm b/3$ prompted Ben Brahim et al. (1984) to consider the simultaneous presence of an identical change in the y coordinate of the A_1 sites, resulting in the formation of two enantiomorphic sites called A_2 and A_3 (Fig. 8.23b and c). The theoretical patterns calculated from these different models led to the following conclusions:

1. The (20, 13) domain does not permit a distinction to be made among the A_1, A_2 and A_3 sites, or between the B and C sites. In this domain, the profiles and intensities of the reflexions are only sensitive to the ratio $(A_1+A_2+A_3)/(B+C)$, where $(A_1+A_2+A_3)$ and $(B+C)$ are respectively the total amounts of water molecules in the A and (B+C) sites (cf. Fig. 8.24).
2. In the (02, 11) domain, the diagram is sensitive to the value of the B/C ratio but the pattern is quite insensitive to the distribution of water molecules among the different A sites, as seen in Fig. 8.25.
3. On the contrary, the (04, 22) domain is significantly modified depending on whether it is the A_1 or the A_2 sites which are occupied by the water molecules. It is, however, not yet possible to distinguish between the occupancy of the two enantiomorphic A_2 and A_3 sites.

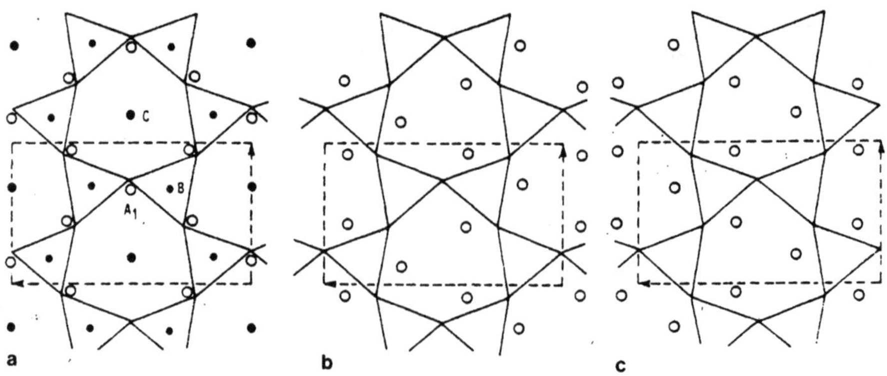

Fig. 8.23. Projection on the (ā, b̄) plane of possible water molecule sites in a two-water-layer homogeneous hydration state of smectite

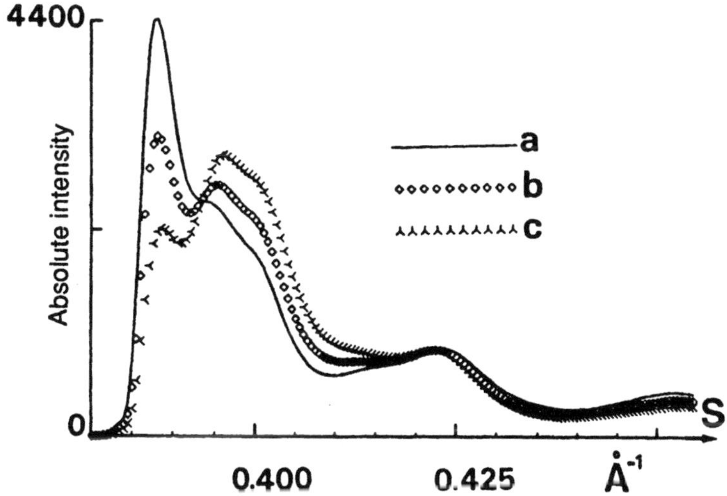

Fig. 8.24. Evolution of the (20, 13) band profiles calculated for the sample E_2 as a function of the ratio $Q = (A_1 + A_2 + A_3)/(B + C)$ describing the relative occupancy of the different sites by the ten molecules of water present per unit cell; **a** $Q = 4/6$; **b** $Q = 6/4$; **c** $Q = 8/2$

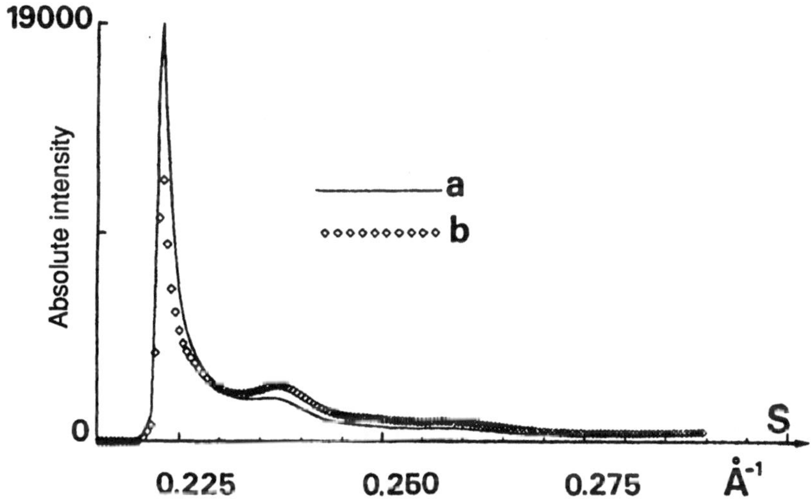

Fig. 8.25. Evolution of the (02, 11) band profiles calculated for the sample E_2 as a function of the relative occupancy of sites B and C, six of the ten water molecules are distributed among the A_1 sites, with the last four **a** all on B sites, or **b** equally distributed among B and C sites

8.3.5 The Different Possible Stackings of Layers in Two-Water-Layer Na-Beidellite

Let $\vec{t}_0$ be the translation vector between any given point on a layer and its crystallographically equivalent point in the adjacent layer of a defect-free 1 M polytype. When all the layers are identical, and with ideal unit cells, $\vec{t}_0 = -\vec{a}/3 + \vec{u}_z d_{001}$, where d_{001} is the basal distance. This implies that the projections of the centers of the ditrigonal cavities which are opposite each other in two adjacent layers must lie on the same point of the (a, b) plane. Ben Brahim et al. (1984) showed that in two-water-layer Na-beidellite, the stacking faults which are found can only be supplementary translations $\vec{T}$ and not rotation defects. Four supplementary translations were examined:

1. Random translations, with the probability P_A of existing between two adjacent layers;
2. well-defined translations $\vec{T} = \pm\vec{b}/3$;
3. well-defined pyrophyllite-like translations $\vec{T} = -\vec{a}/3 \pm \vec{b}/3$;
4. well-defined translations $\vec{T} = +\vec{a}/3 \pm \vec{b}/3$ similar to those observed in Mg-vermiculite placed in very low relative humidity. Whatever the well-defined translation, the total probability of finding it between two adjacent layers is noted P_T.

The influence of the presence of faults of types 1, 3 or 4 on the band profile was analysed. In each case, the values $P_A = 0.75$ and $P_T = 0.25$ were used in the calculations. Depending on the nature of the supplementary translation creating the defects, the position and intensities of the modulations on the (02, 11) and (20, 13) bands are modified, the latter being particularly sensitive to the presence of these defects.

One should also remember that the presence of well-defined supplementary translations $\vec{T} = \pm\vec{b}/3$ in a stack is shown up by a tendency towards a 2D structure only affecting the hk bands in which k is not a multiple of 3 (cf. Sect. 3.2.2). In the present case, this is not observed on the experimental diagrams (as will be further discussed in the next section) so that this type of translation defect can be rejected outright.

8.3.6 Determination of the Structural Characteristics of the Two-Water-Layer Na^+-Beidellite by Fitting the Calculated Pattern to the Experimental XRD Data

The comparison between the calculated and experimental curves are shown in Fig. 8.26a and b (note, in the first, that the ordinates have been multiplied by 5 with respect to the scale in Fig. 8.26b). Residual contribution from neighboring bands which could eventually increase the intensity of the bands considered were taken into account, as well as the 007 reflection superposed on the (04, 22) band. The parameter used in the calculations were:

- symmetry of the layers $C2$;
- unit cell: $a = 5.184 \pm 0.004$ Å;
  $b = 8.975 \pm 0.005$ Å;
  $d_{001} = 15.250 \pm 0.005$ Å;
- translation vector: $\vec{t}_0 = -(0.31 \pm 0.01)\vec{a}$;
- $P_A = 0.74 \pm 0.02$;
- number of water molecules per unit cell: 10.8, distributed over both faces of the layer in sites B, C, A_2 and/or A_3 with the following probabilities of occupancy for each type of site:

$$P_B = 0.24 \ ; \quad P_C = 0.07 \ ; \quad P_A = 0.69 \ .$$

In the course of fitting the calculations to the experimental results, it became clear that, of the two possible solutions which could not be differentiated by the study of the $00l$ reflections, only the one corresponding to 10.8 H_2O per unit cell at the ordinate $z = 6.3 \pm 0.1$ Å could be retained, as indicated by the study of the (20, 13) band, particularly sensitive to variations in the position of the Na^+ cation and to the number of molecules of water. Furthermore, the comparison of the calculated and experimental patterns in the (04, 22) domain showed that the A_1 sites are vacant, that the water molecules in the A sites are situated only in the A_2 *or* A_3 sites, and that the simultaneous occupation of the A_2 *and* A_3 sites in the same water sheet must be excluded. The best fit was obtained with 7.5 molecules in the A_2 (or A_3) sites, 2.6 in the B sites, and 0.7 in the C sites.

A model of the real distribution of water molecules in the 2D-interlayer space was suggested by Ben Brahim et al. (1984) on the basis of all these results averaged statistically. It is consistent with the crystallographic concepts dealing with the organisation of water molecules in the interlamellar space of layer silicates and corresponds to an incomplete hexagonal network.

8.3.7 Structural Characteristics of One-Water Layer Na^+-Beidellite. Comparison with the Two-Water-Layer Hydrate

Figure 8.27a and b presents the experimental profiles of the (02, 11), (20, 13) and (04, 22) reflections bands of one-water-layer Na^+-beidellite. The comparison of these profiles with those given by the hydrates with two layers of water (Fig. 8.26) shows important differences bearing on the relative intensities of the modulations and on their positions, and suggests that the change in hydration results in important modifications of the structure.

An investigation concerning the occupancy of the different sites in the homogeneous one-water-layer Na^+-beidellite was carried out by Ben Brahim (1985) along lines similar to the ones described above for the two-water-layer hydrate. It proved that, once again, the molecules are distributed among the A, B, and C sites so as to form a nonsaturated plane hexagonal network. The sheet of

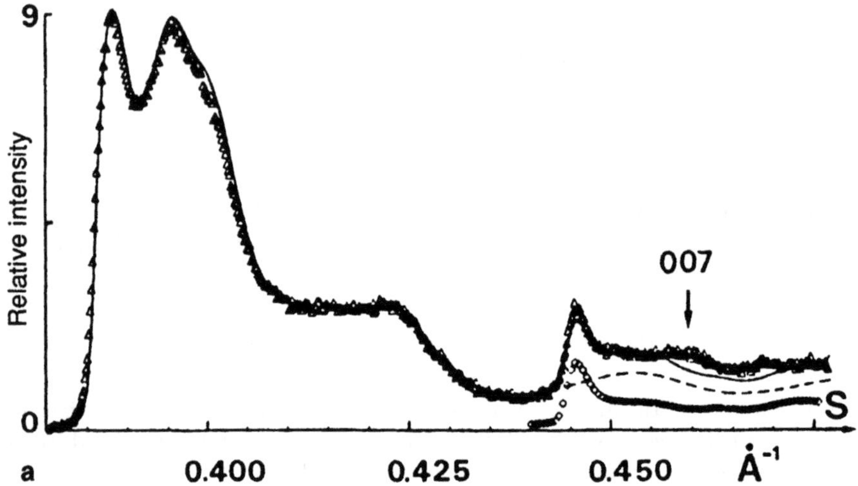

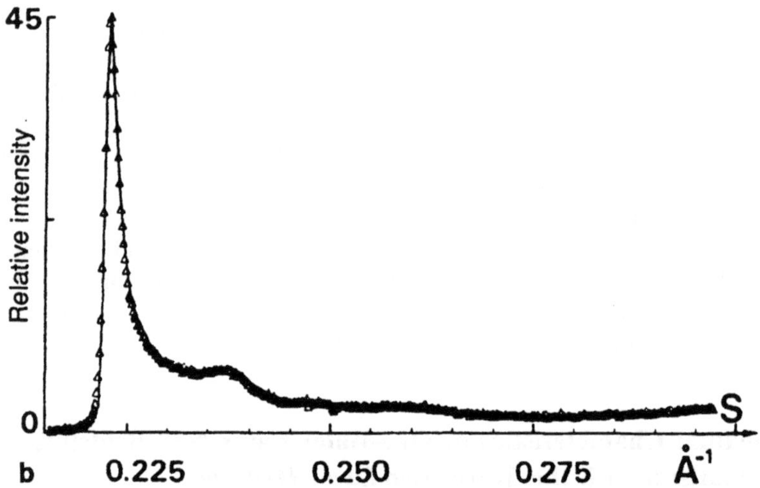

Fig. 8.26. a Agreement between calculated profiles (*full line*) and experimental data from sample E_2. The triangles (△) indicate the profile calculated for the (04, 22) band while the *dashed line* (– – –) gives the calculated (20, 13) band superposed on the (04, 22) band. The contribution of the (02, 11) band in the range (20, 13) to (04, 22) is negligible. **b** Agreement between the calculated profile (*full line*) and experimental data (△) for the (02, 11) band of sample E_2

water molecules is situated in the middle of the interlamellar space while the Na^+ cations remain in the pseudo-hexagonal cavities. This leads to an ordinate $z = +6.2 \pm 0.1$ Å for the six H_2O molecules per unit cell which occupy four A sites, one B, and one C site.

In order to account for the shift in the position of the shoulder observed on the (02, 11) band of the one-water-layer hydrate, it was found necessary to introduce

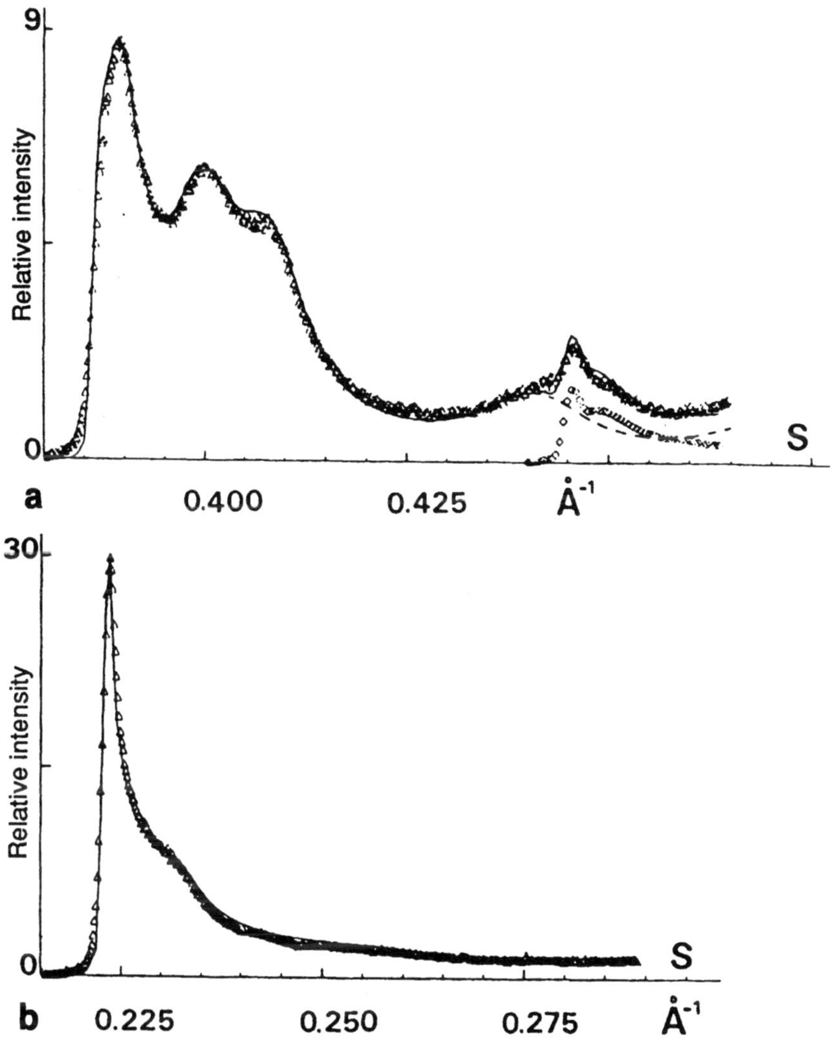

Fig. 8.27. **a** Agreement between the calculated (*full line*) and experimental (△) profiles of the (20, 13) and (04, 32) bands of sample E_1. **b** Agreement between calculated (*full line*) and experimental (△) profiles for the (02, 11) band of sample E_1. The triangles (△) correspond to the calculated (04, 22) profile and the dashed line (– – –) gives the contribution of the (20, 13) band in the range of the (04, 22) band.

in the structural model stacking faults corresponding to supplementary translations $\vec{T} = \pm\vec{b}/3$, with a probability $P_T = 0.30$. Another type of defect, namely random stacking faults with a probability $P_A = 0.70$, was also needed, so that the probability of finding in this hydrate two well-ordered adjacent layers (i.e., with a translation $\vec{t}_0$ between them) is nil. The consequence of the presence of the well-defined translations $\vec{T}$ is that, in each interlamellar space, the basic oxygens and the Na^+ cations situated on either side of the sheet of water molecules show an identical configuration with respect to the water molecules. The latter are therefore in a symmetrical and stable configuration which, otherwise, would not be possible.

The passage from a two-water-layer to a one-water-layer hydrate thus induces not only a modification of the number, the coordinates and the positions of the water molecules, but also a slip of the layers. This process is reversible and the supplementary $\vec{T}$ translations disappear on taking up a second layer of water (Ben Brahim 1985).

8.4 Structural Defects in Glauconites

8.4.1 Structure of the Glauconites

Glauconites are dioctahedral, microdivided, and iron-containing silicates. The globular aggregates of micron-sized particles generally range from a few tenths to some millimeters in size, and the minerals are formed in specific conditions, in recent or ancient oceanic media.

The idealized structure of glauconite may be described as a stacking of 2:1 layers with the same azimuthal orientation. The compensating cations are K^+ and the mutual orientation of adjacent layers is therefore analogous to the one in micas, i.e., corresponding to the 1 M polytype.

The composition of each half-unit cell of glauconite is

$$(Si_{4-x}Al_x)(R^{3+}_{2-y}R^{2+}_y)O_{10}(OH)_2 \cdot K_{x+y}$$

in which $x \geqslant 0.2$, R^{3+} is generally Fe^{3+} or Al^{3+} and R^{2+} represents Mg^{2+} or Fe^{2+}.

In most glauconites, the dominant trivalent ion is iron. However, more ancient (Precambrian) specimens of micas with the same globular morphology as glauconites have dominant Al^{3+} ions; we shall call these minerals Al-glauconites.

The diffractometric and crystallochemical data do not coincide with the idealized structural model described above (Odom 1984). It is therefore necessary to admit the existence of the following defects compatible with the actual structure of glauconites:

- defects connected with the formation of interstratification of layers with different chemical composition and/or structure;
- stacking faults due to the rotation of layers or fluctuation of their relative positions.

For many years, the structural disorder in glauconites was only attributed to interstratification, i.e., to a random stacking of mica-type and smectite layers in proportions which could vary widely in different samples. This was sufficient to explain *qualitatively* the different diffractograms assumed to be given by structures evolving gradually from a 1 M to a 1 Md polytype. More recently, it was proved that some samples of glauconites give diffractograms with important differences in widths and relative intensities of *hkl* reflections in spite of the fact that the minerals only contain very small proportions (less than 5%) of smectite layers (the principal papers giving accounts of these concepts are reviewed by Odom 1984). It is therefore highly likely that the defects correspond to stacking faults of identical layers. Nevertheless, only in the last years were the exact nature of the defects determined and their distribution in the stacks ascertained, by comparing the experimental diffractograms with calculated XRD patterns (Sakharov et al. 1987). These results form the basis of Section 8.4, in which we shall examine the structural models with different types of stacking faults compatible with the crystallochemical characteristics of glauconites before describing the latest results obtained by the modelization method, and end by the updated structure of glauconites.

8.4.2 Choicc of the Samples, Experimental Conditions and Description of the Experimental Diffractograms

Among an extensive set of glauconite samples studied by Nicolaeva (1977) and Kameneva (1985), Sakharov et al. (1987) chose five minerals (whose chemical composition are given in Table 8.13) with a very low proportion of smectite (swelling) layers, differing from each other by their Fe^{3+}/Al^{3+} ratio and showing wide variations of their band profiles.

The XRD intensities were measured with an automatic powder diffractometer equipped with a horizontal goniometer and a vertical object-plane, using Cu Kα radiation and a flat graphite monochromator. Intensities were measured every 0.01° as number of counts registered over a period of 100–200 s. Checks were made to verify that the width of the X-ray beam and of the slits did not induce convolution effects on the reflections. All the samples were run under the same experimental conditions to avoid spurious artefacts resulting from changes in the diffractometer.

Table 8.13. Cristallochemical formulas of glauconites studied

Samples	
60	$K_{0.74}Na_{0.01}Ca_{0.06}[Si_{3.54}Al_{0.46}][Al_{1.24}Fe^{3+}_{0.24}Fe^{2+}_{0.12}Mg_{0.44}]O_{10}(OH)_2$
223	$K_{0.74}Na_{0.01}Ca_{0.02}[Si_{3.44}Al_{0.56}][Al_{1.27}Fe^{3+}_{0.40}Fe^{2+}_{0.13}Mg_{0.24}]O_{10}(OH)_2$
B.P.	$K_{0.74}Na_{0.01}Ca_{0.03}[Si_{3.46}Al_{0.54}][Al_{1.11}Fe^{3+}_{0.41}Fe^{2+}_{0.13}Mg_{0.42}]O_{10}(OH)_2$
314	$K_{0.59}Na_{0.02}Ca_{0.03}[Si_{3.63}Al_{0.37}][Al_{0.90}Fe^{3+}_{0.67}Fe^{2+}_{0.13}Mg_{0.37}]O_{10}(OH)_2$
402	$K_{0.71}Na_{0.01}Ca_{0.04}[Si_{3.78}Al_{0.22}][Al_{0.53}Fe^{3+}_{0.75}Fe^{2+}_{0.31}Mg_{0.48}]O_{10}(OH)_2$

The globular samples were not subjected to grinding to obtain disoriented preparations. A special specimen holder was prepared by covering the 5 mm-wide holder window by 0.1 mm thick beryllium foil so that the goniometer axis coincided with the inner surface of the foil.

The four experimental XRD patterns are shown in Fig. 8.28. The diffracting domains correspond to the $00l$, $02l$, $11l$, $20l$, and $13l$ reflections. Depending on the specimen, there are modifications of the $02l$ and $11l$ reflections whereas the $20l$ and $13l$ reflections remain practically unchanged.

8.4.3 Determination of the Unit Cell Parameters and of the Atomic Coordinates

As the position of the reflections in the diffractograms of glauconites are very close to those given by 1M micas, indexation is very easy. However, since the effect of the presence of faults on the position of the reflections from 1M polytypes has not been examined so far, it is impossible to predict the inaccuracy attached to the use of traditional methods of determination of the unit cell parameters from the positions of the reflections in the diffractograms. This led Sakharov et al. (1987) to use the following method to get at the parameters of the unit cell of glauconites:

- using $00l$ reflections with high values of $l\,(l \geqslant 5)$, $c \sin \beta$ is obtained from $c \sin \beta = l d_{00l}$;
- the 006 reflection (upon which the reflection $\bar{3}31$ is superposed) yields parameter b;
- from the d-spacings corresponding to the reflections $\bar{1}12$ and 112, which can be found on practically all the diffractograms of glauconites, the parameter a and the angle β are obtained by solving the simultaneous equations

$$\frac{1}{d_{\bar{1}12}^2} = \frac{1}{(a \sin \beta)^2} + \frac{1}{b^2} + \frac{4}{(c \sin \beta)^2} + \frac{4 \cos \beta}{a c \sin^2 \beta} \tag{8.9}$$

$$\frac{1}{d_{112}^2} = \frac{1}{(a \sin \beta)^2} + \frac{1}{b^2} + \frac{4}{(c \sin \beta)^2} - \frac{4 \cos \beta}{a c \sin^2 \beta} \;. \tag{8.10}$$

These values of the unit cell parameters are first approximations which are modified during the subsequent calculations to yield the best agreement between the experimental and calculated XRD patterns. Experience shows that most benefit is obtained by adjusting the parameter $t = c \cos \beta$. The final values of the unit cell parameters are found in Table 8.14.

Once the chemical composition and the unit cell parameters are known, the regression formulas of Tsipursky and Drits (1984) lead to atomic coordinates closer to the actual values than to those deduced from the ideal structure. This step is particularly necessary when the structure under study only contains a small con-

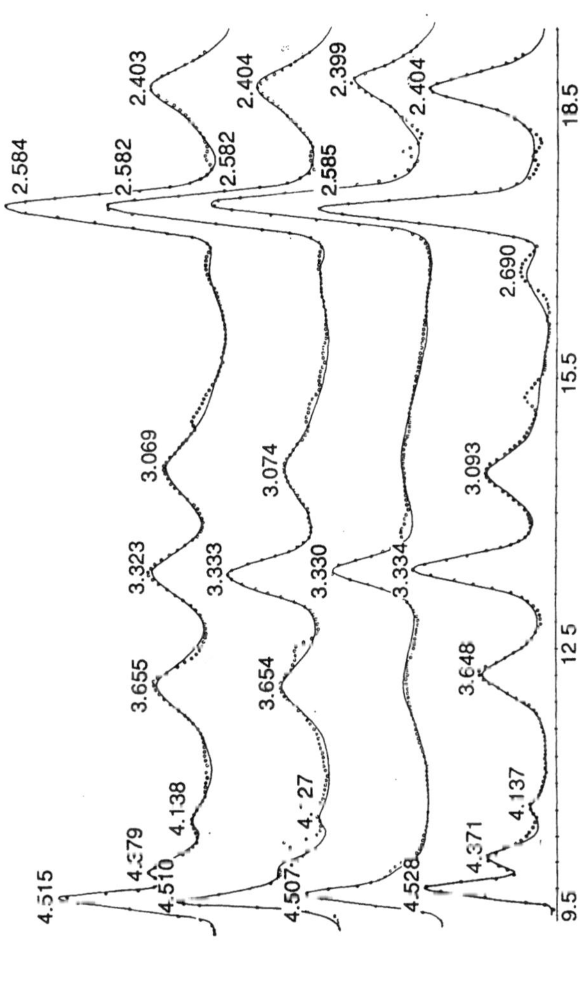

Fig. 8.28. Comparison between the experimental (○) and calculated (*full line*) X-ray diffraction patterns for glauconites with different degrees of structural order. From top: sample 60, 223, 314, 402

Table 8.14. Characteristic structural parameters of glauconites

Fitting parameters \ Samples	60	223	B.P.	314	402
a	5.235	5.228	5.218	5.220	5.235
b	9.038	9.046	9.038	9.041	9.082
$c \sin \beta$	9.96	9.99	9.98	9.99	10.00
$t_{kk} = c \cos \beta / a$	−0.380	−0.380	−0.384	−0.380	−0.366
α(M)	3−10	5−13	3−14	6−15	5−15
D(Å)	240	140−350	160	240	240
W_i	0.167	0.167	0.167	0.167	0.167
P_{ii}	0.59	0.48	0.55	0.30	0.65
trans sites	Vacant	Vacant	0.06	Vacant	0.20
α	5°	7°	5°	5°	4°
R(%)	4.2	5.3	7.5	6.1	5.9

centration of defects since the relative intensities are then highly sensitive even to small modifications of the structure factor. Thus, to characterize 1 M dioctahedral micas with *trans*-vacant octahedra, Tsipursky and Drits (1984) found that the determination of the following parameters was sufficient for the first approximation:

d_o, d_t, the mean cation-anion bond lengths respectively in octahedra and tetrahedra;
t_r, the mean length of the roof edges in *cis*-octahedra;
t_n and t_m, the mean lengths of the edges shared by two *cis*-octahedra and by a *cis*- and a *trans*-octahedron, respectively;
h_o, h_t, the mean thicknesses of the octahedral and tetrahedral sheets, respectively;
ε, the rotation angle of the upper and lower oxygen triads of the octahedra;
ψ, the octahedral flattening angle;
l_b, the mean basal tetrahedral edge length;
l_a, the mean apical tetrahedral edge length;
α, the tetrahedral rotation angle;
b_t, the periodicity of the tetrahedral sheet along y when $\alpha = 0$;
r(K−O), the mean distance between K and the surrounding oxygen atoms in the interlayer;
η, the interlayer thickness.

To determine these parameters, they made use of the following relationships:

$$d_o = \sum_i C_i d_i(R_i - O);$$

$$d_t = 1.61\,x + 1.75\,(1-x) \quad \text{for cation composition } (Si_x Al_{1-x});$$

$$t_n = 0.86\,t_m = (4d_o^2 - b^2/9)^{0.5};$$

$$4\sqrt{2}\,d_o = 2t_r + t_n + t_m;$$

$$b = 2\sqrt{3}\,t_r \cos(30° - \varepsilon);$$

$$d_o^2 = \frac{t_r^2}{3} + \frac{h_o^2}{4} = \frac{t_r^2 \cos^2(30° + \varepsilon)}{3} + \frac{t_m^2}{4};$$

$$\sin\psi = \frac{t_r}{\sqrt{3}\,d_o};$$

$$2\sqrt{2/3}\,d_t = \frac{l_a + l_b}{2} = 1.01\,l_b;$$

$$b_t = 2\sqrt{3}\,l_b;$$

$$\alpha = \arccos(b/b_t);$$

$$\eta = c \cdot \sin\beta - 2h_t - h_o;$$

$$r(K-O) = \left[\frac{\eta^2}{4} + \frac{b^2}{36}(\sqrt{3} - \tan\alpha)^2\right]^{0.5}$$

where C_i is the relative content of the cation R_i in the *cis*-octahedron, with a length d_i for the (R_i-O) bond. It was assumed that the mean bond lengths R−(O, OH) were 1.93, 1.99, 2.08 and 2.12 Å for Al, Fe^{3+}, Mg and Fe^{2+}, respectively. Most of the formulas listed above were known (Donnay et al. 1964; Radoslovich and Norrish 1964; Drits 1969, 1975) and the relations between t_n, t_m and l_b, l_a were obtained empirically from comparison of results from structure refinements of dioctahedral (Al, Mg)-micas. Table 8.15 (column 2) shows that the basic features of the real (Al, Mg)-mica structures could be satisfactorily predicted with the help of the formulas listed above. Similarly Sakharov et al. (1987) used these formulas to determine the bond lengths and, subsequently, the atomic coordinates of glauconites with vacant *trans*-octahedra. Table 8.16 gives the atomic coordinates used by these authors in their calculations.

8.4.4 Structural Models for Glauconites Devoid of Stacking Defects

Starting from ideal structural models of glauconites with different chemical compositions, unit cell parameters, and distribution of cations in the octahedral sites, theoretical powder patterns were calculated. Figure 8.29 presents two of these synthetic diagrams for 1M glauconites: the chemical composition of one of the models corresponds to the Bolshoi Patom (B.P.) Al glauconite, whereas the other corresponds to the iron-containing glauconite (sample 402). The chemical compositions, unit cell parameters and atomic coordinates used in the calculation of the diagrams are given respectively in Tables 8.13, 8.14 and 8.16.

Table 8.15. Main structural parameters of dioctahedral micas. (1): Refined structure. (2): Model

	Phengite 1 M (Tsipursky and Drits 1977)		Muscovite 2 M_1 (Güven 1971)		Muscovite 1 M (Zvyagin et al. 1979)		Phengite 2 M_1 (Güven 1971)		Paragonite 2 M_1 (Radoslovich and Burnham 1964)	
	1	2	1	2	1	2	1	2	1	2
d_t	1.626	1.63	1.643	1.64	1.624	1.62	1.622	1.62	1.651	1.65
l_a	2.677	2.69	2.711	2.70	2.671	2.67	2.686	2.67	2.713	2.72
l_b	2.631	2.63	2.657	2.65	2.623	2.62	2.625	2.62	2.680	2.67
$\alpha°$	8.63	8.7	11.7	11.0	9.33	9.4	6.0	5.1		15.6
h_t	2.207	2.22	2.245	2.24	2.196	2.21	2.22	2.22		2.26
d_o	1.947	1.95	1.930	1.93	1.939	1.94	1.956	1.95	1.913	1.91
t_r	2.808	2.81	2.810	2.82	2.815	2.80	2.842	2.84		2.78
t_{η}	2.481	2.48	2.427	2.42	2.469	2.48	2.497	2.48	2.417	2.40
t_m	2.932	2.91	2.881	2.85	2.879	2.88	2.885	2.88	2.807	2.82
h_o	2.157	2.16	2.087	2.07	2.113	2.14	2.12	2.11		2.07
$\psi°$	56.4	56.3	57.2	57.5	56.95	56.4	57.0	57.1		57.2
$\varepsilon°$	7.76	7.6	7.84	7.2	6.9	7.3	6.6	6.7		7.6
η	3.395	3.37	3.393	3.42	3.399	3.36	3.36	3.37		3.06
r(K–O)	2.917	2.90	2.857	2.87	2.897	2.87	2.97	2.98	2.641	2.64

Table 8.16. Atomic coordinates of the unit cells of two glauconites

Atoms	Sample 402			Sample B.P.		
	x	y	z	x	y	z
O	0.638	0.319	0.441	0.633	0.318	0.432
O	0.638	0.681	0.441	0.633	0.682	0.432
O	0.993	0.319	0.221	0.997	0.318	0.219
O	0.993	0.681	0.221	0.997	0.682	0.219
O	0.379	0.260	0.665	0.371	0.262	0.658
O	0.879	0.240	0.665	0.871	0.238	0.658
O	0.251	0.260	−0.004	0.259	0.262	−0.006
O	0.751	0.240	−0.004	0.759	0.238	−0.006
O	0.971	0.500	0.007	0.974	0.500	0.006
O	0.660	0.500	0.654	0.657	0.500	0.646
OH	0.688	0.000	0.437	0.677	0.000	0.431
OH	0.943	0.000	0.224	0.953	0.000	0.221
Si, Al	0.639	0.330	0.602	0.633	0.330	0.594
Si, Al	0.639	0.669	0.602	0.633	0.670	0.594
Si, Al	0.991	0.330	0.059	0.997	0.330	0.058
Si, Al	0.991	0.669	0.059	0.997	0.670	0.058
Al, Fe, Mg	0.315	0.333	0.331	0.315	0.333	0.326
Al, Fe, Mg	0.315	0.667	0.331	0.315	0.667	0.326
Al, Fe, Mg	0.315	0.000	0.331	0.315	0.000	0.326
K	0.631	0.000	0.831	0.630	0.000	0.826

As with smectites, the general shape of the diagram does not depend much on the chemical composition of the samples and the most important factor, as far as the relative intensities of the reflections are concerned, is the nature of the octahedral sites occupied by the cations. In Fig. 8.29, corresponding to a sample with *trans*-vacant sites, the most intense reflections are the $\bar{1}11$, $\bar{1}12$, and 112.

Kameneva (1985) has calculated the diffraction spectra corresponding to all the possible defect-free structural variants of glauconites compatible with the crystallochemistry of micas. This systematic study showed that none of the models considered gave a synthetic diagram analogous to any of the experimental spectra, either in relative intensities, or profiles, or bandwidths, and resolution. The conclusion of the author was that glauconites are systems in which the vacant octahedral sites of mica-type layers are preferentially *trans* but that glauconites corresponding to the true 1M polytype, i.e., with no stacking defects, probably do not exist naturally.

It is nevertheless possible to reject completely the possibility of existence of translation defects in the layers for the following reasons. In micas which do not contain swelling layers and which have an important number of interlamellar cations (as in glauconites), the probability of finding supplementary well-defined translation faults is nil since the presence of these defects would automatically modify some interlamellar spaces in such a way that the ditrigonal cavities on either side would no longer face each other, and the environment which is normal-

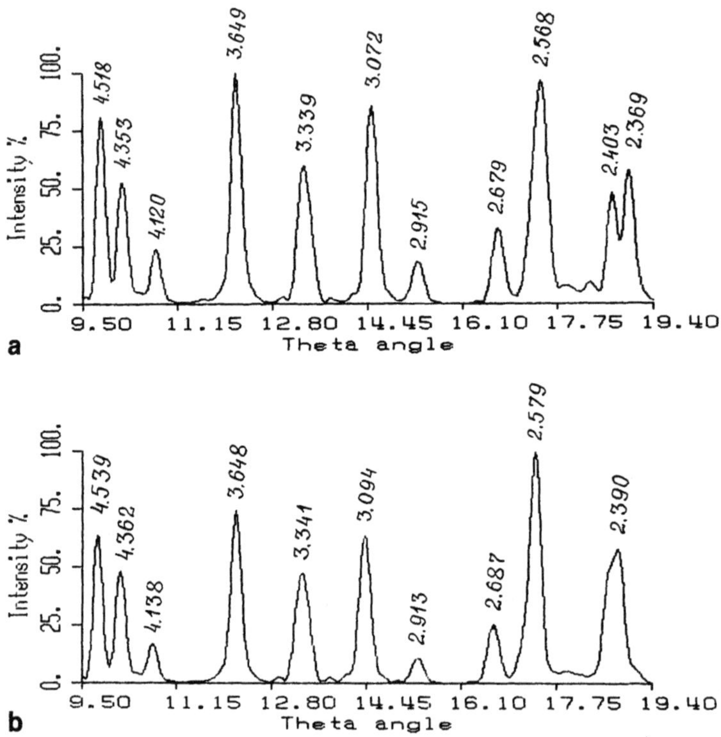

Fig. 8.29. Calculated X-ray curves for defect-free glauconite models; **a** model with $a = 5.218$ Å, $b = 9.038$ Å, $c \sin\beta = 9.98$ Å, $\gamma = 90°$, $t_0 = -0.392\,a$, $\bar{M} = 15$, and chemical composition of Bolshoi Patom glauconite; **b**: model with $a = 5.235$ Å, $b = 9.082$ Å, $c \sin\beta = 10.00$ Å, $t_0 = -0.366\,a$, $\bar{M} = 15$ and chemical composition of sample 402

ly found around the K^+ cations would not be obtained. Similarly, the possible existence of arbitrary translation defects among the layers can be excluded, since the important background which their presence would create between the $20l$ and $13l$ reflections is not observed.

The only stacking defects which need be considered, compatible with the anionic environment of the interlamellar K^+ cation and the qualitative characteristics of the experimental diagrams, are rotation faults among the layers, or the presence of enantiomorphs. The effects of the presence of these categories of defects on the calculated diagrams will be considered in the following sections.

8.4.5 Models with Well-Defined ±120° Rotational Stacking Faults

In fault-free 1M micas, the tetrahedra which form the two tetrahedral sheets of a 2:1 layer are placed in such a way that the ditrigonal surface cavities on one side of a layer are rotated by 180° with respect to the ditrigonal cavities on the opposite side of the layer. As a consequence, each K^+ cation in the interlamellar space is in an octahedral anionic environment. On crystallochemical criteria, it can be argued that the most likely defect is a rotation by ±120°, as this preserves the octahedral environment of the K^+ cations. Figure 8.30 gives three diagrams calculated on this basis. The calculations were done using the chemical composition, unit cell parameters and atomic coordinates of defect-free B.P. glauconite. The proportions of layers turned by +120° and by −120° were taken to be the same ($W_{+120°} = W_{-120°}$) and it was assumed that there was no interaction between the defects (R = 0). The three diagrams differ by the total concentration $W_{\pm 120°}$ of rotation faults taken to be 20, 30 and 60% respectively. The comparison of the calculated spectra shows that:

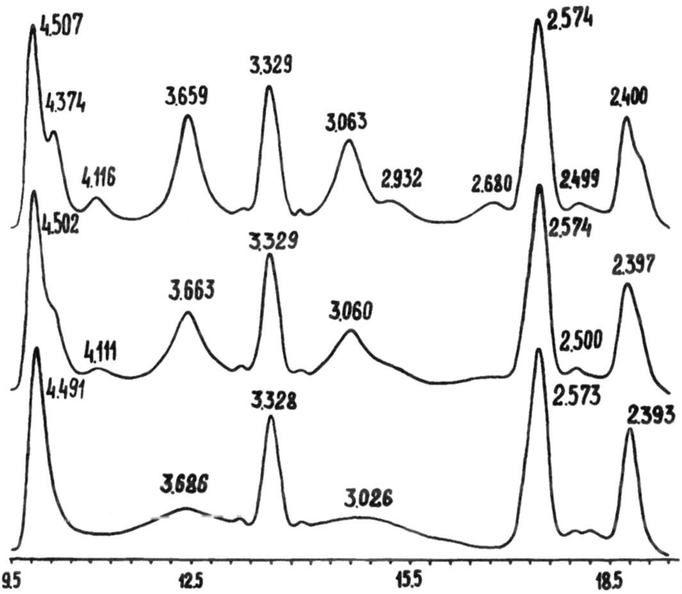

Fig. 8.30. Influence of ±120° rotation stacking faults on the modification of the profile of the calculated X-ray diffraction patterns for a model with parameters and chemical composition of the Bolshoi Patom sample, with proportion $W_{0°}$: $W_{+120°}$: $W_{-120°}$ of stacking faults equal to 0.8:0.1:0.1 (curve *a*), or 0.7:0.15:0.15 (curve *b*), or 0.4:0.3:0.3 (curve *c*)

- when the concentration of defects increases, the intensities of the reflections $11l$ and $02l$ decrease but their relative proportions are maintained. The nature of the vacant octahedral sites can therefore be determined even in disordered glauconites;
- since an increase of $W_{\pm 120°}$ displaces the maxima of the reflections (cf. Table 8.17), it is clear that traditional methods of determination of unit cell parameters, based on the exploitation of the position of reflections, must lead to errors when the structures of the glauconites are disordered;
- the presence of $\pm 120°$ rotation faults shows up by the appearance of a diffuse arc-like background in the region of the $02l$ and $11l$ reflections;
- for the structural models which have $b/a = \sqrt{3}$ and $t = c \cos \beta = -a/3$, the presence of $\pm 120°$ rotation faults does not affect the $(20l, 13l)$ reflection domain, but if $t \neq -a/3$ (as in the Al-glauconites such as the B.P. sample, in which $t = -0.392\,a$), calculations show that the second part of the $(20l, 13l)$ domain becomes narrower when $W_{\pm 120°}$ increases (cf. Fig. 8.30), owing to the fact that the reflections 201 and $\bar{1}32$ come closer to each other as the projections along $\vec{a}$ of the translation vectors $\vec{t}$ between adjacent layers become averaged.

In spite of this, the XRD curves calculated on the basis of a model with $\pm 120°$ rotation defects do not tally with the experimental diagrams and the model must be discarded, mainly because of the total lack of agreement between the calculated and experimental profiles in the $20l, 13l$ domain.

8.4.6 Structural Models with Enantiomorphic Layers

When the parameters of the 2:1 layers are such that $b/a \neq \sqrt{3}$, it is possible to describe the fault-free structure using identical layers with an angle $\gamma \neq 90°$ (see Sect. 1.6.3). In such a structure, stacking disorder consists in the formation of pairs of enantiomorphic layers connected by a glide plane. In other words, the structural defects constitute inclusions of fragments of $2M_1$ polytype in the 1M polytype. In such a model, the octahedral coordination of the K^+ cations is preserved in all interlamellar spaces, whether it be in the 1M polytype or in the $2M_1$ fragments.

The three diagrams in Fig. 8.31 were calculated on the basis of this model, respectively with 20, 30, and 40% concentration of "defective" layers (called type B), enantiomorphic with the initial (type A) layers and randomly distributed (R = 0), with the following crystallographic parameters for the A-layers: $a = 5.202$ Å; $b = 9.100$ Å; $c \sin \beta = 9.99$ Å; $b/a = \sqrt{3.06}$; $\gamma = 89°$ and $t = -0.365\,a$. The analysis of the three calculated spectra indicates that:

- as long as the concentration of B-layers is less than 20% ($W_B < 0.20$), the band profiles are analogous to the ones obtained with the model containing defects caused by $\pm 120°$ rotations of the layers;
- when the concentration of defects increases, the diagrams given by each category of model are well differentiated; for high concentrations of B-layers,

Table 8.17. Influence of the concentration of ±120° rotation faults of layers on the intensities (I) and positions of the reflections calculated on the basis of a structural model corresponding to the chemical composition of B.P. glauconite, with the parameters: $a = 5.218$ Å; $b = 9.038$ Å; $c \sin\beta = 9.98$ Å, $t_k = -0.392a$, $\gamma = 90°$, M = 14 layers

hkl	$W_0 : W_{-120°} : W_{+120°}$		$W_0 : W_{-120°} : W_{+120°}$		$W_0 : W_{-120°} : W_{+120°}$		$W_0 : W_{-120°} : W_{+120°}$		$W_0 : W_{-120°} : W_{+120°}$		$W_0 : W_{-120°} : W_{+120°}$	
	0.95 : 0.05 : 0.05		0.8 : 0.1 : 0.1		0.7 : 0.15 : 0.15		0.6 : 0.2 : 0.2		0.5 : 0.25 : 0.25		0.4 : 0.3 : 0.3 (a) 0.333 : 0.333 : 0.333 (b)	
	d (Å)	I	d (Å)	I	d (Å)	I	d Å)	I	d (Å)	I	d (Å)	I
110	4 506	77	4.497	80	4.489	82	4.485	84	4.483	86	4.482	87 (a)
											4.482	88 (b)
$\bar{1}11$	4.123	20	4.120	16	4.117	14						
$\bar{1}12$	3.657	18	3.659	54	3.662	40	3.667	32	3.675	27	3.685	25 (a)
											3.687	25 (b)
022	3.351	18	3.352	16	3.359	15						
112	3.065	69	3.062	47	3.058	34	3.050	27	3.034	24	3.017	23 (a)
											3.014	22 (b)
$\bar{1}13$	2.925	18	2.939	17								
023	2.678	25	2.676	17	2.3							
200, $\bar{1}31$	2.576	100	2.576	100	2.575	100	2.575	100	2.575	100	2.575	100 (a)
130, $\bar{2}01$											2.575	100 (b)
$\bar{2}02$, 131	2.461	18	2.462	18	2.463	18	2.463	19	2.464	19	2.464	19 (a)
											2.464	19 (b)
$\bar{1}32$, 201	2.399	48	2.396	50	2.394	52	2.393	54	2.393	55	2.393	56 (a)
											2.393	56 (b)

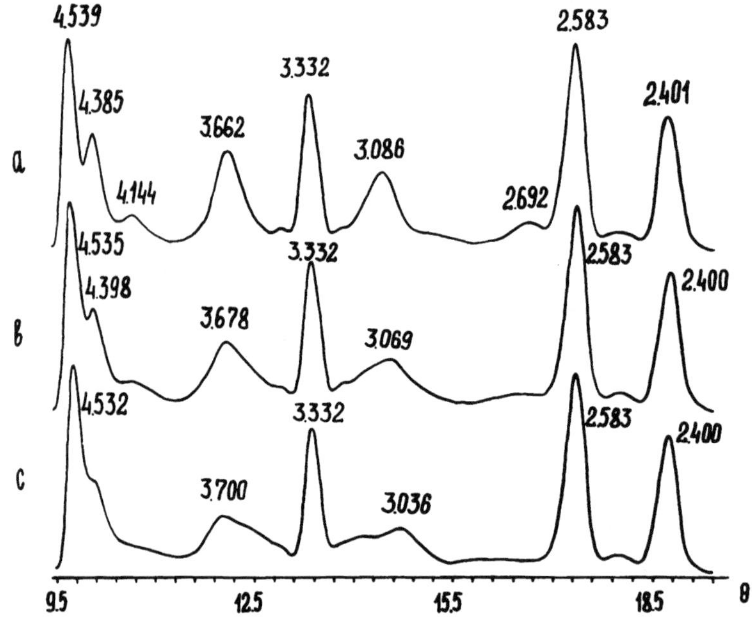

Fig. 8.31. Influence of the concentrations W_A, W_B of enantiomorphic layer pairs on the profile of the calculated X-ray diffraction pattern for a model with parameters and chemical composition sample 402 with $W_A : W_B = 0.8 : 0.2$ (curve *a*), or 0.7 : 0.3 (curve *b*), or 0.6 : 0.4 (curve *c*)

the $11l$ reflections become asymmetrical so that, for instance, the 112 reflection splits when $W_B = 0.40$; furthermore, the decrease in width of the 201 and $\bar{1}32$ reflections is not observed;

– the increase of W_B induces a large displacement of the maximums of the $02l$ and $11l$ reflections; the parameters obtained by standard methods from the positions of the maximums in the diagrams calculated with $W_B = 0.30$ are $a = 5.317$ Å; $b/a = \sqrt{2.93}$ and $t = -0.397\,a$, i.e., quite different from the values used in the initial calculations of the diagram.

Once again, this structural model assumed for glauconites gives synthetic diagrams incompatible with the experimental results and must therefore be rejected.

8.4.7 Structural Model with n60° Rotational Stacking Faults and R = 0

In opposition with the previous structural models involving rotations by ±120° or the presence of enantiomorphic layers, the present model, in which some layers are rotated by multiples of 60°(n60°, with n = integer), creates two types of coordination for the interlamellar K^+: unchanged octahedral coordination (for n = 0, 2 or 4) and prismatic coordination (for n = 1, 3 or 5). The prismatic coordination of K^+ does not exist in 1M micas but has been found in the $2M_2$ polytype. Even if this polytype is rather scarce in nature, its existence justifies the need to examine the present structural model.

The three XRD patterns shown in Fig. 8.32 were calculated with the following assumptions:

- all values of n are equiprobable;
- the concentration W_0 of fault-free layers (which have n = 0) is such that

$$W_0 > \sum_{n=1}^{5} W_{n60^\circ}$$

- the rotated layers are randomly dispersed in the stack (R = 0).

There is an important difference between the three diagrams in Fig. 8.32 and those in Figs. 8.29 to 8.31: an increase in the concentration of defects results in a broadening of the 201 and $\bar{1}32$ bands and a decrease of the relative intensity of this domain, so that the profiles now resemble those of the experimental tracings.

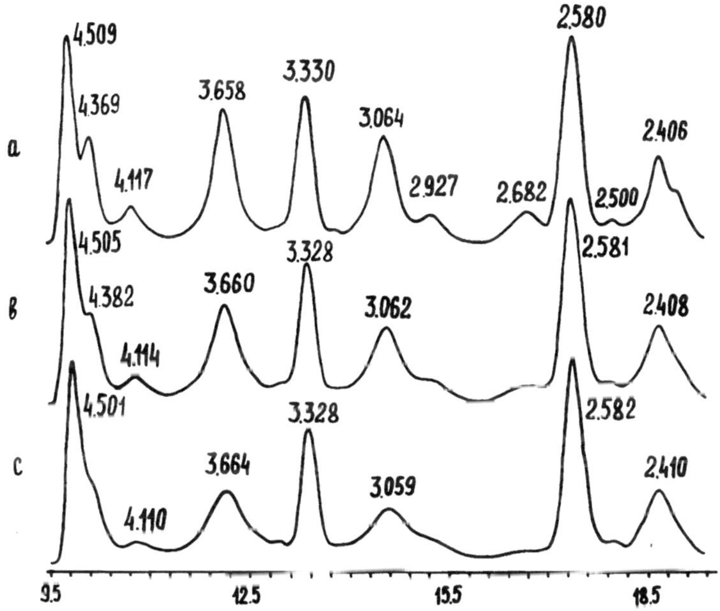

Fig. 8.32. Influence of n 60° rotation stacking faults on the profile of the calculated X-ray diffraction patterns for a model with the parameters and chemical composition of sample 402, and model parameters S = 0, W_{0°; W_{120°; ... W_{300° = 0.8 : 0.04 : 0.04 : ... : 0.04 (curve *a*), or 0.7 : 0.06 : 0.06 : ... : 0.06 (curve *b*), or 0.6 : 0.08 : ... : 0.08 (curve *c*)

Although the model with uncorrelated n 60° rotations of the layers does therefore give an evolution of the profiles in the 20*l*, 13*l* domain compatible with the experimental observations, it was abandoned, since all attempts to obtain a satisfactory agreement between the calculated and experimental curves in the other regions were unsuccessful.

8.4.8 Structural Model with n 60° Rotational Stacking Faults and R = 1

In this model, the concentrations W_i of layers in each of the six possible positions are taken to be equal ($W_i = 0.167$, with $0 \leqslant i \leqslant 5$) and it is also assumed that the layers in each of the orientations are distributed in the stack with a more or less important degree of segregation corresponding to an interaction parameter R = 1. The variable which characterizes the amount of segregation is the probability P_{ii} of finding two identical adjacent layers, i.e., layers rotated by the same angle with respect to a reference direction. The limiting values for this probability are $P_{ii} = 1$ (total segregation) and $P_{ii} = W_i$ (random distribution). All P_{ii} are taken to be equal and independent of the value of i.

The three curves in Fig. 8.33 correspond respectively to $P_{ii} = 0.7$, 0.4 and 0.167, and it can be seen that the calculated profiles in the 02*l*, 11*l* domain are analogous to the ones obtained with the model in which the defect was chosen to be ±120° rotation of layers, whereas the reflections in the 20*l*, 13*l* domain are analogous to the ones observed with the previous model (uncorrelated rotations by n 60°).

Since the qualitative comparison shows good similarity between the calculated and experimental XRD curves, the final step is the adjustment of the specific parameters of this model to obtain the best possible fit with the experimental diagrams.

8.4.9 Determination of the Structural Parameters Characteristic of Glauconites

Table 8.14 shows, for each glauconite, the values which give the best fit for the model in which the defects are rotations of layers by n 60° correlated by first-neighbor interaction (R = 1), and the results thus obtained are shown in Fig. 8.24. In Table 8.14, $\alpha(M)$ is the distribution function of the thicknesses, and the maximum and minimum values are given for each sample. D is the diameter (in Å) of the interferential coherence domain, assumed to be circular in samples 60, 314, 402, and B.P. For sample 223, it was found necessary to use a rectangular coherence domain, with sides parallel to $\vec{a}$ and $\vec{b}$ and respectively equal to 140 and 350 Å. The coefficient P_{ii} is the proportion of segregation of layers with identical orientations, and trans-site is the filling coefficient of *trans*-octahedral cavities by cations. Lastly, α is the angle of rotation of the tetrahedra.

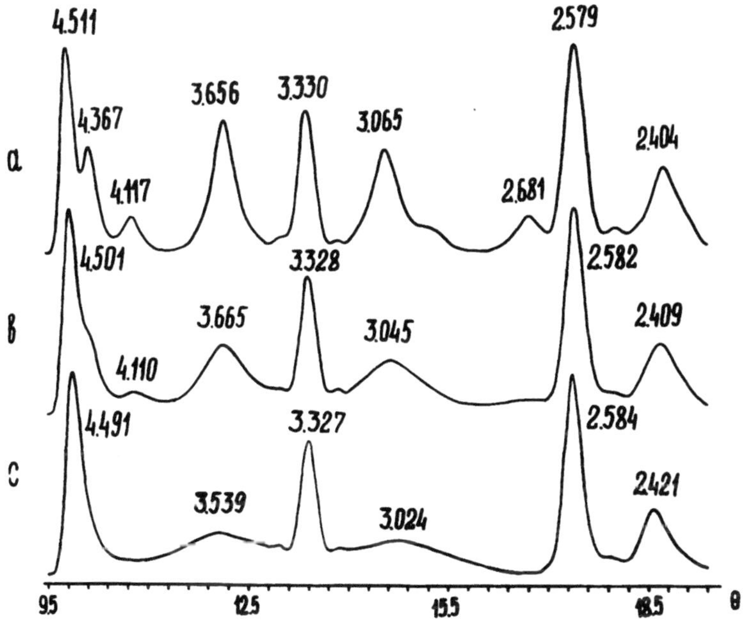

Fig. 8.33. Influence of n60° rotation stacking faults on the profile of the calculated X-ray diffraction pattern for a model with the parameters and chemical composition of the Bolshoi Patom sample, model parameters $S = 1$, $W_{0°} = W_{60°} = \ldots W_{300°} = 0.167$ and degree of structural disorder depending on the degree of segregation of the layers with the same azimuthal orientation characterized by the probability parameter P_{11} equal to 0.7 (curve *a*), or 0.4 (curve *b*), or 0.167 (curve *c*)

From Table 8.14 and Fig. 8.24, the following conclusions can be drawn:

- for all samples, excellent agreement was obtained between the experimental and calculated results, as shown by the reliability factor:

$$R = \frac{\sum |I_{calc.} - I_{exp.}|}{\sum I_{exp.}}$$

which never exceeds 8%.

- regardless of the degree of order or of the chemical composition of the glauconites, it was possible to describe all the samples in terms of the same model involving equally probable occurrence of layers rotated by multiples of 60°. The layer types alternate with a definite tendency to segregation, and the structural order is largely controlled by a single parameter, P_{ii}, which describes this degree of segregation for layers with the given orientation;
- the model is characteristic for all glauconites, irrespective of composition, age, and type of rock which contains them; the alternation of layers rotated by 60° in the structures leads to the presence of both octahedral and prismatic coor-

dination for the interlayer K^+ cations within the same microcrystal, and it is noteworthy that this alternation of different coordination for K^+ is observed in Al-rich globular micas with a relatively large angle of rotation of the tetrahedra;

- the influence of different types of stacking faults on the profiles and intensity distributions was found to be most evident on the (20*l*, 13*l*) bands.

One can deduce from these results and conclusions that, regardless of their age or rock type, globular micaceous minerals were formed by the same mechanism which, most probably, consisted in a solid phase transformation of smectites. In this case, the fixation of K was accompanied by inheriting the orientation of the initial smectite matrix with consequent identical stacking faults in all globular glauconites.

References

Adams JM, Hewat AW (1981) Hydrogen atom positions in dickite. Clays Clay Mineral 29:316–319

Alcover JF, Gatineau L (1980) Structure de l'espace interlamellaire des vermiculites Ba monocouches. Clay Mineral 15:25–35 and 193–203

Bailey SW (1963) Polymorphism of the kaolin minerals. Am Mineral 48:1186–1209

Bailey SW (1980) Structures of layer silicates. In: Brindley GW, Brown G (eds) Crystal structures of clay minerals and their X-ray identification. Mineral Soc, London, pp 1–124

Barshad I (1949) The nature of lattice expansion and its relation to hydration in montmorillonite and vermiculite. Am Mineral 34:675–684

Ben Brahim J (1985) Contribution à l'étude des systèmes eau-argile par diffraction des rayons X. Structure des couches insérées et mode d'empilement des feuillets dans les hydrates homogènes à une et deux couches d'eau de la beidellite-Na. Thèse de Doctorat, Orléans Univ, Fr

Ben Brahim J, Armagan N, Besson G, Tchoubar C (1983) X-ray diffraction studies on the arrangement of water molecules in a smectite: homogeneous two-water-layer Na-beidellite. J Appl Cryst 16:264–269

Ben Brahim J, Besson G, Tchoubar C (1984) Etude des profiles des bandes de diffraction X d'une beidellite-Na hydratée à deux couches d'eau. Détermination du mode d'empilement des feuillets et des sites occupés par l'eau. J Appl Cryst 17:179–188

Besson G (1980) Structure des smectites dioctaédriques – paramètres conditionnant les fautes d'empilement des feuillets. Thèse de Doctorat, Orléans Univ, Fr

Besson G, Tchoubar C (1972) Détermination du groupe de symétrie du feuillet élémentaire de beidellite. CR Acad Sci Paris 275:633–636

Besson G, Bookin AS, Dainyak LG, Rautureau M, Tsipursky SI, Tchoubar C, Drits VA (1983a) Use of diffraction and Mössbauer methods for the structural and crystallochemical characterization of nontronites. J Appl Crystallogr 16:374–383

Besson G, Glaeser R, Tchoubar C (1983b) Le césium, révélateur de structure des smectites. Clay Mineral 18:11–19

Besson G, de la Calle C, Rautureau M, Tchoubar C, Tsipursky SI, Drits VA (1982) X-Ray and electron diffraction study of the structure of the Garfield nontronite. Proc Int Clay Conf, Bologne, vol 1, pp 29–40

Blount AM, Threadgold IM, Bailey SW (1969) Refinement of the crystal structure of nacrite. Clays Clay Mineral 17:185–194

Bookin AS, Dainyak LG, Drits VA (1978) Interpretation of the Mössbauer spectra of layer silicates on the basis of the structural modelling. Phys Chem Miner 3:58–59

Bookin AS, Dainyak LG, Drits VA (1979) Interpretation of the Mössbauer spectre of Fe^{3+}-containing layer silicate on the basis of structural modelling (in Russian). In: Lapides LI (ed) Cation ordering in structure of minerals. Nauka, Novosibirsk

Bookin AS, Drits VA, Plançon A, Tchoubar C (1989a) Stacking faults in kaolin minerals in the light of real structural features. Clay Mineral 37:297–307

Bookin AS, Drits VA, Tcherkashin VI, Salyn AL (1989b) Comparison of kaolinite and dickite unit cell parameters (in Russian). Mineral J 11:13–20

Boutouzova G Yu, Drits VA, Lisitsina NA, Tsipursky SY (1979) The dynamics of the formation of clay minerals in ore-bearing sediments of the Atlantis II (Red-Sea) (in Russian). Lithol Polesn Iskop 1:20–42

Bradley WF, Serratosa JM (1960) A discussion of water content of vermiculite. Clays Clay Mineral 7:260–270

Brindley GW (1980) Chap. 2: Order-disorder in clay mineral structures. In: Brindley GW, Brown G (eds) Crystal structures of clay minerals and their X-ray identification. Mineral Soc, London, 125–196

Brindley GW, Kao Chich-Chun, Harrison JL, Lipsicas M, Raythatha R (1986) Relation between structural disorder and other characteristics of kaolinites and dickites. Clays Clay Mineral 34:239–249

Brindley GW, Robinson K (1946) Randomness in the structures of kaolinitic clay minerals. Trans Faraday Soc 42B:198–205

De la Calle C (1977) Structure des vermiculites. Facteurs conditionnant les mouvements des feuillets. Thèse de Doctorat, Univ Paris IV, Fr

Cardille CM, Johnson JH, Dickson DPE (1986) Magnetic ordering at 4.2 and 1.3 K in nontronites of different iron contents: a ^{57}Fe Mössbauer spectroscopic study. Clays Clay Mineral 34:233–238

Dainyak LG (1980) The interpretation of the Mössbauer spectra of some Fe^{3+}-containing silicates on the basis of structural modelling. Ph D Thesis, Geol Inst Academy of Sci, Moscow

Deluca S, Slaughter M (1985) Existence of multiple kaolinite phases and their relationship to disorder in kaolin minerals. Amer Miner 70:149–158

Donnay G, Donnay JDH, Takeda H (1964) Trioctahedral one-layer micas. II. Prediction of the structure from composition and cell dimension. Acta Cryst 17:1374–1377

Drits VA (1969) Some general remarks on the structure of trioctahedral micas. Proc Intern Clay Conf, Tokyo, 1:51–59

Drits VA (1971) Crystallochemical features of trioctahedral mica (in Russian). In: Kossovskaya AG (ed) Epigenesis and mineral indicators. Nauka, Moscow, pp 96–110

Drits VA (1975) Structure and crystallochemical features of layer silicates (in Russian). In: Kossovskaya AG (ed) Crystallochemistry of minerals and geological problems. Nauka, Moscow, pp 35–52

Drits VA (1987) Mixed-layer minerals: diffraction methods and structural features. Proc Intern Clay Conf, Denver, 1:33–45

Drits VA, Kashaev A (1960) An X-ray study of a kaolinite single crystal (in Russian). Kristallografia 5:207–210

Drits VA, Kossovskaya AG (1980) Geological crystal chemistry of rock-forming dioctahedral smectites (in Russian). Lithol Polesn Iskop 1:84–112

Giese RF (1982) Theoretical studies of the kaolin minerals: electrostatic calculations. Bull Mineral 105:417–424

Glaeser R, Méring J (1954) Isothermes d'hydratation des montmorillonites biioniques (Na, Ca). Clay Miner Bull 2:188–190

Goodman BA, Russell JD, Fraser AR, Woodhams FWD (1976) A Mössbauer and I.R. spectroscopic study of the structure of nontronite. Clays Clay Mineral 24:53–59

Goodyear B, Duffin MA (1961) An X-ray examination of an exceptionally well crystallized kaolinite. Mineral Mag 32:902–907

Grim R (1968) Clay mineralogy. McGraw-Hill Book Company, New York (ed), 506 pp

Güven N (1971) The crystal structure of 2M-phengite and 2M-muscovite. Z Krist 134:487–490

Güven N, Pease RW, Murr LE (1977) Fine structure in the selected area diffraction patterns of beidellite. Clay Mineral 12:67–74

Heller-Kallai L, Rozenson I (1980) Dehydroxylation of dioctahedral phyllosilicates. Clays Clay Miner 28:355–368

Joswig W, Drits VA (1986) The orientation of the hydroxyl groups in dickite by X-ray diffraction. N Jb Miner Mh 1:19–22

Kameneva M Yu (1985) Crystallochemical features of glauconite minerals. Ph D Inst Geology and Geophysics. Academy of Sci of USSR, Novosibirsk

Mamy J, Gaultier JP (1976) Les phénomènes de diffraction des rayonnements X et électroniques par les réseaux atomiques. Application à l'étude de l'ordre cristallin dans les minéraux argileux. Evolution structurelle de la montmorillonite associée au phénomène de fixation irréversible du potassium. Anales Agro 27-I:1–16

Mathieson AM, Walker GF (1954) Crystal structure of magnesium vermiculite. Amer Mineral 39:231–235

Méring J, Glaeser R (1954) Sur le rôle de la valeur des cations échangeables dans la montmorillonite. Bull Soc Fr Miner Crist 77:519–522

Méring J, Oberlin A (1971) Chap 6: The smectites. In: Gard JA (ed) The electron-optical investigation of clays. Mineral Soc, London, pp 193–229

Mitra GB, Bhattacherjee S (1970) X-ray diffraction studies on the transformation of kaolinite into metakaolin: study of layer shift. Acta Cryst B26:2124–2128

Murray HH (1954) Structural variations of some kaolinites in relation to dehydrated halloysite. Amer Mineral 39:97–108

Newnham RE (1961) A refinement of the dickite structure and some remarks on polymorphism in kaolin minerals. Miner Mag 32:683–704

Nicolaeva IV (1977) Glauconite minerals in sediment formations. In: Yanshin AL (ed) Sediment formations (in Russian). Nauka, Moscow, 321 pp

Noble FR (1971) A study of disorder in kaolinite. Clay Mineral 9:71–80

Odom IE (1984) Glauconite and celadonite minerals. In: Bailey SW (ed) Review in Mineralogy, Mineral Soc, London, 13:545–572

Osthaus BB (1954) Chemical determination of tetrahedral ions in nontronite and montmorillonite. Clays Clay Mineral, NRC-NAS Pub. no 327:404–416

Plançon A, Giese RF, Snyder R, Bookin AS, Drits VA (1989a) Stacking faults in the kaolin minerals. I. General review. Clays Clay Mineral 37(3):203–210

Plançon A, Giese RF, Snyder R (1989b) The Hinckley index for kaolinite. Clays Clay Mineral 24:249–260

Plançon A, Tchoubar C (1975) Etude des fautes d'empilement dans les kaolinites partiellement désordonnées. I. Modèle ne comportant que des fautes par translation. J Appl Cryst 8:582–588

Plançon A, Tchoubar C (1976) Etude des fautes d'empilement dans les kaolinites partiellement désordonées. II. Modèle comportant des fautes par rotation. J Appl Cryst 9:279–285

Plançon A, Tchoubar C (1977a) Determination of structural defects in phyllosilicates by X-ray powder diffraction. I. Principle of calculation of the diffraction phenomena. Clays Clay Mineral 25:430–435

Plançon A, Tchoubar C (1977b) Determination of structural defects in phyllosilicates by X-ray diffraction. II. Nature and proportion of defects in natural kaolinites. Clays Clay Miner 25:436–450

Pons CH, Tchoubar C, Tchoubar D (1980) Organisation des molécules d'eau à la surface des feuillets dans un gel de montmorillonite-Na. Bull Mineral 103:452–456

Radoslovich EW, Burnham CW (1964) Crystal structure of coexisting muscovite $2M_1$ and paragonite $2M_1$. Carnegie Inst Year Book 63:232–234

Radolovich EW, Norrish K (1964) The cell dimensions and symmetry of layer-lattice silicates. I. Some structural considerations. Amer Mineral 47:599–616

Rozenson I, Heller-Kallai L (1977) Mössbauer spectra of dioctahedral smectites. Clays Clay Mineral 25:94–101

Russel JD, Goodman BA, Fraser AR (1979) Infra-red and Mössbauer studies of reduced nontronites. Clays Clay Mineral 27:63–71

Sakharov BA, Besson G, Kameneva M Yu, Smolyar BB, Drits VA (1987) Studies of stacking faults nature in the structure of Fe^{3+}-containing mica by X-ray analysis. In: Book of summaries; Proc of the 6th meeting of the Europ Clay Group, Sevilla, Spain

Sakharov BA, Besson G, Kameneva M Yu, Smolyar BB, Drits VA (1989) Studies of stacking faults in glauconite structures. Clay Mienral (submitted)

Samotoin ND (1966) Study of surfaces of kaolinite and dickite monocrystals by decoration method (in Russian). Zapisky Usesojusn Miner Obschest 95:390–399

Samotoin ND, Chekin SS, Finko VI (1980) Spiral growth and polytypism of kaolin group minerals. Proc VIth Intern Congr Crystal Growth, Moscow, 4:230–236

Shirozu H, Bailey SW (1966) Chlorite polytypism: III. Crystal structure of an orthohexagonal iron chlorite. Amer Mineral 51:1124–1143

Suitch PR, Young RA (1983) Atom position in highly ordered kaolinite. Clays Clay Miner 31:357–366

Tchoubar C, Plançon A, Ben Brahim J, Clinard C, Sow C (1982) Caractéristiques structurales des kaolinites désordonnées. Bull Mineral 105:477–491

Tsipursky SI, Drits VA (1977) Effectiveness of the electronometric method of measuring the intensities at the electron diffraction structural studies (in Russian). Izv Akad Nauk SSSR, Phys Ser, 2263–2271

Tsipursky SI, Drits VA (1984) The distribution of octahedral cations in the 2:1 layers of dioctahedral smectites studied by oblique texture electron diffraction. Clay Miner 19:177–193

Tsipursky SI, Drits VA, Chekin SS (1978) Revealing of the structural ordering of nontronites by oblique texture electron diffraction (in Russian). Izv Akad Nauk SSSR, Ser Geol 10:105–113

Zvyagin BB (1960) Electrodiffraction determination of the structure of kaolinite (in Russian). Soviet Phys Crystallogr 5:32–42

Zvyagin BB (1967) Electron diffraction analysis of clay mineral structure (translated from Russian). Plenum Press, New York, 364 pp

Zvyagin BB, Vrublevskaya ZV, Zhoukhlistov AP, Sidorenko SV, Fedotov AF (1979) In: Drits VA (ed) High voltage electron diffraction study of layer minerals. Nauka, Moscow, 215 pp

CHAPTER 9

Determination of the Structural Characteristics of Mixed-Layer Minerals

Mixed-layer minerals (MLM) occur quite frequently in different geological surroundings. They are met in soils, in weathering crusts, in hydrothermal formations, in modern sediments, in rock altered by post-sedimentary transformations etc. They have been used as indicators of the dynamics of the physicochemical processes of various geological environments, and the study of these minerals is important for the comprehension of the mechanism of structural phase transitions: for example, the well-known solid-phase transformation of phlogopite into vermiculite occurs through a series of intermediary interstratified phases. On the contrary, the present knowledge of the structural mechanism of the transformation of dioctahedral smectites to illites by diagenesis is still incomplete. These few examples show the necessity of being able to identify without ambiguity mixed layer minerals and the need to determine at least their main structural characteristics.

It was shown in Chapter 5 that the determination of the structural characteristics of MLM required the development of specific direct and indirect methods of analysis of the XRD patterns. In practice, the study of the diffractograms given by MLM is limited to the analysis of the basal reflections since their position, profile, and intensity constitute the criteria which are most sensitive to the nature of the layers, their chemical composition and internal structure, as well as their sequence in the stack. All these parameters are specific characteristics needed to obtain the actual structure of the MLM.

As was done in the preceding chapter, we shall not attempt to present here all the practical and methodological approaches described in the literature, nor shall we try to give all the structural characteristics of all the known MLM. We shall first detail the possibilities, limits, and practical use of the direct method of analysis of D'yakonov, especially as this useful method is practically unknown, but the largest part of this chapter will deal with the methodological difficulties associated with the interpretation of XRD patterns using indirect methods. The approach is identical to the one given in Chapter 8, i.e., we shall strive to determine all the factors likely to influence the characteristics of the basal reflections.

9.1 The Method of D'yakonov

It will be recalled that the direct method of analysis of the basal reflections of D'yakonov first requires the computation of the function $\varphi'(z)$ from the relationship in Eq. (5.6):

$$\varphi'(z) = \sum_{K=1}^{n} I_K \cos(2\pi Z_K z)$$

in which I_K is the integrated intensity of the basal reflection K found at the distance Z_K of the origin in reciprocal space and corresponding to the interlayer distance d_K through $Z_K = 1/d_K$. This function has principal maxima (PM) at values of z equal to the distances between the layers, and the amplitude of these maxima increases when the frequency of occurrence of the corresponding distances itself increases in the stack. Let us, for instance, suppose that layers of type A and B, respectively of thickness d_A and d_B, occur in a stack. The probability of finding a substack composed of m type-A and n type-B layers is

$$(W_{A^m B^n})_{exp} = \frac{h_{A^m B^n}}{H_s} \quad \text{with} \quad m, n = 0, 1, 2 \ldots ,$$

where $h_{A^m B^n}$ is the amplitude of the PM of the function $\varphi'(z)$ for $z = m d_A + n d_B$ and where

$$H_s = \sum_{m=0}^{q} h_{A^m B^{(q-m)}} \tag{9.1}$$

with $q = m+n$.

The set of probalities $(W_{A^m B^n})_{exp}$ corresponding to the different values of q contains, at least in principle, all the information needed to determine the probability parameters which characterize the interstratified structure composed of A and B layers. However, one of the major difficulties in this analysis lies in the fact that each PM is generally surrounded by secondary peaks reflecting the inner structure of the layers, with relative amplitudes proportional to that of the PM beside which they lie, and which may be superposed on other principal maxima, thus giving the latter an overlarge apparent intensity. D'yakonov (1978) suggested that their contribution to the amplitude of "composite" peaks could be eliminated by the following procedure: since the PM at $z = 0$ is the only one upon which no secondary maximum will ever be superposed, it can be used to determine the coefficients K_i by the relationship $K_i = h_i/H_0$, in which h_i is the amplitude of the i-th secondary maximum at the distance z_i of the origin and H_0 is that of the PM at $z = 0$. As will be seen in the next section, these coefficients will be used to determine the contributions of the secondary maxima to each of the composite principal peaks.

9.1.1 Practical Example of the Use of the D'yakonov Method

Figure 9.1 shows the function $\varphi'(z)$ obtained from the integrated intensities of the basal reflections given by a chlorite-montmorillonite interstratified mineral

(D'yakonov 1978) with basal spacings (in Å) and relative integrated intensities (in brackets, in arbitrary units) equal to 31.9 (78); 15.3 (100); 7.91 (25); 5.36 (2); 4.68 (22); 3.58 (10); 2.886 (5); 2.026 (1); 1.828 (0.1); 1.569 (0.3). In Fig. 9.1, A and B refer respectively to layers of chlorite ($d_A = 14.25$ Å) and montmorillonite after absorption of glycerine ($d_B = 18.0$ Å). The succession of letters A, B, A^2, AB, etc. gives the positions of the principal maxima corresponding to the isolated layers and to substacks AA, AB and BA, etc. The secondary maxima (two, in the present case) due to the internal structure of the layers are indicated by "a" and "b".

It is possible to determine the coeffients K_i from Fig. 9.1, namely

$$K_a = h_a/H_0 \quad \text{and} \quad K_b = h_b/H_0 \ .$$

It is evident from Fig. 9.1 that the true amplitude of the principal maximum A is increased by a contribution from the secondary "a" maximum due to the neighboring B PM. Similarly, the true amplitude of the latter is increased by the

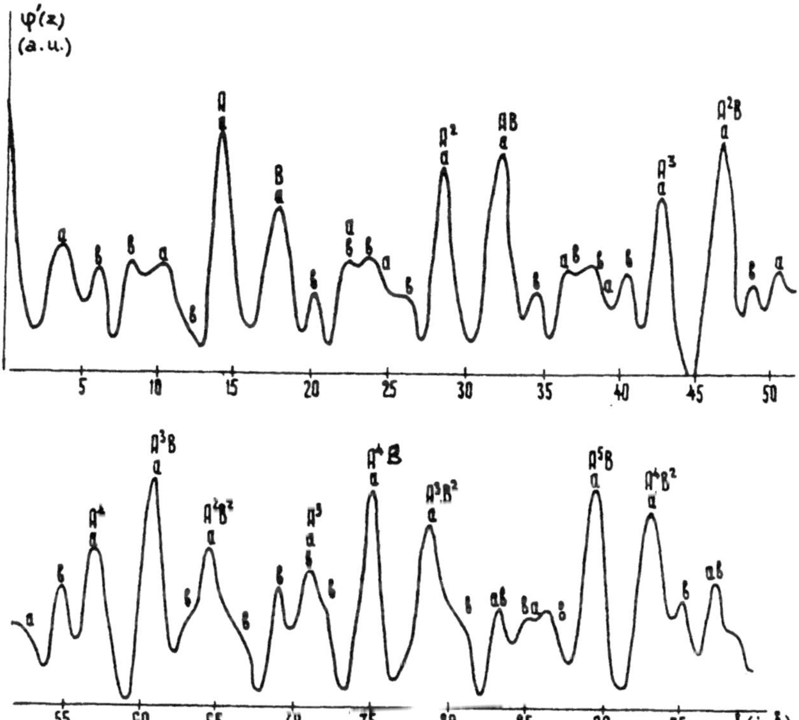

Fig. 9.1. φ'(z) curve corresponding to an interstratified chlorite-montmorillonite mineral; *A* and *B* respectively identify chlorite and montmorillonite layers, $A^m B^n$ represents a combination of m chlorite and n montmorillonite layers, *a* and *b* correspond to successive secondary maxima (D'yakonov 1978)

contribution from a secondary "a" maximum due to A. Writing now h_A and h_B for the respective amplitudes of the composite principal maxima A and B, we have:

$$\begin{aligned} h_A + K_a h_B &= H_A \\ K_a h_A + h_B &= H_B \end{aligned} \tag{9.2}$$

in which h_A and h_B are the true amplitudes of the A and B principal maxima in the absence of contributions from secondary maxima. Since H_A, H_B, K_a and K_b are known, the system of Eqs. (9.2) yields h_A and h_B, from which the probabilities $(W_A)_{exp}$ and $(W_B)_{exp}$ can be calculated through:

$$(W_A)_{exp} = \frac{h_A}{h_A + h_B} \quad \text{and} \quad (W_B)_{exp} = \frac{h_B}{h_A + h_B} . \tag{9.3}$$

Insofar as substacks of two layers are concerned, we find, from Fig. 9.1, that the sequence BB is absent, but that principal maxima corresponding to the pairs AA and AB (and BA) exist. By following the same type of derivation as the one used above for the singletons A and B, we can write the following set of simultaneous equations:

$$\begin{aligned} h_{A^2} + K_a h_{AB} &= H_{A^2} \\ K_a h_{A^2} + h_{AB} &= H_{AB} \end{aligned} \tag{9.4}$$

from which h_{A^2} and h_{AB} are easily obtained, which can then be combined to yield:

$$(W_{A^2})_{exp} = \frac{h_{A^2}}{h_{A^2} + h_{AB}} \quad \text{and} \quad (W_{AB})_{exp} = \frac{h_{AB}}{h_{A^2} + h_{AB}} .$$

The other composite principal maxima are dealt with in exactly the same way, unless several secondary maxima are superposed on the same PM, in which case a slightly different procedure must be adopted. Consider, for instance, the PM corresponding to the substack A^3B upon which two "a" secondary maxima are superposed, one corresponding to the A^4PM and the other to A^2B^2. The system of equations is

$$\begin{aligned} h_{A^4} + K_a h_{A^3B} &= H_{A^4} \\ K_a h_{A^4} + h_{A^3B} + K_a h_{A^2B^2} &= H_{A^3B} \\ K_a h_{A^3B} + h_{A^2B^2} &= H_{A^2B^2} \end{aligned} \tag{9.5}$$

whereas, for the composite principal maximum A^5, which includes contributions from "a" (from A^4B) and "b" (from A^2B^2) secondary maxima, the three simultaneous equations to be solved are:

$$h_{A^5} + K_a h_{A^4B} + K_b h_{A^2B^2} = H_{A^5}$$
$$K_a h_{A^5} + h_{A^4B} + K_a h_{A^3B^2} = H_{A^4B} \quad (9.6)$$
$$K_a h_{A^4B} + h_{A^3B^2} = H_{A^3B^2} \ .$$

Table 9.1 gives the results of the calculations performed on the MLM considered above. The number of layers in each category of substacks is indicated in column 1, with the nature and number of each type of layer detailed in column 2. Columns 3 and 4 respectively give the calculated values of $h_{A^mB^n}$ and the sums H_S obtained from Eq. (9.1). Lastly, column 5 gives the proportion $(W_{A^mB^n})_{exp}$ of each substack of q layers such that its thickness corresponds to that of the sequence A^mB^n.

In this connection, it should be noted that the *global* proportion of each type of layer (A and B) in *all* the substacks with q layers is obtained from

$$W_A = \frac{1}{q} \sum_{m=1}^{q} m\,[W_{A^mB^{(q-m)}}]_{exp} \quad \text{and} \quad W_B = 1 - W_A \ . \quad (9.7)$$

Both equalities must be satisfied for each value of q. Column 6 in Table 9.1 gives the values of W_A calculated using Eq. (9.7). The mean value $\overline{W}_A$ of the entries in

Table 9.1. Characteristic parameters for the sequence of chlorite (A) and montmorillonite (B) layers in an interstratified mineral. (After D'yakonov 1978)

q	Substack types	h	H_S	$(W_{A^mB^n})_{exp}$	W_A	$(W_{A^mB^n})_{cal}$ S = 1	$(W_{A^mB^n})_{cal}$ S = 2
1	A	222		0.75		0.75	0.75
			295		0.750		
	B	73		0.25		0.25	0.25
2	A^2	139		0.45		0.50	0.50
			311		0.725		
	AB	172		0.55		0.50	0.50
3	A^3	91		0.30		0.33	0.30
	A^2B	208	299	0.70	0.767	0.58	0.65
	AB^2	0		0		0.08	0.03
4	A^4	95		0.27		0.22	0.18
	A^3B	155	347	0.45	0.748	0.56	0.64
	A^2B^2	97		0.28		0.22	0.18
5	A^5	32		0.10		0.15	0.11
	A^4B	165	317	0.52	0.744	0.48	0.54
	A^3B^2	120		0.38		0.37	0.34
6	A^6	0		0		0.10	0.07
	A^5B	173	305	0.57	0.762	0.39	0.42
	A^4B^2	132		0.43		0.45	0.45
	A^3B^3	0		0		0.06	0.04

Notes: The mineral species is the one to which the function φ′(z) in Fig. 9.1 refers column 6 is 0.75, with a mean deviation $\Delta W_A = 0.01$ essentially due to the inaccuracy in the determination of the amplitude of the various maxima.

Mere inspection of the function $\varphi'(z)$ is sometimes sufficient to eliminate *a priori* some of the possible sequences: in the example shown in Fig. 9.1, the absence of PM corresponding to BB pairs implies that $P_{BB} = 0$ and therefore that the stacking of A and B layers cannot be random (S = 0) since one should then have $P_{BB} = W_B$. On the contrary, the elaboration of a model in which layer interaction would extend to first neighbors (S = 1) requires the choice of two independent parameters (cf. Chap. 4), which, in the present case, would be $W_A = (W_A)_{exp} = 0.75$ and $P_{BB} = (P_{BB})_{exp} = 0$.

To establish the real mode of distribution of the layer types in the mineral under study, it is necessary to calcule the global values $(W_{A^mB^n})_{cal}$ corresponding to the *models* of substacks with different values of the interaction range parameter: S = 1,2, ... For instance $(W_{A^2B^2})_{cal}$ is given by

$$(W_{A^2B^2})_{cal} = W_{AABB} + W_{ABAB} + W_{BAAB} + W_{BABA} + W_{ABBA} + W_{BBAA} \ , \quad (9.8)$$

in which $W_{AABB}, \ldots$ must be calculated with the help of the different relationships corresponding to S = 1,2, ... given in Chapter 4.

When S = 1, the values of $(W_{A^mB^n})_{cal}$ calculated are shown in column 7 of Table 9.1. It can be seen that the agreement between these values and the ones given in column 5 deteriorates as q increases. We are therefore led to try a model with S = 2. This choice, with $W_A = 0.75$, $P_{BB} = 0$ and $P_{AAA} = (P_{AAA})_{exp} = 0.60$ yields a second set of values for $(W_{A^mB^n})_{cal}$ shown in column 8. The parameter RP (equivalent to a reliability parameter) calculated from

$$RP = \frac{\sum |(W_{A^mB^n})_{exp} - (W_{A^mB^n})_{cal}|}{\sum (W_{A^mB^n})_{exp}} \quad (9.9)$$

shows that there is indeed a better agreement between the calculated and experimental W values, though the improvement is not spectacular: S = 2 yields RP = 0.15, instead of RP = 0.17 obtained with S = 1.

9.1.2 Appraisal of the D'yakonov Method

D'yakonov used the method which he had elaborated to study a number of interstratified illite-smectite, kaolinite-smectite, and phlogopite-vermiculite minerals (D'yakonov 1981, 1983; D'yakonov and Volosnykh 1979). This is illustrated by the results on three illite-montmorillonite minerals (samples 1, 2 and 3) in Table 9.2. The values of $(W_{A^mB^n})_{exp}$ determined for each sample by analyzing its $\varphi'(z)$ function are presented in column 1, while column 2 gives $(W_{A^mB^n})_{cal}$ calculated with the help of the relationships in Chapter 4. For each sample, the lowest values of RP were obtained with the set of parameters S, W_i, P_{ij} and P_{ijk} shown in Table 9.3, in which the thicknesses d_A and d_B of the layers of mica (A) and smectite (B) are also given, as well as the uncertainties Δd_A, Δd_B, ΔW_B and RP.

Table 9.2. Comparison between experimental $(W_{A^mB^n})_{exp}$ and calculated $(W_{A^mB^n})_{cal}$ values for three samples of mixed illite (A) – montmorillonite (B). (After D'yakonov and Volosnykh 1979)

q	Substack types	Sample 1		Sample 2		Sample 3	
		$(W_{A^mB^n})_{exp}$	$(W_{A^mB^n})_{cal}$	$(W_{A^mB^n})_{exp}$	$(W_{A^mB^n})_{cal}$	$(W_{A^mB^n})_{exp}$	$(W_{A^mB^n})_{cal}$
1	A	0.64	0.62	0.66	0.71	0.86	0.82
	B	0.36	0.38	0.34	0.29	0.14	0.18
2	A^2	0.18	0.24	0.40	0.42	0.61	0.64
	AB	0.82	0.76	0.60	0.58	0.39	0.36
3	A^3	0	0	0.19	0.16	0.77	0.46
	A^2B	0.89	0.86	0.77	0.81	0.53	0.54
	AB^2	0.11	0.14	0.04	0.03	0	0
4	A^4	0	0	0.11	0.10	0.26	0.28
	A^3B	0.45	0.48	0.62	0.64	0.74	0.72
	A^2B^2	0.55	0.52	0.27	0.26	0	0
5	A^5	0	0	0.07	0.06	0.21	0.21
	A^4B	0.12	0.10	0.41	0.42	0.67	0.68
	A^3B^2	0.88	0.90	0.52	0.51	0.12	0.11
6	A^6	0	0	0.05	0.04	0.19	0.16
	A^5B	0	0	0.28	0.23	0.56	0.60
	A^4B^2	0.76	0.72	0.64	0.68	0.25	0.24
	A^3B^3	0.24	0.28	0.03	0.05	0	0
7	A^7	0	0	0	0.02	0.09	0.12
	A^6B	0	0	0.21	0.15	0.49	0.50
	A^5B^2	0.34	0.34	0.56	0.58	0.42	0.38
	A^4B^3	0.66	0.66	0.23	0.24	0	0
8	A^8	0	0	0	0	0.06	0.09
	A^7B	0	0	0	0	0.43	0.38
	A^6B^2	0	0	0	0	0.51	0.53

Although the values of RP are low, one may ask to what extend the "best values" of the probability coefficients adopted in the D'yakonov method and detailed in Table 9.2 depict the actual structure of the minerals. Attempting to answer this question, Sakharov et al. (1989) compared the experimental diffractograms with the diagrams calculated using the probability coefficients deduced from the analysis of the function $\varphi'(z)$ for the three samples presented above and for which D'yakonov had found $S \geqslant 2$. The modelization method which they used is described in Section 9.2, and Table 9.4 gives, for each sample, the positions of the basal reflections in the experimental diffractograms (d_{exp}) and in the diagrams calculated (d_{cal}) on the basis of the model derived from the D'yakonov method. For the three samples, the agreement between d_{exp} and d_{cal} is only good when the diffraction angle is relatively large, but the main discrepancy is the presence, in the synthetic diagrams, of a non-negligible reflection which does not exist in the experimental diffractogram for small values of the angle of diffraction. The presence of this supplementary reflection was traced to the fact that, for $S \geqslant 2$, the propor-

Table 9.3. Characteristic parameters of the structural models used in Table 9.2 for the three samples of interstratified illite (A) – montmorillonite (B). (After D'yakonov and Volosnykh 1979)

Samples	d_A (in A)	$\lvert\Delta d_A\rvert$ (in A)	d_B (in A)	$\lvert\Delta d_B\rvert$ (in A)	$\overline{W_B}$	$\lvert\Delta W_B\rvert$	$\lvert\Delta W_{A^mB^n}\rvert$	S	Independant parameters used to obtain the $(W_{A^mB^n})_{cal}$ values in Table 9.2	RP
1	9.8	0.1	17.9	0.1	0.38	0.02	0.03	2	$W_B = 0.38$ $P_{BB} = P_{AAA} = 0$	0.06
2	9.97	0.02	17.9	0.1	0.29	0.01	0.02	3	$W_B = 0.29$ $P_{BB} = 0$ $P_{BAB} = 0.103$ $P_{BAAB} = 0.769$	0.07
3	10.0	0.02	17.69	0.06	0.18	0.01	0.04	4	$W_B = 0.18$ $P_{BB} = P_{BAB} = P_{BAAB} = 0$ $P_{BAAAB} = 0.611$	0.05

Table 9.4. Comparison between the experimental (d_{exp}) and calculated (d_{cal}) values of the basal spacings for the three illite (A) – montmorillonite (B) samples described in Tables 9.2 and 9.3 (Sakharov et al. 1989).

S.1	Model with $S = 2$; $W_B = 0.38$ $P_{BB} = P_{AAA} = 0$	Model with $S = 1$; $W_B = 0.38$ $P_{BB} = 0$	S.2	Model with $S = 3$; $W_B = 0.29$ $P_{BB} = 0$; $P_{BAB} = 0.103$ $P_{BAAB} = 0.769$	S.3	Model with $S = 4$; $W_B = 0.18$ $P_{BB} = P_{BAB} = P_{BAAB} = 0$ $P_{BAAAB} = 0.611$
d_{exp}	d_{cal}	d_{cal}	d_{exp}	d_{cal}	d_{exp}	d_{cal}
29.6	34.68	29.9	33.1	37.94	34.9	51.4
	19.59			19.21		24.2
13.58	13.00	13.58	13.37	12.78	13.26	16.01
9.36	9.40	9.36	9.56	9.58		11.93
5.42	5.42	5.43	5.36	5.41	9.58	9.71
4.68	4.67	4.66	4.74	4.74	5.37	5.27
3.42	3.426	3.430	3.42	3.435	4.76	4.80
3.11	3.110	3.102	3.14	3.155	3.38	3.393
2.53	2.509	2.508	2.53	2.523	3.18	3.189
1.979	1.975	1.975	1.991	1.992	2.53	
			1.66	1.646	1.989	

Notes: The values d_{cal} are calculated using the structural characteristics obtained by the D'yakonov method. In the case of sample 1, another simpler model is given in column 3. The d-values are given in Angström units; S.1, S.2, S.3 are the samples 1, 2, 3 respectively tions of the sequence AABAAB... (in samples 1 and 2) or AAABAAAB... (in sample 3) can no longer be neglected. However, Sakharov et al. (1989) noted that agreement between the experimental and calculated diagrams, insofar as the number and position of the reflections were concerned, was obtained if a value of S lower than the one deduced by D'yakonov was adopted. This is illustrated in Table 9.4, column 3, which gives the values of d_{cal} calculated with S = 1 for sample 1. One can therefore conclude that the actual *disorder* in an interstratified mineral is greater than the one deduced from the application of the D'yakonov method.

A first hypothesis which can be suggested is that the difference between the actual structure and the model resulting from the D'yakonov method could be due to loose conditions of recording and of exploitation of the experimental diffractograms. This possibility was tested by Sakharov et al. (1989), who calculated the synthetic XRD diagrams expected from stacking models consisting of illite (A) and montmorillonite (B) layers with $W_A:W_B = 0.7:0.3$, $d_A = 10.0$ Å, $d_B = 17.78$ Å and S = 0, 1, 2 or 3. Two categories of models were studied for S > 0, namely one in which the sequence of layers showed the MPDO, and one in which the layers had a tendency to segregate. Using then the synthetic XRD patterns as if they were experimental diffractograms, the authors calculated the corresponding $\varphi'(z)$ function and obtained from it the different parameters by the D'yakonov method. The conclusions which they reached were:

1. The value of d_A and d_B obtained by the D'yakonov-method are quite accurate (to within 0.25%) for all the structural models considered.
2. The relative proportions W_A and W_B are also correct to within 2 to 3% for the models in which the degree of order is highest, but the error is more important when there is segregation in the mode of stacking of each layer type: for a model with nominal values S = 1, $W_B = 0.3$ and $P_{BB} = 0.5$, the value obtained from $\varphi'(z)$ for $\overline{W}_B$ is 0.23.
3. There is a clear difference between the values of $(W_{A^mB^n})_{mod}$ introduced in the initial model and the $(W_{A^mB^n})_{cal}$ calculated using a model giving the best fit with the "experimentally observed" values of $(W_{A^mB^n})_{exp}$ yielded by the analysis of $\varphi'(z)$. The latter values differ all the more as S decreases. Thus, for a model with S = 2, the parameter RP relative to the experimental and model values of $(W_{A^mB^n})$ is 5.5%, but RP increases to 8.1 and 14.4% for models with S = 1 and S = 0 respectively.
4. Agreement between the experimental and calculated values of $(W_{A^mB^n})$ is always improved when S is increased by one unit with respect to the actual S value used in the starting model. Thus, using $(W_{A^mB^n})_{mod}$ (obtained from a model with S = 1 and $P_{BB} = 0$) and $(W_{A^mB^n})_{cal}$ (obtained from a model with S = 2, $P_{BB} = 0$ and $P_{BAB} = 0.2$), the value of RP is 8.1% if we compare $(W_{A^mB^n})_{mod}$ with $(W_{A^mB^m})_{exp}$ and 5.5% when $(W_{A^mB^n})_{cal}$ and $(W_{A^mB^n})_{exp}$ are compared. Similarly, RP decreases from 14.4% (from values of W_{mod} given by a model with S = 0 and $P_{BB} = W_B$) to 7.6% (from W_{cal} values given by a model with S = 1 and $P_{BB} = 0$) when these values are compared with W_{exp}.

5. The lack of agreement among the values of W_{exp}, W_{mod}, and W_{cal} is much more important when the structural models include some segregation in the layers of each type. For example, an initial model with $S = 1$ and $P_{BB} = 0.5$ leads to RP = 20% when W_{mod} and W_{exp} are compared, and no model obtained by the analysis of $\varphi'(z)$ function could yield a better value of RP for the agreement between W_{exp} and W_{cal}.

The method of D'yakonov involving the Fourier transformation of the intensities of the basal reflections given by an interstratified mineral can therefore be expected to yield the following results:

- good accuracy regarding the determination of the thickness of each type of layer, whatever the sequence in the stack;
- good accuracy in the determination of the concentration of the types of layers only if there is no tendency to segregation in the stacking sequence;
- best results when the interaction parameter S is large and when the sequence of layers is close to the one with MPDO;
- on overestimation of the degree of order in the sequence of layers and a value of S larger than its actual value.

These conclusions are quite general and remain valid even if strict control of all the experimental parameters is respected, and whatever the method used to calculate the function $\varphi'(z)$, i.e., whether it be from the integrated intensity of the basal reflections or from their profile. Strictly speaking, perfect agreement between the model worked out from $\varphi'(z)$ and the actual structure can only be expected if $\varphi'(z)$ were reduced to a succession of Dirac functions, an impossible occurrence since all layers have their own internal structure. It should also be noted that the D'yakonov method becomes inapplicable when there are less than five or six reflections or when the sample is composed of a mixture of phases such that there is even only one superposition of a basal reflection from each phase.

9.2 General Guidelines for the Use of Modelization of X-Ray Diffractograms in the Study of Mixed-Layer Minerals

The indirect methods of analysis of X-ray diffractograms exploiting a given structural model to calculate a synthetic powder pattern which can be compared to the experimental tracing have been extensively developed during the last 20 years and, in Chapter 5, we examined the theoretical bases of some of these methods. More specifically, we described a matrix formalism particularly well adapted to the solution of practical problems met in disordered systems. In the present section, we shall describe the variables which need to be closely controlled for the practical use of this method in the determination of structural characteristics, with particular emphasis on the difficulties which may arise in the interpretation of the experimental diffractograms given by MLM.

The elaboration of a calculated XRD powder pattern given by an interstratified system starts from a structural model, with explicit reference to the following parameters:

1. the number of types of layers;
2. the chemical composition of each type of layer, its thickness, and the coordinates of the atoms in it;
3. the concentration of each type and the mode of succession of the different types of layers;
4. the mean number of layers, i.e., the mean thickness of the interferential coherence domains, as well as the law of distribution of these domains;
5. the experimental function which may perturb or distort the diffracted waves.

Each of these points will be examined, to obtain a full understanding to the way the structural models needed to calculate the synthetic XRD diagrams are built. The different possibilities of interpreting experimental diffractograms by comparison with synthetic diagrams will then be reviewed and, lastly, we shall give practical examples of the determination of structural characteristics for some naturally occurring MLM.

9.2.1 Determination of the Nature of the Layer Types

The discussion will be limited to models for the most common natural interstratified layer silicates such as illite-montmorillonite, chlorite-smectite, kaolinite-montmorillonite, etc.

Since most mixed-layer silicates contain expandable interlamellar spaces similar to those in smectites or vermiculites and with exchangeable cations, one of the first steps consists in obtaining the diffractograms of the cation-exchanged interstratified mineral after submitting it to the action of organic compounds (such as glycerine or ethylene glycol), cation exchange, heating, etc. Any modification in the intensity and position of the basal reflections is proof that the sample contains solvatable interlamellar spaces. The fundamental principle at the basis of these treatments is that the interlamellar spaces associated with each type of layer will have the same composition and properties as the interlamellar spaces in the noninterstratified minerals (containing only one type of layer) whose behavior, when submitted to the same treatment, is well-known. Table 9.5 gives the effect of these different treatments on the noninterstratified minerals. Using the data in this Table, it is possible to predict the most important characteristics of the XRD patterns obtained from MLM after various treatments. For example, one of the following three situations may be found after heating:

Table 9.5. Changes in the basal spacings consecutive to various treatments on some non-interstratified minerals

Minerals	Basal spacings (in Å) consecutive to different treatments
Smectites: montmorillonite, nontronite, saponite, etc.	With glycerin, $d_{001} = 17.8$; with ethylene glycol, $d_{001} = 16.8$; after heating (400 °C $\leqslant$ T $\leqslant$ 500 °C), $9.6 \leqslant d_{001} \leqslant 10$ (depending on the nature of the mineral); with H_2O, d_{001} depends on the humidity and on the exchangeable cations
Vermiculites	With glycerin, $d_{001} = 14.3$ (for Mg-vermiculite); after K^+ exchange, $d_{001} = 10$; after heating (600 °C $\leqslant$ T $\leqslant$ 700 °C), the Mg-vermiculite shows similar behaviour as smectites
Chlorites	$14.2 \leqslant d_{001} \leqslant 14.4$ depending on the Al-content in tetrahedra; heating and organic solvation do not modify the basal spacings
Micas	$d_{001} = 10.0$ for K-micas and $d_{001} = 9.6$ for Na- and Ca-micas; no modification by heating or by organic solvation
Kaolinites	$d_{001} = 7.15$; the layer is destroyed when T $\geqslant$ 600 °C
Talcs Pyrophyllites	$(d_{001})_{talc} = 9.35$; $(d_{001})_{pyrophyl.} = 9.16$; basal spacings are very stable when the mineral is heated or solvated

1. a first reflection corresponding to 7.15 Å $\leqslant d \leqslant$ 9.6 Å; the layers are kaolinite and smectite, or kaolinite, smectite, and illite;
2. a first reflection corresponding to a d-spacing between 9.6 and 10 Å; the interstratified mineral then contains smectite and mica-like or talc-pyrophyllite-like layers;
3. a first reflection corresponding to 10 Å $\leqslant d \leqslant$ 14 Å; the interstratified mineral contains smectite-chlorite layers.

These conclusions can be confirmed by heating the mineral to different ranges of temperature: between 250 and 350 °C, only the smectite type interlayer spaces are dehydrated, whereas heating to 500–600 °C dehydroxylates the kaolinic layers and simultaneously destroys the brucitic layers.

In some interstratified species, the rational indices of the basal reflections in the diffractograms of the natural mineral become non-rational after the treatments described above: this happens, for instance, with chlorite-swelling chlorite after reaction with ethylene glycol, or with Mg-vermiculite-chlorite in which the exchangeable cations have been replaced by K^+ ions. The opposite situation can also be met: interstratified biotite-vermiculite layers only give rational indices of the basal reflections after K^+-exchange, while interstratified montmorillonite-chlorite requires Mg^{2+}-exchange to yield rational indices for these reflections.

Although, as described here, the determination of the types of layers present in a given sample is fairly straightforward, it should be remembered that the problem becomes much more complex when the MLM contains more than two types of components.

9.2.2 Chemical Composition and Structure of the Layers and of the Interlayer Spaces

The number, the nature, and the z ordinates of the atoms present in the silicate layers and the interlayer spaces are easily obtained from the study of the appropriate pure single-type layer minerals, thanks to the fact that, apart from very few exceptions, the 2:1 and 1:1 layers retain the same anionic framework, irrespective of the nature of the cation in the octahedral and tetrahedral sites. Furthermore, since the X-ray scattering power of Si, Al or Mg is nearly the same, it is possible to neglect any tetrahedral Si-Al or octahedral Al-Mg substitution, if present. Similarly, because of the clear difference in the proportions of octahedral cations of di- and trioctahedral layers, the representation of the layers may be simplified by admitting that they contain four octahedral cations per unit cell when they are dioctahedral, and six cations when they are trioctahedral.

It is, however, much more difficult to describe the layer when the mineral contains iron – particularly when the chemical composition is unknown – or when the interstratified species is mixed with other phases. The difficulty is even larger when it comes to describing the interlayer spaces as these vary considerably in composition as a function of the mineral and of the kind of treatment it may have undergone.

In view of these difficulties, the description of the layers and of the interlayer spaces has been done in terms of rather loose parameters, this being one of the reasons for the lack of agreement often observed between experimental and calculated values of the basal reflection intensities. However, although a modification of the chemical composition of the layers or of the interlayer spaces does bear upon the intensities of the reflections, it has little or no effect on the position of the peaks, thereby considerably simplifying the identification of the structure of interstratified minerals.

Table 9.6 gives the z ordinates and the number of atoms (or molecules) per unit cell for the various components most commonly found in the interstratified lamellar silicates.

9.2.3 Choice of the Origin for the z Ordinates of Atoms in the Scattering Units

For interstratified minerals composed of 2:1 and/or 1:1 layers, the origin for the z atomic ordinates is placed on the sheet of surface oxygen atoms of the layers. *The interlayer space situated just above a layer is always associated with it and, together, they constitute the "scattering unit".* Obviously, the distances between the basal oxygen of successive silicate layers are equal to the thicknesses of these scattering units. In MLM containing 1:1 and 2:1 layers, the successive layers are stacked in such a way that the surface OH groups of the 1:1 layers can form hydrogen bonds with the basal oxygens of both 2:1 and 1:1 adjacent layers.

Table 9.6. Species of atoms, their number and *z*-ordinates for scattering units of different layered silicates

16.86 Å-smectite			17.78 Å-smectite			13.5 Å-smectite with one layer of glycol molecules		
Species	*z*-Ordinates	Number	Species	*z*-Ordinates	Number	Species	*z*-Ordinates	Number
O	0.0000	6.0000	O	0.0000	6.0000	O	0.0000	6.0000
Si	0.0355	4.0000	Si	0.0340	4.0000	Si	0.0400	4.0000
O	0.1308	6.0000	O	0.1240	6.0000	O	0.1570	6.0000
Al	0.1941	4.0000	Al	0.1840	4.0000	Al	0.2300	4.0000
O	0.2574	6.0000	O	0.2450	6.0000	O	0.3040	6.0000
Si	0.3527	4.0000	Si	0.3350	4.0000	Si	0.4210	4.0000
O	0.3882	6.0000	O	0.3690	6.0000	O	0.4610	6.0000
O	0.5562	1.7000	O	0.5470	2.2500	O	0.6640	1.7000
C	0.5562	1.7000	C	0.5470	2.2500	C	0.6640	1.7000
O	0.6124	1.7000	O	0.5920	2.2500	O	0.7970	1.7000
C	0.6124	1.7000	C	0.5920	2.2500	C	0.7970	1.7000
Ca, O	0.6639	1.3000	Na	0.6840	0.6000			
Ca, O	0.7243	1.3000	O	0.7770	2.2500			
O	0.7757	1.7000	C	0.7770	2.2500			
C	0.7757	1.7000	O	0.8220	2.2500			
O	0.8320	1.7000	C	0.8820	2.2500			
C	0.8320	1.7000						

9.6 Å-smectite			9.38 Å-talc			10 Å-mica		
Species	*z*-Ordinates	Number	Species	*z*-Ordinates	Number	Species	*z*-Ordinates	Number
O	0.0000	6.0000	O	0.0000	6.0000	O	0.0000	6.0000
Si	0.0600	4.0000	Si	0.0580	4.0000	Si	0.0600	4.0000
O	0.2330	6.0000	O	0.2340	6.0000	O	0.2210	6.0000
Al	0.3430	4.0000	Mg	0.3540	6.0000	Al	0.3280	4.0000
O	0.4520	6.0000	O	0.4740	6.0000	O	0.4350	6.0000
Si	0.6250	4.0000	Si	0.6510	4.0000	Si	0.5960	4.0000
O	0.6580	6.0000	O	0.7080	6.0000	O	0.6560	6.0000
						K	0.8280	1.8000

Furthermore, with reference to a direction parallel to $\vec{c}^*$, the tetrahedral sheets of the 1 : 1 layers are all on the same side with respect to the octahedral sheets of the same layers. Note that the choice of a sense for the $\vec{c}^*$ direction determines the nature of the scattering units. For example, the interlayer space associated with the surface OH groups of a 1 : 1 layer is always kaolin-type and the 1 : 1 layer itself plus this interlayer space form a *kaolin-type scattering unit.* On the contrary, the interlayer space next to a 2 : 1 layer may or may not be solvatable: the set "2 : 1 layer + interlamellar space" is a *smectite-type scattering unit* in the first case and a *mica-type scattering unit* in the second case. Figure 9.2 gives a diagram of an interstratified kaolinite-smectite with irregular sequences of kaolinite-, mica-, and smectite-type scattering units (k-, m- and s-units).

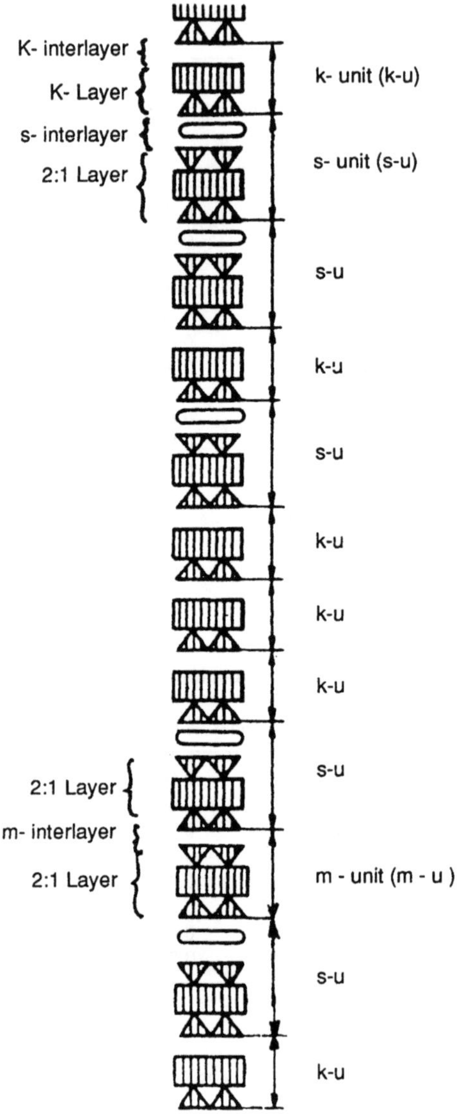

Fig. 9.2. Schematic representation of the interlayer spaces and of the scattering units in a kaolinite-smectite-mica mixed-layer sequence, $k-u$, $s-u$ and $m-u$ correspond respectively to kaolinite-, smectite- and mica-type scattering units

9.3 Calculation of the Reference X-Ray Diffractogram for Quasi-Homogeneous Interstratified Minerals

The method of models is deemed capable of giving all the synthetic XRD patterns associated with the variations of the different parameters which describe each type of statistical sequence in a stack but such systematic studies have been mainly applied to quasi-homogeneous systems, i.e., those which follow Markovian statistics (Sect. 1.7.2). However, investigations in this field only became efficient when the mathematical formalism was developed which not only permits the calculation of the intensity distribution along the (00) rod but also takes into account the powder effect. This last formalism therefore allows a direct comparison between the synthetic and experimental XRD curves.

These comparisons were first performed by Reynold and Hower (1970), who examined interstratified illite-smectite (i-s) minerals, and by Sakharov and Drits (1973) working on kaolinite-smectite (k-s). Drits and Sakharov (1976) were also the first to show the correlation between the specific features of the X-ray diffractograms from systems like i-s, k-s, ch-s (chlorite-smectite), k-s-i etc. and the structural characteristic of these mixed-layer minerals.

In order to establish these correlations for an interstratified system of two components (A and B), the last authors first chose a value for S (the range of interaction parameter), then the values of W_A and W_B, after which the synthetic spectra of the basal reflections were calculated using different sets of values for the other independent parameters involved in the structural models. Thus, for $S = 1$ and a given value of W_A, the probability P_{BB} (if $W_A > W_B$) or P_{AA} (if $W_B > W_A$) was changed from 0 to 1 by increments of 0.1 to determine the whole set of synthetic spectra corresponding to the values of P_{ii}, and the procedure repeated for other values of W_A between 0 and 1. As stated in Chapter 4, this study, carried out with $S = 1$, includes ipso facto the case $S = 0$. For $S > 1$, Drits and Sakharov limited their systematic study to interstratified i-m systems which, for a structural model with $S = 1$, would have corresponded to a system with MPDO in the succession of scattering units, as detailed in Section 4.3.2. For instance, with $S = 2$, having chosen $W_A > W_B$ and $P_{BB} = 0$, the parameters BAB (if $1/2 < W_A < 2/3$) or P_{AAA} (if $W_A > 2/3$) are varied between 0 and 1. Similarly, for $S = 3$, the only spectra calculated were those of an interstratified system which would show MPDO in a model with $S = 2$: for each fixed value of W_A, the independent parameters were varied and the different spectra calculated with the help of the relationships in Chapter 5, with the angle of diffraction θ varying in the domain 0 to 20° by increments of 0.1° (for Cu$K\alpha$). The number of scattering units in the mean interferential coherence domain was taken to be 20. For each model considered and each set of values of the independent probability parameters, separated studies involving different states of solvation were performed: the interstratified i-m (m – montmorillonite) contained ethylene glycol or glycerol, while interstratified k-m was studied with organic molecules or dehydrated. The rationale behind this procedure lies in the following facts:

1. organic solvents stabilize the interlamellar space of smectite layers, whereas simple hydratation leads to interlamellar spaces whose thickness varies with ambient relative humidity and the quantity or nature of the exchangeable cations;
2. for a wide angular diffractions domain, the intensity of the basal reflections are more intense when the cations are solvated by organic molecules than when solvated by water;

3. the comparison between the synthetic and experimental spectra yields firmer conclusions regarding the structural characteristics of the MLM when the latter is submitted to the treatments described above.

The systematic analysis of all the calculated spectra led to a number of criteria which may be used to determine the main structural features of the natural MLM.

9.3.1 Specific Features of the X-Ray Diffractograms Given by Interstratified i-m Systems with S = 0 or 1

The detailed description and the specific features of all the models considered in which the interaction parameter S is 0 or 1 are given by Drits and Sakharov (1976). We shall therefore discuss here only a few diagrams relative to interstratified i-m systems solvated by ethylene glycol, in which the thickness of the smectite scattering units is 16.86 Å, or 17.78 Å when they are solvated by glycerol.

Figure 9.3 presents the synthetic diagrams yielded by the structural models with S = 0 and different concentrations of illite and montmorillonite scattering units. Similar diagrams obtained with S = 1 and MPDO are shown in Fig. 9.4. Also shown (in Fig. 9.5) are some synthetic diagrams calculated with S = 1, $W_i : W_m = 0.7 : 0.3$ and different values of P_{mm}, i.e., models which do not have MPDO in the distribution of the scattering units.

The description of the results is given in terms of two parameters for each reflection in the diagrams:

- the *position* of the maximum of the peak will be simply given as "the maximum at d";
- the *indexation* of the maxima indicating between which $00l$ reflections of the pure minerals (illite or montmorillonite) each reflection from the interstratified MLM is found: for instance, the reflection at $d = 9.5$ Å in the diffractogram of the i-m treated with glycerol is indexed as $(001)_i/(002)_m$ since it lies between d_{001} of pure illite (at 10.05 Å) and d_{002} of glycerine-treated montmorillonite (at 8.89 Å). In some cases, we shall make use of a notation such as $(002)_{im}/(003)_{iim}$, which indicates that the maximum is found between the 002 reflection from an ordered imim...im sequence and the 003 reflection from the ordered iimiim...iim sequence.

Examining the spectra in Figs. 9.3 and 9.4, we see that the position of certain peaks found at $d < 10$ Å (such as the maxima $(001)_i/(002)_m$ and $(003)_i/(005)_m$) are not very sensitive to the degree of order-disorder in the sequence of scattering units of each type or when S changes from 0 to 1. This is clearly seen in Fig. 9.6 which gives, for six reflections, d as a function of W_m for the models with S = 0, and for the models with S = 1 and MPDO.

The existence of interaction between the i and m scattering units in the sequences is mainly revealed by the number and position of the reflections in the range

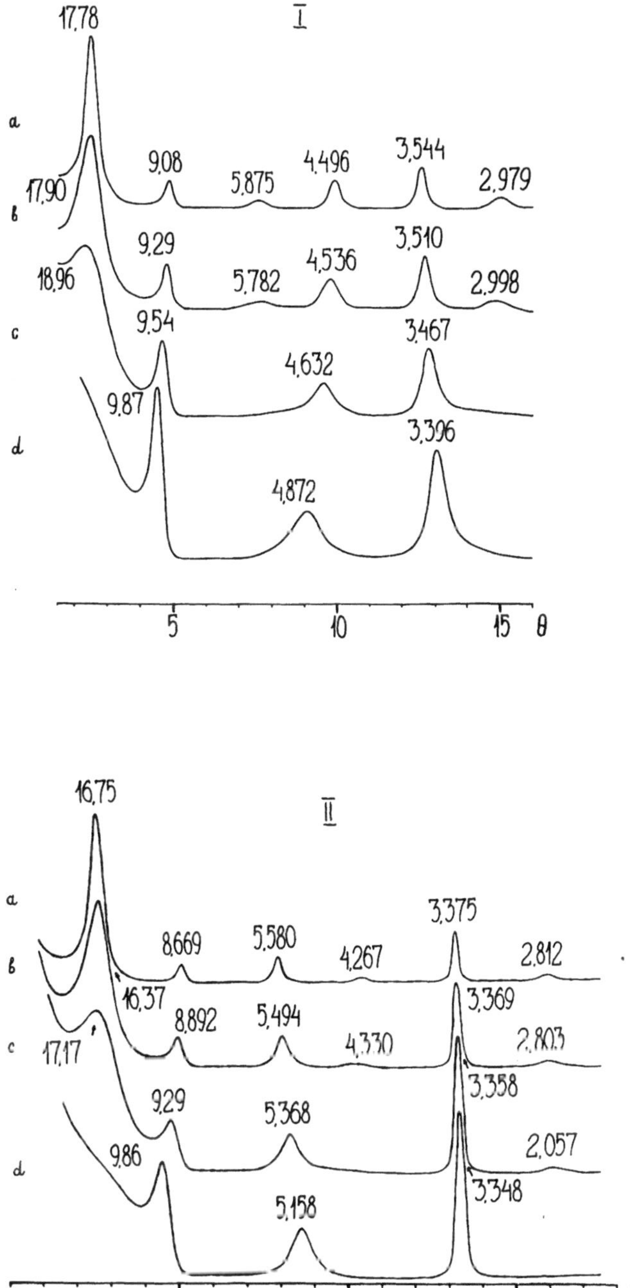

Fig. 9.3. Calculated XRD diagrams for an interstratified illite-montmorillonite mineral with S = 0, for various concentrations of glycerol solvated (**I**) or ethylene glycol-solvated (**II**) montmorillonite layers. The label *a*, *b*, *c* or *d* refer to concentrations $W_m = 0.8, 0.6, 0.4$ or 0.2, respectively

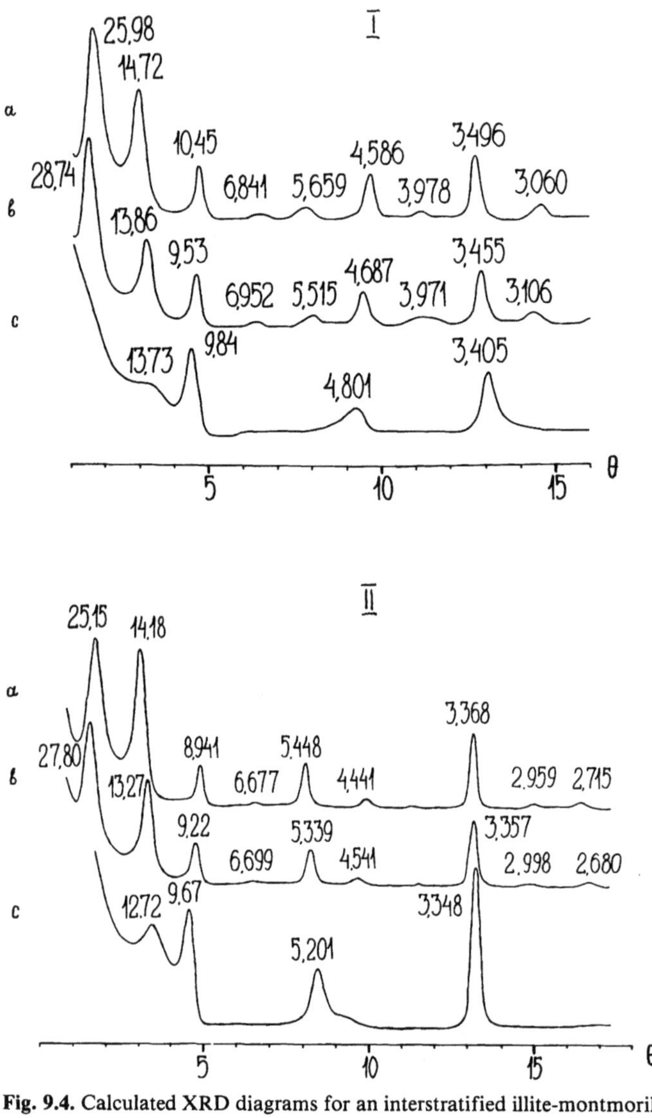

Fig. 9.4. Calculated XRD diagrams for an interstratified illite-montmorillonite mineral with S = 1 and MPDO, for various concentrations of glycerol-solvated (**I**) or ethylene glycol-solvated (**II**) montmorillonite layers. The label *a*, *b*, or *c* refer to concentrations $W_m = 0.6, 0.4$ or 0.2, respectively

$d > 10$ Å. In the spectra of Fig. 9.3 calculated with $S = 0$ and $W_m > 0.3$, there is a maximum with d between 16.4 Å and 17.2 Å when the solvating molecule is ethylene glycol, or between 17.8 Å and 19 Å when the solvating molecule is glycerol; for the models with $W_m < 0.3$, all reflections have $d \leqslant 10$ Å. The situation is different in the spectra calculated on the basis of models with $S = 1$ and

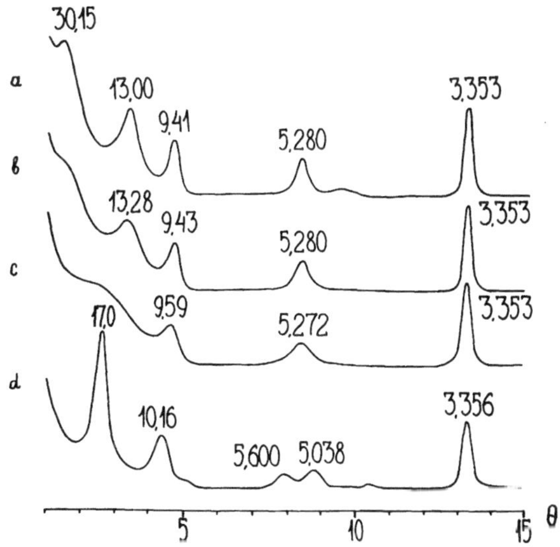

Fig. 9.5. Calculated XRD diagrams for an ethylene glycol-solvated interstratified illite-montmorillonite mineral with $S = 1$, $W_i : W_m = 0.7 : 0.3$. The letters *a, b, c,* or *d* respectively correspond to $P_{mm} = 0, 0.1, 0.3$ or 0.7

MPDO shown in Fig. 9.4: with $0.2 < W_m < 0.8$, the maximum is now between 12.7 Å and 16.2 Å (ethylene glycol) or between 13.6 Å and 17.5 Å (glycerol), i.e., in regions where the former models with $S = 0$ do not show a maximum. Furthermore, for models with $0.3 \leqslant W_m \leqslant 0.7$, a new peak appears whose position depends on the value of W_m: it is found at 22 Å $< d <$ 30 Å (ethylene glycol) or at 26 Å $< d <$ 28.8 Å (glycerol). Note that the distinction between the models with $S = 0$ and $S = 1$ becomes practically impossible if $W_m > 0.8$ as the diffractograms then become quite alike.

When $S = 1$, and with a fixed $W_i : W_m$ ratio, models which differ from each other by the tendency to segregation of the i and m scattering units give diagrams which are fairly similar: with $W_i > 0.5$ and $P_{mm} > 0.5$, the calculated spectra of interstratified i-m are hardly distinguishable from those given by a mechanical mixture of illite and montmorillonite, especially if the system was treated with ethylene glycol (see Fig. 9.5). It is then necessary to exploit the experimental spectra recorded under conditions yielding optimum resolution of the reflections and allowing precise measurements of their positions and intensities.

9.3.2 Characteristics of the X-Ray Diffractograms Given by i-m System with $S = 2$ and $W_i > W_m$

1. We shall first consider i-m systems with $S = 2$ and MPDO.

It will be recalled that structures of this type can be subdivided into two groups, depending on whether $W_i \leq 0.67$ or $W_i > 0.67$ (Sect. 4.4.2). For both groups, the

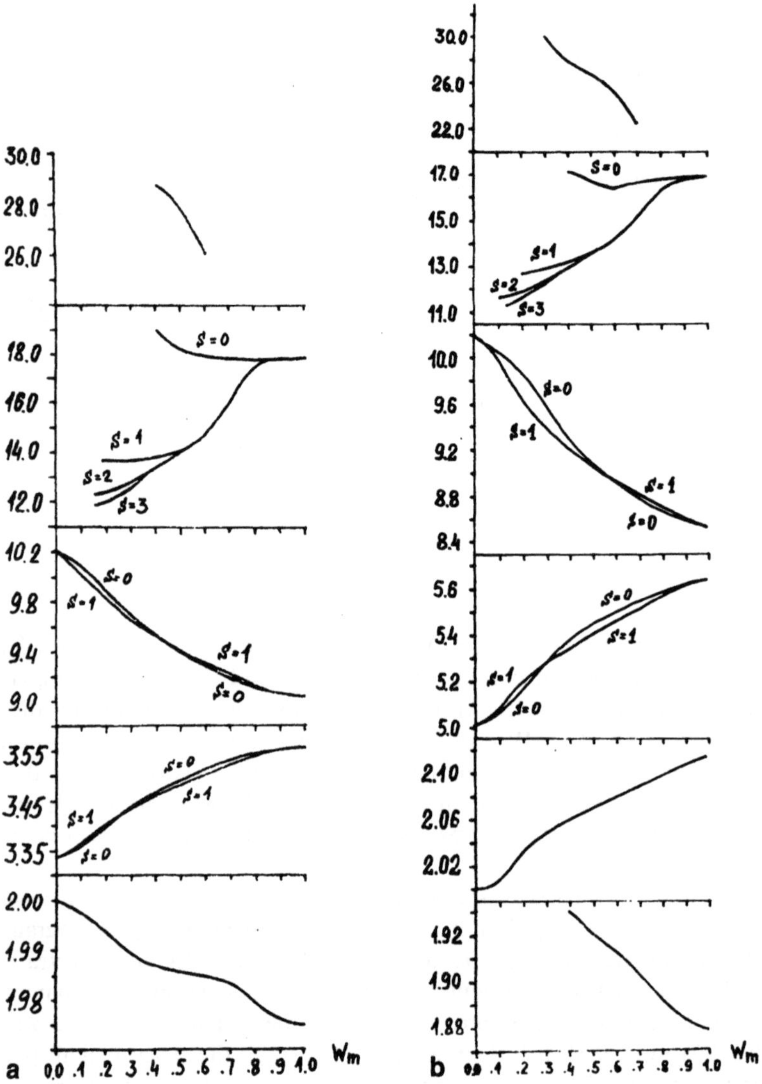

Fig. 9.6. Nomograms giving the dependence of the apparent *d*-spacings on W_m for a glycerol-solvated (**a**) or ethylene glycol-solvated (**b**) interstratified illite-montmorillonite minerals with S = 0, 1, 2 or 3 and MPDO

independent parameters chosen to draw up the structural models are the conditional probabilities P_{iii} and P_{mim}. It was also shown (Sect. 4.4.2) that increasing P_{mim} from 0 to W_m/W_i was equivalent to changing from the MPDO structure with S = 2 to a MPDO structure with S = 1. The same result is obtained when P_{iii} changes from 0 to $(W_i - W_m)/W_i$.

Figures 9.7 and 9.8 give the spectra calculated for structural models with S = 2 and different values of the probabilities P_{mim}, P_{iii} and of the ratio $W_i : W_m$. By

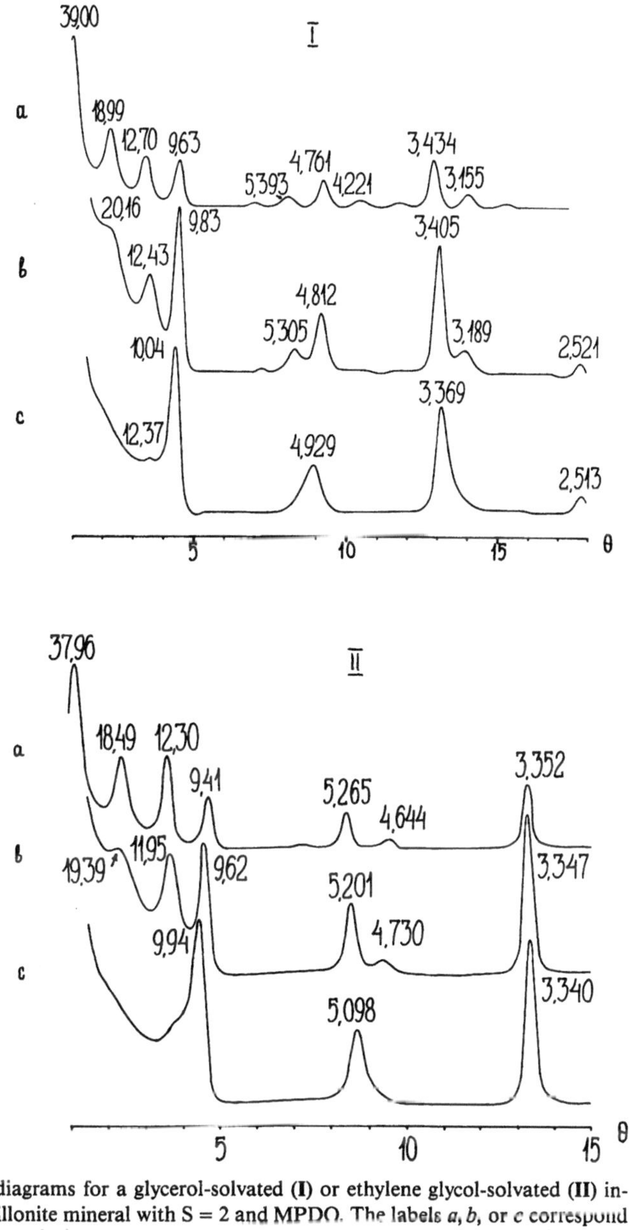

Fig. 9.7. Calculated XRD diagrams for a glycerol-solvated (**I**) or ethylene glycol-solvated (**II**) interstratified illite-montmorillonite mineral with S = 2 and MPDO. The labels *a*, *b*, or *c* correspond to $W_m = 0.3, 0.2$ or 0.1, respectively

comparing these spectra with those corresponding to models with S = 1 and MPDO (cf. particularly Fig. 9.4), we are led to the following conclusions:

a) in the angular domain corresponding to $d < 10$ Å, all the spectra calculated with different values of W_i, P_{mim} and P_{iii} show the same number of intense basal reflections with positions mainly dependent on W_i. For any given value of W_i, spectra calculated with S = 2 and MPDO or with S = 1 and MPDO only differ in minor details.
b) The most important differences in these two categories of spectra are found in the angular domain corresponding to $d > 10$ Å.
c) Increasing W_i from 0.5 to 0.67 shifts the reflections at small values of the scattering angle from positions characteristic of rectorites (with structure imim...im) to those found in the ordered structure iimiim...iim: curve a of Fig. 9.7 shows a peak at $d = 12.7$ Å indexed $(002)_{im}/(003)_{iim}$ placed between the 002 reflection of rectorite ($d_{002} = 13.43$ Å) and the 003 reflection which would be given by an iimiim... structure ($d_{003} = 12.28$ Å). As W_i increases, the positions of the basal reflections constitute a series which first tends to become more nonrational but finally settles down to rational values.
d) the basal reflection found at 11.5 Å $< d <$ 13.6 Å gives the possibility of distinguishing between MPDO structures with S = 2 and those with S = 1. For instance, in the former case and with a glycerol-treated structure, this reflection shifts from $d = 12.84$ Å to $d = 12.36$ Å when W_i changes from 0.67 to 0.85, whereas the reflection stays at $d = 13.6$ Å when the MPDO structure has S = 1. When the same system is treated with ethylene glycol, and for the same values of W_i, the reflection is found between 11.71 Å and 12.43 Å in the diagram for the first structural model and between 12.7 Å and 13.1 Å if the MPDO structure has S = 1 (cf. Figs. 9.7 and 9.4). The relation between the position of this reflection and the value of W_m is indicated in Fig. 9.6.

2. Consider now the models for interstratified i-m structures with S = 2 but without MPDO in the sequence of scattering units. Depending upon whether W_i was chosen to be larger or smaller than 0.67, the parameter P_{mim} or P_{iii} can be varied, with the following results on the calculated spectra, as seen in Fig. 9.8:

a) The increase of P_{mim} (or P_{iii}) from 0 to the value corresponding to a MPDO structure with S = 1 results in important modifications of the spectra, limited, however, to the region $d > 10$ Å: the principal modifications are shifts of the $(001)_{im}/(001)_{iim}$ and $(002)_{im}/(003)_{iim}$ reflections towards the positions of $(001)_{im}$ and $(002)_{im}$.
b) The increase of P_{mim} beyond this value and up to $P_{mim} = 1$ implies increasing segregation of the imim and iii substacks, with the appearance, in the region $d > 10$ Å, of two intense reflections very close to the position of the 001 and 002 reflections of rectorites.

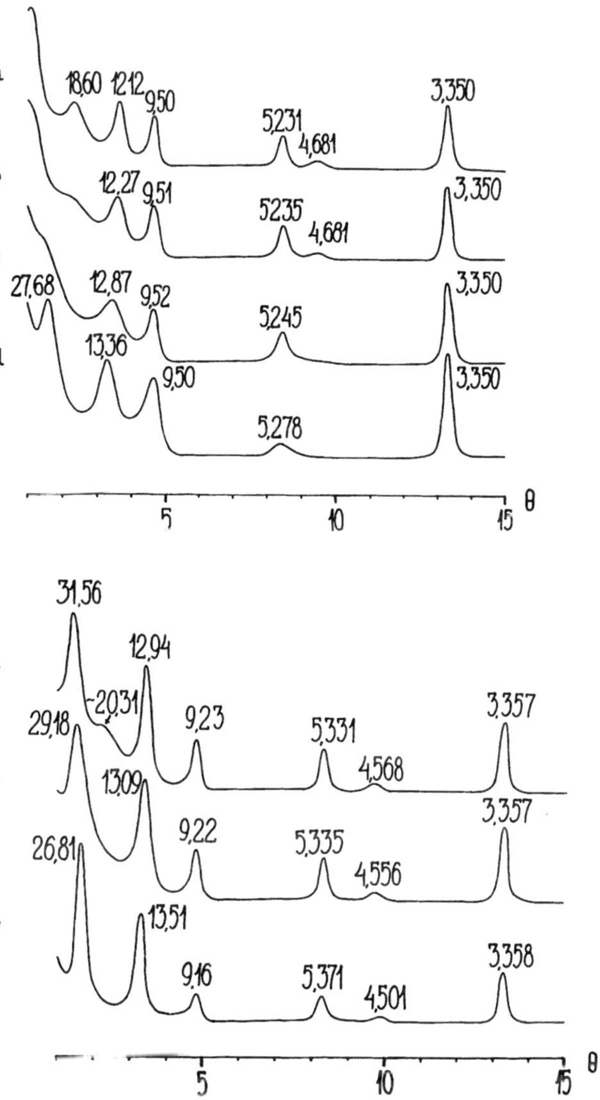

Fig. 9.8. Calculated XRD diagrams for an ethylene glycol-solvated interstratified illite-montmorillonite mineral with S = 2. The labels *a, b, c,* or *d* respectively correspond to $W_i : W_m = 0.75 : 0.25$, $P_{mm} = 0$ and $P_{mim} = 0, 0.15$, or 0.5, whereas *e, f,* or *g* respectively correspond to $W_i : W_m = 0.6 : 0.4$, $P_{mm} = 0$ and $P_{mim} = 0$, 0.2 or 0.7

9.3.3 Specific Features of the X-Ray Diffractograms from Interstratified i-m Structures with S = 3

Figure 9.9 shows three spectra based on structural models with S = 3 and MPDO, with W_i taken as 0.80, 0.85 or 0.9 respectively. There is an obvious change in the number and position of the basal reflections in the region of small angles of diffraction. For $W_i = 0.8$, four intense and narrow reflections are found at $d = 24.2$ Å, 15.6 Å, 11.66 Å, and 9.62 Å. Their presence is connected with the high proportion of iiim substacks with thickness 46.9 Å. An increase of W_i when S = 3 implies a greater proportion of substacks such as iiiim, iiiiim, ... causing the gradual disappearance of the reflections at $d > 12$ Å in the diagram.

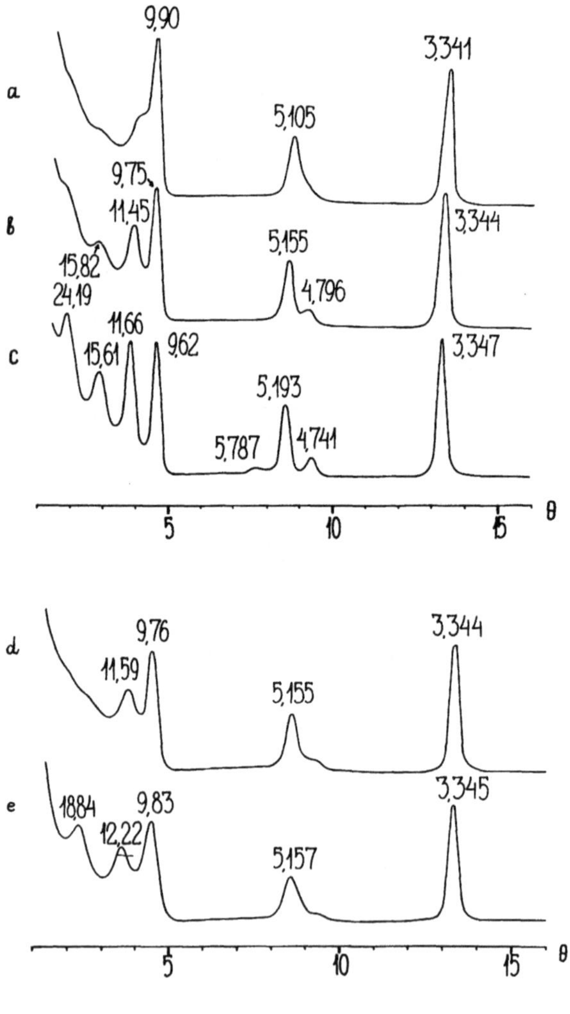

Fig. 9.9. Calculated XRD diagrams for a glycerol-solvated interstratified illite-montmorillonite mineral with S = 3. The labels *a*, *b*, or *c* correspond to structures with MPDO and $W_m = 0.1$, 0.15 or 0.20 respectively, whereas *d* and *e* correspond to structures with $W_i : W_m = 0.85 : 0.15$, $P_{mm} = P_{mim} = 0$, $P_{miim} = 0.15$ or 0.40

A behavior similar to that of i-m systems with S = 1 or S = 2 can be noted: in structural models with S = 3, the peaks at $d < 10$ Å are at positions close to the ones expected from a model with S = 2. Systems with S = 3 and S = 2 can be differentiated by the reflection at 11 Å $< d <$ 13 Å: with identical values of W_i, a model with S = 3 gives a $(004)_{iiim}/(001)_i$ reflection at $d \approx 11-12$ Å, but the same reflection is found at $d > 12$ Å and is indexed $(003)_{iim}/(001)_i$ in the diagram calculated for a model with S = 2. The position of this reflection, characteristic for S = 3, is shown in Fig. 9.6 as a function of W_m.

In systems with S = 3 and without MPDO in the sequence of scattering i and m units, the variation of P_{miim} from 0 to the value of the parameter P_{iim} is seen as an evolution of the MPDO structure with S = 3 to one with S = 2 and MPDO: in the diagram calculated for a system with S = 3, $W_i = 0.85$ and $P_{miim} = 0.15$, the basal reflection at $d > 12$ Å is absent (curve d, Fig. 9.9), but the reflections at $d < 10$ Å are approximately at the same positions as the ones found in the diagrams given by systems with S = 3 and MPDO, or S = 2 and MPDO.

Lastly, when S = 3, $P_{miim} > P_{iim}$ and P_{miim} increases, the structural systems show a tendency towards segregation of the iim substacks and i scattering units resulting in the presence of basal reflections whose positions get closer to those of the 00*l* reflections given by a regular iimiim...iim structure (curve e, Fig. 9.9).

9.3.4 Comparison of the Specific Features of the Diagrams Given by Systems with S = 0, 1, 2, 3

The descriptions given in the preceding sections seem to show that it is easy to determine both the relative proportions of the scattering i and m units and their sequence, i.e., the value of the parameter S, thanks to the fact that the set of reflections found at $d \leqslant 10$ Å is sensitive to changes in the concentration of each type of scattering unit, whereas the reflections found at $d > 10$ Å are most sensitive to variations of S. Using different values for the independent probabilistic parameters, it is possible to draw up nomograms giving the position of the reflections as a function of W, analogous to those presented in Fig. 9.6 and convenient for the determination of the structural characteristics of natural i-m systems. This was done by Watanabe (1981), who proposed an easy and fast method for the determination of W_i and S with the help of the curves in Fig. 9.10. Using the angular difference between the positions of the three successive reflections between 4.5 Å $< d <$ 18 Å in systems with S = 0, 1, 2 or 3, he showed that the variations of these differences can lead to the characteristics of each interstratified i-m system. Other authors (Drits and Sakharov 1976; Reynolds 1980; Srodon 1980, 1984) have also given curves useful for the identification of MLM, apparently without the need to calculate the synthetic diagrams.

In actual fact, extreme caution must be exercised when using these curves as the profiles, intensities, and positions of the peaks do not depend only on the concentrations W_i and W_m and on the sequence of scattering units but also on a number

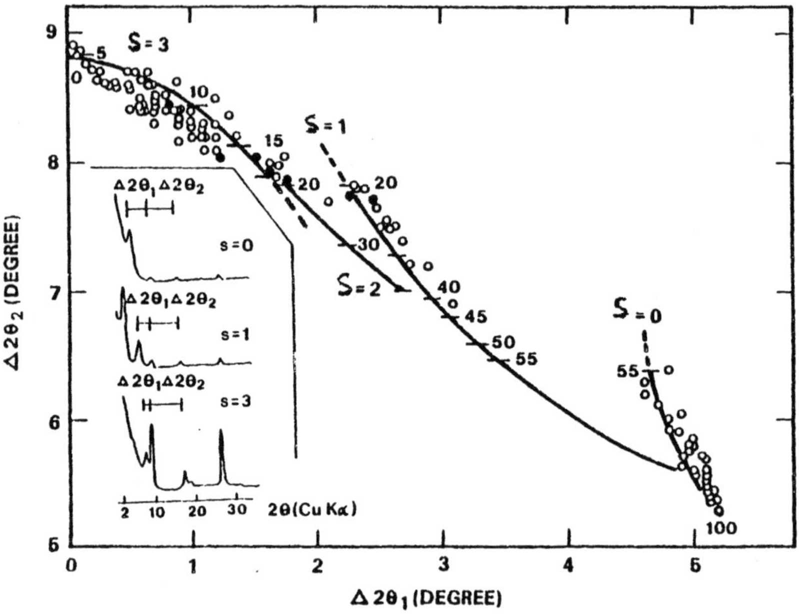

Fig. 9.10. Dependence of the position of the first three low-angle reflections, of the relative proportion, and pattern of layer sequence in interstratified illite-montmorillonite systems. (After Watanabe 1981)

of other parameters which can only be taken into account during the modelization of the synthetic X-ray diffractograms, as will be seen in the next section.

9.4 Parameters Other than W and S which Influence the Profile of the Calculated X-Ray Diagrams of Two-Component Interstratified Systems

9.4.1 Influence of the Thickness of the Scattering Units

In a study of dioctahedral smectites solvated by ethylene glycol, Srodon (1980) showed that the thickness of the scattering units varied from 17.3 Å to 16.5 Å, depending on the charge of the layer and the nature of the interlamellar cation. D'yakonov and Volosnykh (1974) had previously described similar effects in various interstratified i-m minerals solvated by glycerol, with variations between 17.5 Å and 18.1 Å of the scattering m-unit thickness. Similarly, the thickness of mica-type scattering units with interlamellar K^+ cations depends on the cationic content of the silicate layer and varies from 10 Å, in illite or muscovite, to 9.85 Å in leucophillite (Sokolova et al. 1976, 1978).

This is illustrated in Fig. 9.11; the six spectra depicted were calculated for an i-m system with S = 1, $W_i : W_m = 0.7 : 0.3$ and $P_{mm} = 0$, with the thicknesses of the mica-type scattering units taken as 9.8 Å and 10 Å, and those of the glycerol solvated montmorillonite-type units as 17.4 Å, 17.8 Å or 18.2 Å. It is clear from the diagrams that there are shifts in the positions of the basal reflections as well as important variations of their intensity, so that the thickness of the scattering units is a parameter which must be taken into account in the modelization of the XRD patterns.

9.4.2 Influence of the Thickness of the Interferential Coherence Domains

It was shown in Chapter 4 that when the nominal distribution of sequences in an interstratified system followed Markovian statistics, a small thickness of the interferential coherence domains was conducive to powders with lack of homogeneity in chemical composition of the domains as well as in the sequence of scattering units. It is also clear that the smaller the thickness of the domains, the lower will the intensity and the greater will the smearing of the basal reflections be, because of the small number of layers in each domain, and because of the heterogeneity in the proportion of the different scattering units in the undersized domains.

Another consequence of the presence of small-sized interferential coherence domains, namely a shift in the position of the maxima of the basal reflections, was described and discussed in Section 2.2.2.

Lastly, we saw in Section 3.1.2 that an efficient comparison of experimental and calculated patterns must take into account the distribution functions $\alpha(M)$ of the interferential coherence domain thicknesses. Experience shows that naturally occuring interstratified two-component microdivided systems contain 5 to 15 scattering units per domain. We note that the existence of the distribution function $\alpha(M)$ results in an increase of heterogeneity of composition in the domains.

Calculations show, however, that when all these parameters are taken into account, and even if the structural models examined have maximum disorder in the sequence of scattering units (S = 0), the basal reflections are relatively narrow and intense. We are therefore led to conclude that the calculated diffraction pattern weakly reflects the heterogeneity of the system, be it in the composition of the interferential coherence domains or in the mode of succession of scattering units. The main information brought by the patterns is rather the type of domains most frequently occurring in the powder which is examined.

Figure 9.12 illustrates the influence of the thickness of the coherence domains on calculated spectra. The model used was an interstratified i-m structure solvated by ethylene glycol, with S = 1, $W_i = 0.8$ and $P_{mm} = 0$. Curve *a* corresponds to domains with the same mean thickness $\overline{M} = 15$ while curves *b* and *c* correspond to two distributions of thicknesses, one with $5 \leqslant M \leqslant 15$ and the other with $11 \leqslant M \leqslant 10$. The curves show that the modification of the thickness is mostly felt

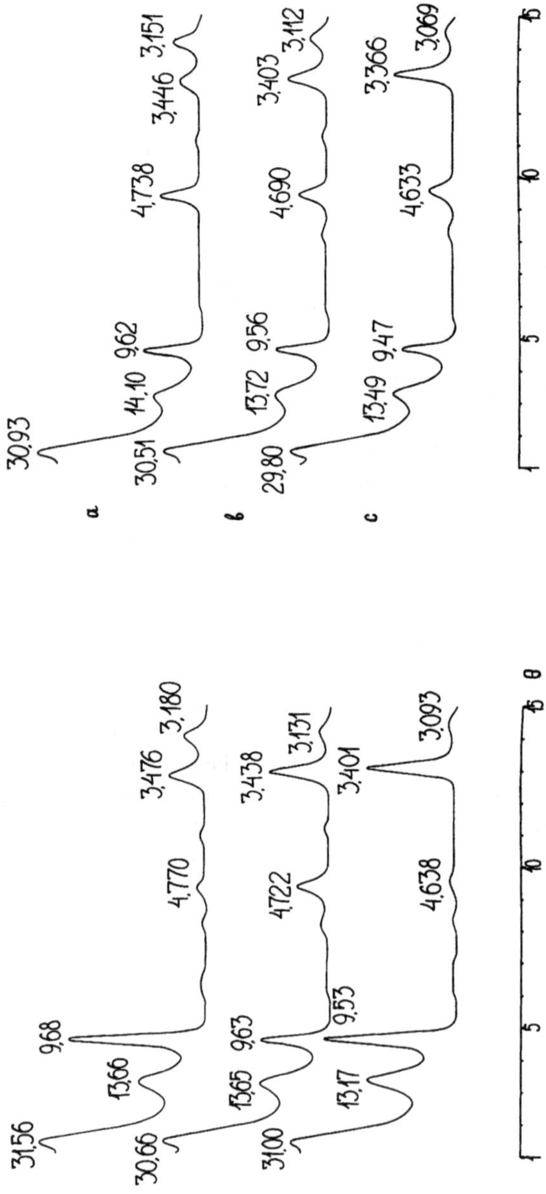

Fig. 9.11. Calculated XRD diagrams for a glycerol-solvated interstratified illite-montmorillonite mineral with $W_i = 0.7$, $S = 1$, MPDO and two thicknesses for the mica layer: *left-hand curves* $d_{mica} = 10$ Å; *right-hand curves* $d_{mica} = 9.8$ Å; the labels ***a***, ***b*** or ***c*** refer to montmorillonite layer thicknesses respectively equal to 18.2 Å, 17.78 Å or 17.4 Å

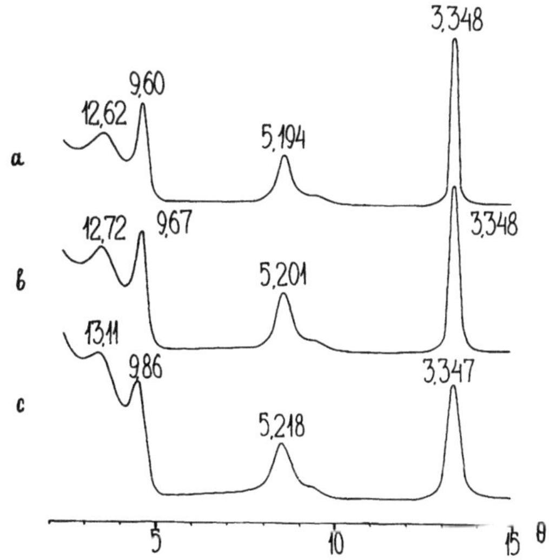

Fig. 9.12. Calculated XRD diagrams for an ethylene glycol-solvated interstratified illite-montmorillonite minerals with $W_i = 0.8$, $S = 1$ and MPDO for different mean thicknesses and distributions of thicknesses of interferential coherence domains: $\overline{M} = 15$ (curve *a*), $5 \leqslant M \leqslant 15$ (curve *b*) or $1 \leqslant M \leqslant 10$ (curve *c*); the proportions of the various domain thicknesses are assumed to be equal

in the region of small angles of diffraction. It will later be shown that this enables the basal reflections found at larger angles of diffraction to be used for the determination of the concentration of each type of scattering unit in the interferential coherence domains.

9.4.3 Physical Mixtures of Quasi-Homogeneous Mixed-Layer Systems

The degree of heterogeneity of MLM obviously reflects that of the micro-environment at the time of their formation or at the time of their deposition. For example, mixing of MLM particles may take place during their transportation or their sedimentation. It is therefore useful to analyze the diffraction effects that would be observed from a physical mixture of quasi-homogeneous mixed-layer species[16] but it is first necessary to know whether it is always possible to distinguish between a physical mixture of two quasi-homogeneous mixed-layer systems and a unique quasi-homogeneous system.

Calculations of XRD curves were made for i-m systems ($d_i = 10$ Å, $d_m = 17.8$ Å) saturated with glycerol. Similar results were given by the comparison with i-m systems saturated with ethylene glycol (Drits 1987).

Figure 9.13 shows the histograms and the diffraction patterns of two interstratified i-m systems as well as the powder pattern of a mixture of these

[16] Physical mixtures of minerals consisting of particles having ordered structures and homogeneous compositions, such as smectite, illite, rectorite (ABAB), tarasovite (AABAAB) will not be considered here as they can be unambiguously identified by their XRD pattern.

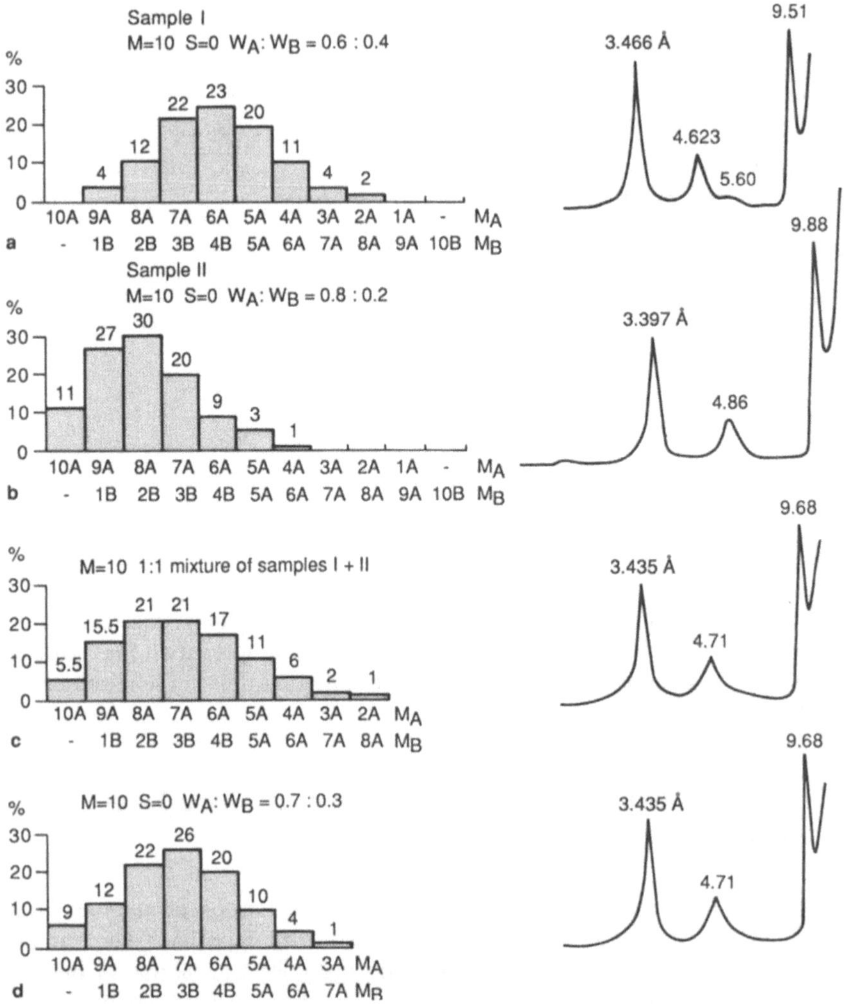

Fig. 9.13. Comparison of the histogram and X-ray powder diffraction curve of the quasi-homogeneous mixed-layer illite-smectite (i-s) and a mixture of sample *I* ($M = 10, S = 0$, $W_A : W_B = 0.6 : 0.4$) and sample *II* ($M = 10$, $S = 0$, $W_A : W_B = 0.8 : 0.2$). The histograms show the distribution of concentrations for coherent domains having M_A layers A and M_B layers B for the constant total number M of layers in the crystal, with $M = M_A + M_B$. Values for M and independent probability parameters are shown above each histogram. Smectite layers in i-s were saturated with glycerol ($d_A = 10.0$ Å and $d_B = 17.8$ Å). For quasi-homogeneous i-s, domains of different thicknesses having $M = 7$ to 13 layers were assumed to be equally distributed. Ordinate axes give the percentage of layers

systems. Parts *a* and *b* correspond to two systems (species I and II) with the same number M = 10 of scattering units and with the same parameters S = 0, but with two different $W_i:W_m$ ratios, i.e., 0.6:0.4 and 0.8:0.2 respectively. Part *c* of the figure is a histogram representing the distribution of compositions in the coherent domains for a physical mixture of equal proportions of species I and II. Part *d* represents a quasi-homogeneous assemblage of coherent domains with $W_i:W_m = 0.7:0.3$ and S = 0. Because the histograms in parts *c* and *d* are quite similar, the XRD curve corresponding to these values of the parameters cannot be distinguished from the mean weighted sum of the curves calculated with $W_i:W_m = 0.6:0.4$ and $W_i:W_m = 0.8:0.2$ respectively, with all other parameters retaining unchanged values.

Curves calculated for samples like those of Fig. 9.13 but having the additional limitation S = 1 imposed on the layer stacking (with $W_i > W_m$, $P_{mm} = 0$) are shown in Fig. 9.14. Comparing *c* and *d*, we see that the peaks, once again, exactly coincide for all the scattering angles but have different intensities in the region of the peak at 13.9 Å. In practice, the curves can, however, hardly be distinguished owing to the number of factors which affect the intensity distribution in the region of small scattering angles and which cannot be included in the calculation, e.g., difference in the degree of orientation of particles with same thickness, the different modes of distribution of thicknesses of the coherent domains, etc. Moreover, it should be remembered that the comparison is carried out under ideal conditions, i.e., with a known degree of heterogeneity of the samples: only two quasi-homogeneous samples were mixed in equal proportions and the parameters characterizing the structural specificities of mixed-layer samples were fixed.

Figure 9.15a shows a curve calculated for a mixture of two quasi-homogeneous samples having the same proportion of i scattering units ($W_i = 0.70$) but differing in the sequence of scattering unit types. One part of the mixture has S = 0, the other S = 1 and $P_{mm} = 0$. Figure 9.15b corresponds to a quasi-homogeneous sample with S = 1, $W_i = 0.70$ and $P_{mm} = 0.15$ (50% of MPDO). Again, the curves coincide in the whole range of angles of diffraction except for a minor difference in the profile of the first low-angle reflection.

Calculations show that a similar diffraction effect appears if the curve simulated for the mixture of quasi-homogeneous i-m samples ($W_m = P_{mm} = 0.4$, S = 0 for the first component, $W_m = 0.4$, $P_{mm} = 0$, S = 1 for the second) is compared with that for a quasi-homogeneous i-m with $W_m = 0.4$, $P_{mm} = 0.2$, S = 1. Only a small difference is observed in the low-angle region when the comparison is carried out for i-m saturated either with glycerol or with ethylene glycol.

If equal quantities of two mixed-layer samples with 60% and 20% of i units respectively but with M = 10 and S = 0 for both are mixed, the histogram of the mixture should be markedly different from that of quasi-homogeneous sample with averaged parameters $W_i = 0.40$ and S = 0. The histogram of the mixture should indeed contain two maxima, one of them corresponding to the composition of the crystals having $M_i = 2$ and $M_m = 8$, the other to $M_i = 6$ and $M_m = 4$. The histogram of the quasi-homogeneous sample having $W_i = 0.40$ and S = 0 contains a single peak corresponding to the composition of the crystals with $M_i = 4$ and

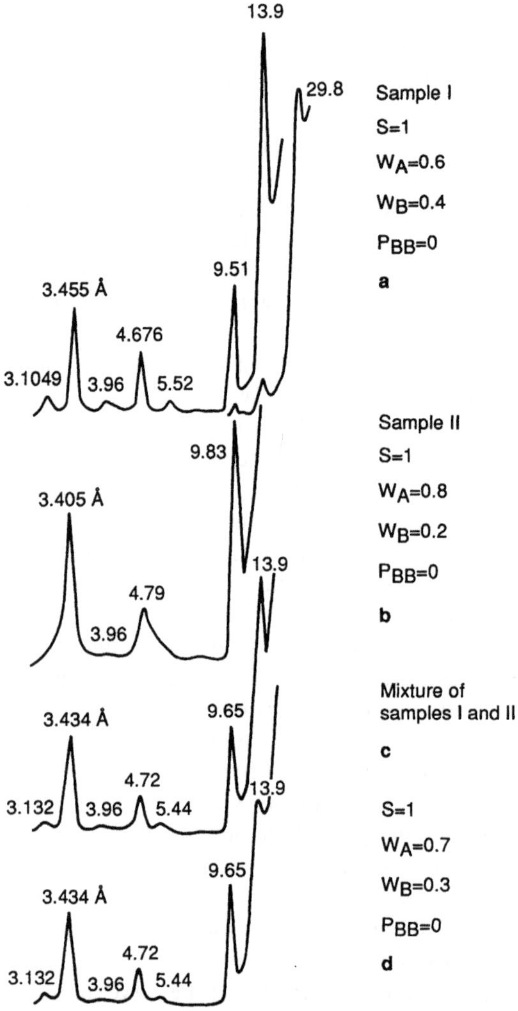

Fig. 9.14. Comparison of X-ray powder diffraction diagrams calculated for the quasi-homogeneous ordered (S = 1) mixed-layer illite-smectite (i-s) and for a mixture of sample I ($M = 10$, $W_A : W_B = 0.6 : 0.4$) and sample II ($M = 10, W_A : W_B = 0.8 : 0.2$). Smectite layers in i-s were saturated by glycerol ($d_A = 10.0$ Å and $d_B = 17.8$ Å). For quasi-homogeneous i-s, domains of different thicknesses with M = 7 to 13 layers were assumed to be equally distributed

$M_m = 6$. On the contrary, a quasi-homogeneous i-m with $W_i = 0.40$ may be described by a histogram with two maxima if different layer-types alternate with a certain degree of segregation. Here, the value S = 1 should be used to describe the sequence.

Figure 9.16 shows the XRD curves for the glycerol-saturated system described above. The lower curve (a) corresponds to a mixture of equal quantities of quasi-

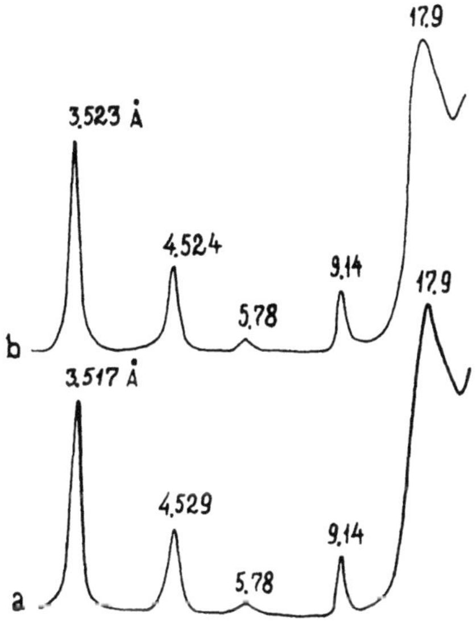

Fig. 9.15. Comparison of X-ray powder diffraction diagrams calculated for glycerol-saturated illite-smectite (i-s) ($d_A = 10.0$ Å and $d_B = 17.8$ Å). The *upper curve* corresponds to quasi-homogeneous i-s having $W_A = 0.70, P_{BB} = 0.15$ and $S = 1$. The *lower curve* corresponds to a 1:1 mixture of quasi-homogeneous i-s having $W_A = 0.70, S = 0$ and $W_A = 0.70, P_{BB} = 0, S = 1$. The quasi-homogeneous samples were assumed to have equally distributed coherent domain thicknesses with $M = 7$ to 13 layers

homogeneous i-m with $S = 0$, $W_i = 0.60$ and $W_i = 0.20$ respectively. The upper curve (b) corresponds to a quasi-homogeneous sample with $S = 1$, $W_i = 0.40$ and $P_{ii} = 0.5$, indicating a tendency to segregation. The intensity distribution and positions of the basal reflections in both curves are similar. Thus, in certain cases, the degree of heterogeneity for a physical mixture of quasi-homogeneous samples may be similar to that of a quasi-homogeneous species. It is therefore evident that, in these cases, using XRD patterns to distinguish between a mixture of samples and a quasi-homogeneous sample with the same average W_i is difficult.

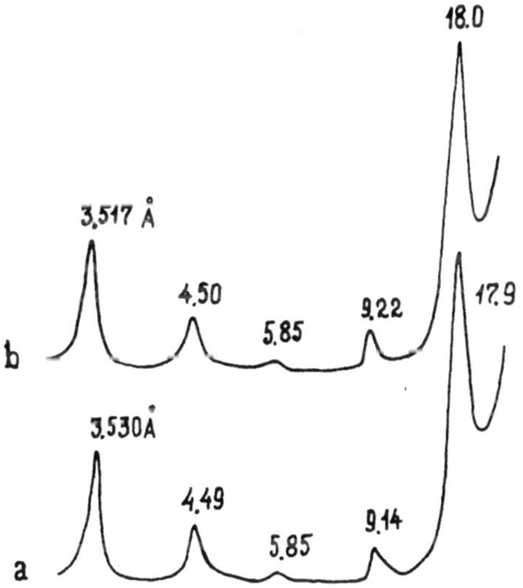

Fig. 9.16. Comparison of X-ray powder diffraction diagrams calculated for glycerol-saturated illite-smectite (i-s) ($d_A = 10.0$ Å and $d_B = 17.8$ Å). The *upper curve* corresponds to quasi-homogeneous i-s having $W_A = 0.40, P_{AA} = 0.5$ and $S = 1$. The *lower curve* corresponds to a physical mixture of random quasi-homogeneous i-s samples having $W_A = 0.60$ and 0.20, respectively, and $S = 0$. For quasi-homogeneous i-s, domains of different thicknesses having $M = 5$ to 15 layers were assumed to be equally distributed

The interpretation of XRD patterns from physical mixtures of quasi-homogeneous samples will be less ambiguous if these patterns contain "split" basal reflections corresponding to the components of the mixture. This will occur if the quasi-homogeneous samples which make up the mixture differ markedly in the proportions of different layer types, and/or the mode of alternation of the layers, and in the intensities of the reflections being mixed. For i-m, such patterns are observed from mixtures of quasi-homogeneous samples having, for instance, $W_i = 0.40$, $S = 0$ and $W_i = 0.40$, $P_{ii} = 0.5$, $S = 1$; or $W_i = 0.20$, $S = 0$ and $W_i = 0.80$, $S = 0$; or $W_i = 0.60$, $P_{mm} = 0$, $S = 1$ and $W_i = 0.20$, $P_{ii} = 0$, $S = 1$, etc. Even then, the interpretation of XRD patterns may be complicated and ambiguous, even for physical mixtures of samples having quite different degrees of heterogeneity.

Figure 9.17 shows curves calculated for a physical mixture of pure montmorillonite and a quasi-homogeneous i-m sample with $W_i = 0.45$, $W_m = 0.55$, $P_{ii} = 0.15$, $S = 1$. The montmorillonite and i-m were mixed in equal quantities but, in drawing up the curves *a* and *b* in Fig. 9.17, it was assumed that the total number of layers in the coherent domains ranged respectively from 4 to 8 and from 1 to 10, with domains of each size equally distributed. From the appearance of these curves, it is difficult to infer that the sample contains 50% of i-m with a high tendency to ordering. This conclusion could perhaps be drawn from the poorly resolved superorder maximum in the low-angle region of the diffraction pattern ($\theta_{CuK\alpha} = 1.5°$) in Fig. 9.17a, but the peak is practically absent in curve *b* and the true structure of the sample cannot therefore be deduced.

To avoid misunderstanding, it should be stressed that the mixture of montmorillonite and rectorite can be unambiguously identified using XRD data, whatever the proportion of these components.

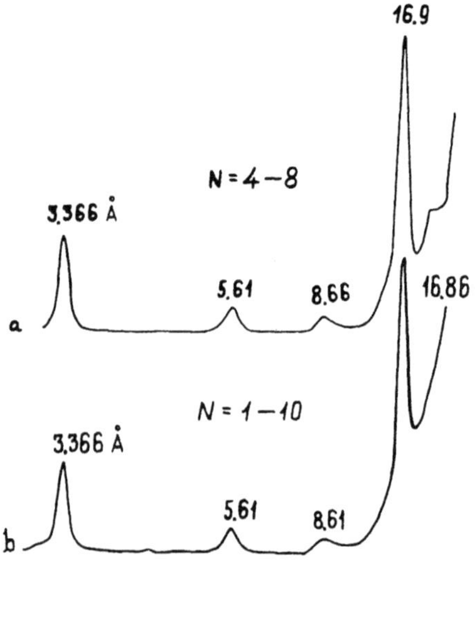

Fig. 9.17. Comparison of X-ray powder diagrams calculated for mixtures of equal quantities of smectite and illite-smectite (i-s) having $W_A = 0.45, P_{AA} = 0.15$ and $S = 1$ with differences in the range of coherent domain thicknesses as marked above each curve. Domain thicknesses were assumed to be equally distributed. Smectite layers in i-s were saturated by ethylene glycol ($d_A = 10$ Å, $d_B = 16.86$ Å)

The considerations above agree with the conclusions drawn by Gradusov (1971) from his experiments on mixtures of interstratified minerals.

9.4.4 Homogeneous Mixed-Layer Models

Statistically homogeneous models (Plançon et al. 1983; Drits et al. 1984), in which all the crystals have the same composition but differ in the distribution of layer types, may also be envisaged. The procedure here is the comparison of the curves calculated on the basis of homogeneous and quasi-homogeneous models with identical contents of layer types and sequence patterns.

Figure 9.18 shows histograms for the distribution of compositions of coherent domains in two random mixed-layer samples with the same mean proportion of i units ($W_i = 0.70$), one of them homogeneous (lower histogram) and the other quasi-homogeneous (upper histogram). The mean number of layers in a coherent domain was assumed to be 10 for the quasi-homogeneous sample. In the same figure, XRD curves calculated for glycerol-saturated i-m samples are shown. The two upper curves respectively correspond to quasi-homogeneous and homogeneous i-m samples, with $W_i = 0.70$, $S = 0$, $d_i = 10$ Å and $d_m = 17.8$ Å. In both samples, the total number M of scattering units in coherent domains was assumed to range from 7 to 13, with domains of different thickness equally distributed. For the homogeneous sample, it was assumed that the proportion M_i/M of i units in different crystals was 0.65 – 0.75 (Fig. 9.18).

Comparison of the curves shows that they are similar with respect to intensity distribution but not identical in terms of shape and position of the basal reflections. Thus, theoretically, these two models may be distinguished by XRD but this, however, is not always easy because of the ambiguity in the precise determination of the thickness of individual layers and of the coherence domains. Figure 9.18c shows, for instance, an XRD curve calculated for a quasi-homogeneous i-m sample with $W_i = 0.65$, $W_m = 0.35$, $S = 0$, $d_i = 9.9$ Å and $d_m = 17.5$ Å. This curve is similar to the one in Fig. 9.18b calculated for a random homogeneous i-m sample with $W_i = 0.70$, $d_i = 10$ Å, and $d_m = 17.8$ Å and the same distribution of thicknesses of coherent domains. Very precise determinations are therefore required to reveal the existence of homogeneous, quasi-homogeneous, and heterogeneous mixed-layer samples.

9.5 Quantitative Determination of the Structural Characteristics of Interstratified Dioctahedral Mica-Smectite Minerals

We shall now present some unpublished results of quantitative determinations of structural characteristics obtained by Sakharov and Drits on three different in-

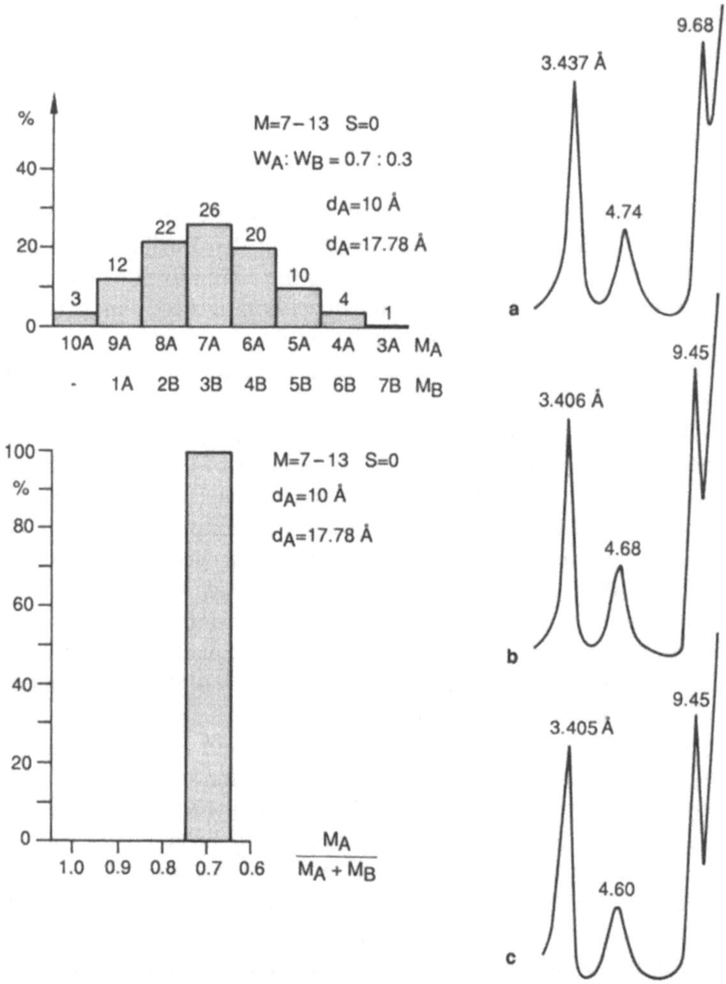

Fig. 9.18. Comparison of histograms and diagrams calculated for random quasi-homogeneous and homogeneous glycerol-saturated illite-smectite (i-s). The *upper histogram* shows the content of coherent domains containing M_AA layers and M_BB layers. The *lower histogram* shows that the homogeneous i-s contains domains in which the content of A layers lies within the interval 0.65−0.75. The *upper diffraction curve* corresponds to a quasi-homogeneous i-s having $W_A = 0.70$, S = 0, the *middle curve* to a homogeneous i-s having $M_A/M = 0.70 \pm 0.05$, the *lower curve* to a quasi-homogeneous i-s having $W_A = 0.65, S = 0$ ($d_A = 9.9$ Å and $d_B = 17.5$ Å). Domain thicknesses ranging from M = 7 to 13 are equally distributed for all the samples

terstratified minerals composed of mica and smectite layers, using the method in which best agreement is sought between experimental and calculated values of the position of the relative intensity and the profile of the peaks, in the XRD diagrams.

The oriented powder samples were sufficiently thick to comply with the conditions for infinite absorption (Chap. 6.1.4) in a reflection mounting. Monochromatized Cu Kα radiation was used and detected by a DRON-UM 1 diffractometer equipped with a stepping motor permitting measurements every 0.1° in θ for 100 s. The experimental conditions were such that convolution effects due to the divergence of the beam could be neglected.

9.5.1 Two-Component Interstratified Minerals: Celadonite-Nontronite

This sample is part of a large series of interstratified celadonite-nontronite minerals fairly widespread in sedimentary hydrothermal volcanic formations from the sea of Japan (Lipkina et al. 1987).

After heat treatment to 300 °C of the mineral, the XRD diagram is reduced to two basal reflections at $d = 9.8$ Å and $d = 3.30$ Å, an indication that this interstratified sample contains mica-type and smectite-type layers (cf. Sect. 9.2.1).

Figure 9.19 shows the tracings (circles) of two nonrational series of basal reflections given by the mineral treated with ethylene glycol (curve a) or glycerol (curve b). The presence of a reflection at d = 17.3 Å or d = 18.2 Å respectively in the two tracings indicates that the sequence of mica-type and smectite-type layers is nearly

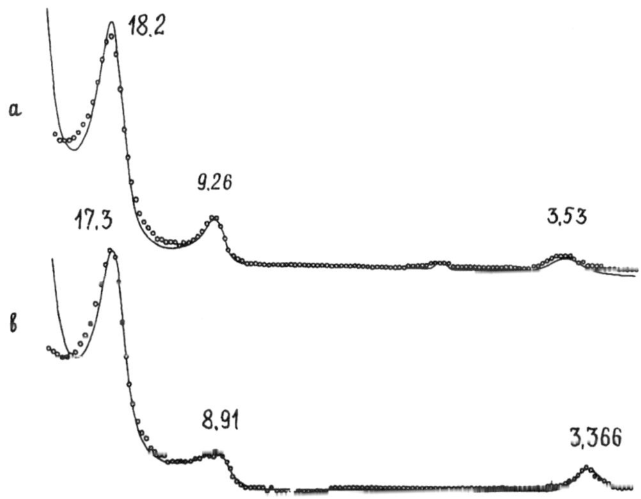

Fig. 9.19. Comparison between experimental (*circles*) and calculated (*full line*) X-ray diffraction curves obtained from a mixed-layer celadonite-nontronite (c-n) mineral saturated with glycerol (**a**) or ethylene glycol (**b**). The parameters of the model used for the calculations were $W_C : W_n = 0.4 : 0.6$, $P_{CC} = 0.55$, $S = 1$, $d_c = 10$ Å, $d_n = 16.90$ Å or 17.78 Å; the distribution of thicknesses of coherence domains ranged from M = 2 to 8

totally disordered. The qualitive comparison of the tracings with the nomograms computed for an interstratified i-m model with $S = 0$ (Fig. 9.6) shows that the mineral contains approximately equal numbers of mica and smectite-type scattering units while its chemical composition ($SiO_2 = 50.69\%$; $Fe_2O_3 = 29.06\%$; $Al_2O_3 = 0.87\%$; $K_2O = 1.64\%$; $MgO = 4.5\%$; $FeO = 0.43\%$; $CaO = 0.44\%$; $Na_2O = 0.50\%$) indicates that the layers are celadonite (or glauconite) and nontronite.

Successive improvements in the calculated XRD curves led to a final structural model of interstratified celadonite-nontronite (c-n) with $S = 1$, $W_c : W_n = 0.40 : 0.60$, and $P_{CC} = 0.55$. The thicknesses of the c and glycerol-solvated or ethylene glycol-solvated n scattering units are 10 Å, 17.78 Å, or 16.9 Å, with a rms deviation of 0.1 Å. The distribution of the interferential coherence domain thicknesses lies between $M = 2$ and $M = 8$. The structural formulas of the layers are:

- for celadonite, $K_2Si_8(FeMg)_2O_{20}(OH)_4$;
- for nontronite, $Na(Si_7Al)Fe_4O_{20}(OH)_4$.

The XRD patterns calculated on the basis of this model are shown in Fig. 9.19 by the continuous lines.

From the values of the structural parameters giving the best agreement between the experimental and calculated diagrams, it is possible to deduce that the interstratified mineral contains 40% c and 60% n layers with a tendency to segregation in the sequence described by the segregation parameter S_{G1} (cf. Chap. 4.3.2) given by Eq. (4.21) as

$$S_{G1} = 1 - \frac{1 - P_{cc}}{1 - W_c} = 0.25 \ .$$

(It will be recalled that $S_{G1} = 0$ describes total disorder in the sequence of c and n layers while $S_{G1} = 1$ describes total segregation of the layers in two distinct phases.)

9.5.2 Three-Component Interstratified Minerals: Leucophyllite-Montmorillonite-"Vermiculite" with S = 1

The basal reflections obtained from this mineral after treatment with ethylene glycol are shown in Fig. 9.20 (broken line). They constitute a nonrational series in which the positions of the maxima are close to the positions calculated for an interstratified i-m system with $S = 1$ and $W_i > W_m$ (cf. Fig. 9.4), particularly for the reflections at $d = 12.72$ Å and $d = 9.67$ Å.

The analytical data ($SiO_2 = 55.4\%$; $Al_2O_3 = 17.13\%$; $Fe_2O_3 = 1.21\%$; $FeO = 0.22\%$: $MgO = 8.30\%$; $CaO = 0.56\%$; $Na_2O = 0.20\%$; $K_2O = 5.04\%$) and

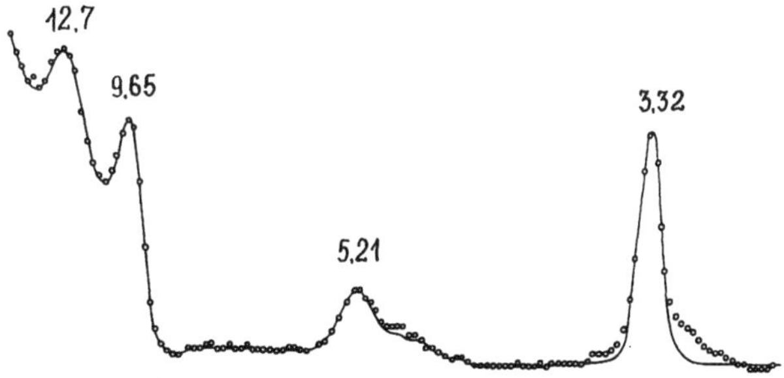

Fig. 9.20. Comparison between experimental (circles) and calculated (full line) X-ray diffraction curves for an interstratified leucophyllite-montmorillonite-"vermiculite" (l-m-v) solvated with ethylene glycol. The parameters of the model used in the calculations were $W_l:W_m:W_v = 0.75:0.21:0.04$, $P_{mm} = P_{mv} = P_{vm} = P_{vv} = 0$, $S = 1$, $d_l = 9.88$ Å, $d_m = 16.82$ Å, $d_v = 13{,}5$ Å, and $5 \leqslant M \leqslant 10$

the diffractogram given by the mineral after dehydration at 300 °C indicate that it is composed of montmorillonite (m) and leucophyllite (l) layers [the ideal formula for leucophyllite is $KSi_4(AlMg)_2O_{10}(OH)_2$]. With this structural formula, the thickness of the leucophyllite scattering unit must be smaller than 10 Å.

The various synthetic diagrams based on two-component l-m systems with $S = 1$ are in fair agreement with the experimental diffractogram, with the exception of a shoulder found on the large-angle side of the reflection at $d = 5.155$ Å and practically nonexistent in the synthetic diagrams. It is therefore necessary to consider a three-component interstratified system in which the solvation of montmorillonite layers could lead to interlamellar spaces with either a single or a double layer of ethylene glycol molecules. This assumption can be justified on the grounds of the relatively large charge of the layers. Conventionally, the scattering units with a single layer of ethylene glycol will be called "vermiculite" (v) units.

This three-component model gives good agreement between the experimental and the calculated (Fig. 9.20, full line) XRD diagrams. The characteristic structural parameters of the final model – and hence of the solvated mineral – are:

- $S = 1$; $W_l:W_m:W_v = 0.75:0.21;0.04$; $P_{mm} = P_{mv} = P_{vm} = P_{vv} = 0$;
- thickness of the scattering units: $d_l = 9.88$ Å; $d_m = 16.82$ Å; $d_v = 13.5$ Å;
- number M of scattering units in the interferential coherence domains: $5 \leqslant M \leqslant 10$.

The fact that the structure does not contain any adjacent mm-, mv, vm- or vv-pairs implies that the mineral has a MPDO structure.

9.5.3 Two-Component Interstratified Minerals: Leucophyllite-Montmorillonite, with S = 3

Figure 9.21 shows a recording of the basal reflections given by the ethylene glycol-treated mineral: the general aspect of the tracing corresponds to a solvated illite except for two differences: a slight nonrationality in the positions of the maxima, and the existence of a shoulder on the low-angle side of the first reflection (at $d \approx 11$ Å).

These two features are sufficient to indicate that the mineral contains sequences of mica- and smectite-type layers, with $S > 2$. Again, the analytical data ($SiO_2 = 63.45\%$; $Al_2O_3 = 20.81\%$; $Fe_2O_3 = 0.92\%$; $FeO = 0.41\%$; $MgO = 9.22\%$; $K_2O = 5.70\%$) and the diffractogram given by the mineral after dehydration at 300 °C lead to the conclusion that the layers are leucophyllite and montmorillonite.

Agreement between the experimental tracing (circles) and the calculated diagram (full line) is best with a structural model having S = 3 and MPDO for the l and m layers, thicknesses $d_l = 9.88$ Å and $d_m = 16.60$ Å, concentrations $W_l : W_m = 0.89 : 0.11$, with $P_{mm} = P_{mlm} = 0$ and $P_{mllm} = 0.15$, and with a number M of layers in the interferential coherence domains distributed between 5 and 15.

9.6 Semi-Quantitative Determinations of the Structural Characteristics of Interstratified Minerals

Numerous studies on the structure of interstratified minerals using the method of modelization of X-ray diffractograms have shown that it is not essential to obtain simultaneous perfect agreement between the calculated and experimental values for the positions, the profiles and the relative intensities of the reflections. Since

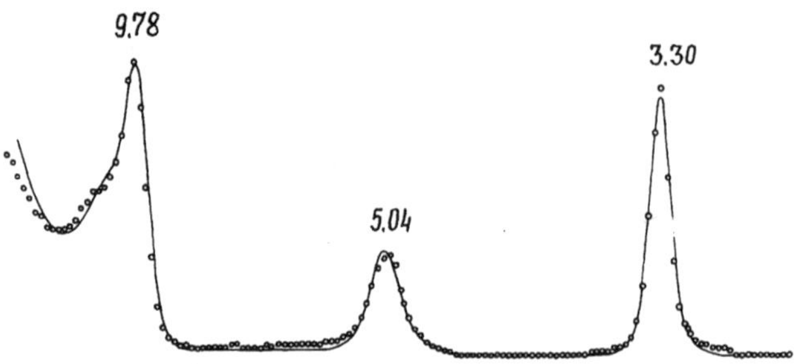

Fig. 9.21. Comparison between experimental (*circles*) and calculated (*full line*) X-ray diffraction curves for an interstratified leucophyllite-montmorillonite (l-m) solvated with ethylene glycol. The parameter of the model used for the calculations were $W_l : W_m = 0.89 : 0.11, P_{mm} = P_{mlm} = 0$, $P_{mllm} = 0.15, S = 3, d_l = 9.88$ Å$, d_m = 16.6$ Å$, 5 \leqslant M \leqslant 10$

the latter (and particularly low-angle reflections) are extremely sensitive to a number of parameters such as:

- the chemical composition of the contents of the interlamellar space and the distribution of the ions and molecules therein,
- the width of the distribution function of the interferential coherence domain thickness,
- the heterogeneity of constitution of these domains, etc.,

and insofar as the accurate determination of the values of these parameters does not usually constitute the principal goal of the structural study, a good agreement for the positions and a semi-quantitative agreement for the profiles and relative intensities my be sufficient, provided this is maintained when the mineral species is submitted to different treatments. Nevertheless, some of the examples to be presented later will show that a good agreement for the profiles may, in some cases, also be necessary to provide an accurate determination of the structural characteristics, particularly when the system is made up from more than two components or when the silicate layers are themselves defective.

9.6.1 Two-Component Interstratified Minerals: Illite-Montmorillonite

Four examples will be given of the determination of the principal structural characteristics of interstratified i-m minerals by the semi-quantitative analysis of their X-ray diffractograms. Two of the minerals (samples 1 and 4) are found in tuffs from Ascana (Georgia, USSR), sample 2 is from Kuli-Kolon (Tajikistan, USSR) and sample 3 is from Sarigukh (Armenia, USSR), (Gradusov 1971).

Figures 9.22 to 9.25 give the experimental diffractograms of the four minerals solvated with glycerol (curves a) or with ethylene glycol (curves b) as well as the diagrams calculated using models corresponding to the same type of solvation. In all cases, the basal reflections in the calculated XRD curves are found at positions and with intensities close to the ones observed experimentally. Considering that this agreement is observed for two different states of solvation of each mineral, this simple qualitative comparison is sufficient to show that the models used for calculating the spectra are good descriptions of the main structural characteristics of the minerals as far as types of layers, relative proportions, and mode of stacking are concerned, namely

- for sample 1: $W_i : W_m = 0.54 : 0.46$, with $S = 0$;
- for sample 2: $W_i : W_m = 0.75 : 0.25$, with $S = 1$ and $P_{mm} = 0$.
- for sample 3: $W_i : W_m = 0.67 : 0.33$, with $S = 2$ and $P_{mm} = 0$, $P_{iii} = 0.4$. Note, in this case, the presence of a reflection at $d \simeq 12$ Å $- 14$ Å characteristic of systems with $S > 1$. These values of the structural parameters place the mineral on the segment OO′ in the square diagram of Fig. 4.3, from which it is deduced that the structure shows a tendency to the order iimiim . . . in the sequence of

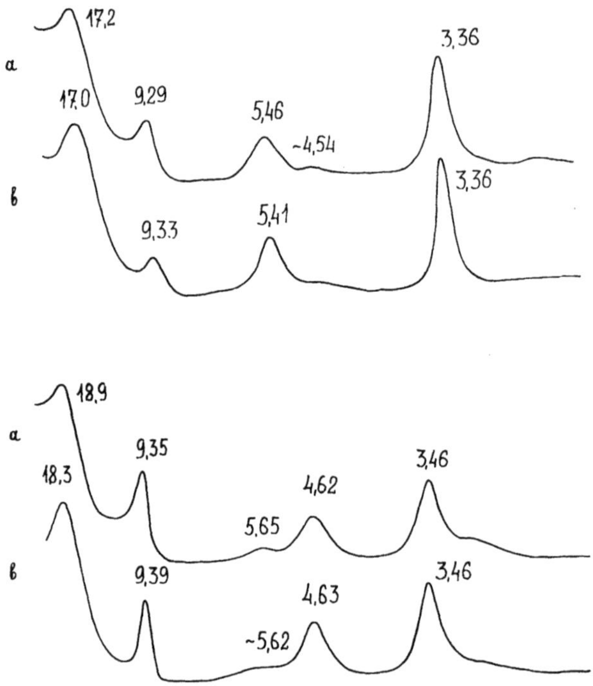

Fig. 9.22. Comparison between experimental (**a**) and calculated (**b**) X-ray diffraction curves for an interstratified illite-montmorillonite (i-m) solvated with ethylene glycol (*upper curves*) or glycerol (*lower curves*). The parameters of the model used for the calculations were $W_i : W_m = 0.55 : 0.45, S = 0, d_i = 10$ Å$, d_m = 16.9$ Å or 17.78 Å, and $5 \leqslant M \leqslant 10$

layers. The value of the coefficient D_{s2} characteristic of the degree of disorder in the sequence of ii-pairs and m layers ($D_{s2} = 0.8$) indicates, however, that the tendency to form the order iimiim... is weak (for MPDO, a structure with $S = 2$ should have $D_{s2} = 0$ while one with $S = 1$ should have MPDO and $D_{s2} = 1$);

- for sample 4: $W_i : W_m = 0.84 : 0.16$, with $S = 3$ and $P_{mm} = P_{mim} = P_{miim} = 0$.

9.6.2 Interstratified Minerals with Kaolinite 1:1 Layers

Drits and Sakharov (1976) have calculated the set of theoretical diagrams corresponding to all possible cases of quasi-homogeneous interstratified k-m with $S = 1$, and k-m-i with $S = 0$. From this systematic study, the following criteria were developed which can be used for the qualitative evaluation of X-ray diffractograms to determine if a given interstratified species contains kaolinite layers.

a) All interstratified minerals containing kaolinite layers give two intense reflections, one of which is at 7.15 Å $\leqslant d \leqslant$ 8.43 Å when ethylene glycol is the

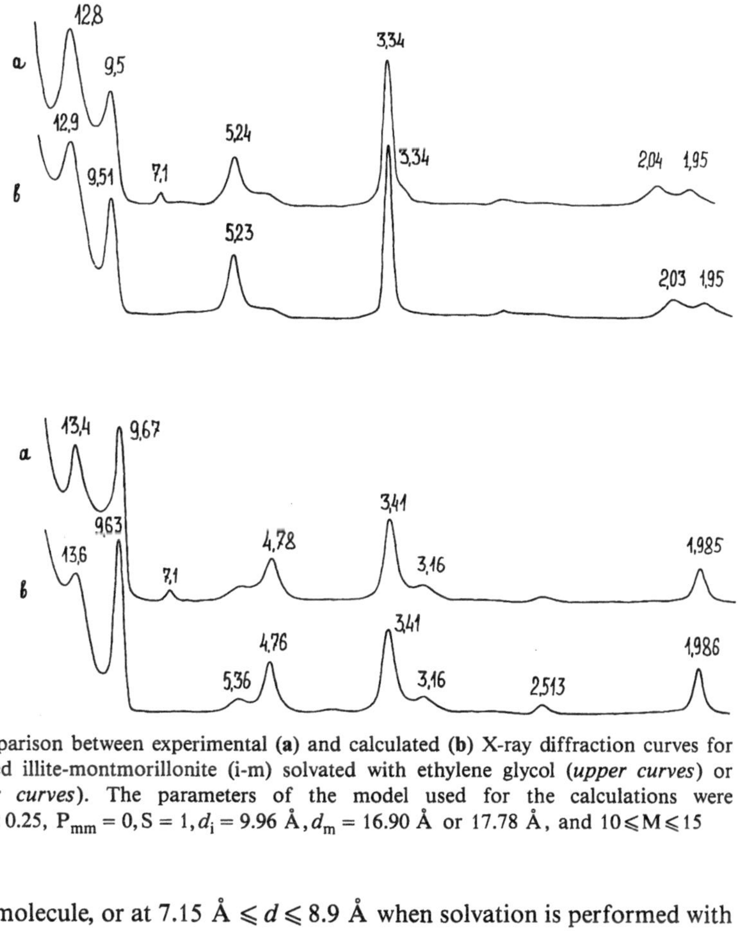

Fig. 9.23. Comparison between experimental (**a**) and calculated (**b**) X-ray diffraction curves for an interstratified illite-montmorillonite (i-m) solvated with ethylene glycol (*upper curves*) or glycerol (*lower curves*). The parameters of the model used for the calculations were $W_i : W_m = 0.75 : 0.25$, $P_{mm} = 0$, $S = 1$, $d_i = 9.96$ Å, $d_m = 16.90$ Å or 17.78 Å, and $10 \leqslant M \leqslant 15$

solvating molecule, or at 7.15 Å $\leqslant d \leqslant$ 8.9 Å when solvation is performed with glycerol. The second reflection is even more characteristic: two-component glycerol-solvated k-m systems give a narrow reflection at $d \approx 3.56$ Å to 3.57 Å, whatever the concentration of k or m layers or their mode of succession. Similarly, glycerol-solvated three-component k-m-i systems give an intense reflection at $d < 3.56$ Å.

b) One way of finding out if the system is composed of k and m layers is to study the mineral after heating it to 350 – 450 °C. For these components, a nonrational series of reflections is observed due to the presence of k layers 7.15 Å thick and of dehydrated m layers 9.5 Å – 9.7 Å thick in the mineral.
c) Another useful criterion for the identification of interstratified k-m minerals is found in the study of the Mg^{2+}-exchanged samples which give a nearly rational series of basal reflections owing to the fact that the k and m scattering units are then respectively 7.15 Å and 14.4 Å thick.

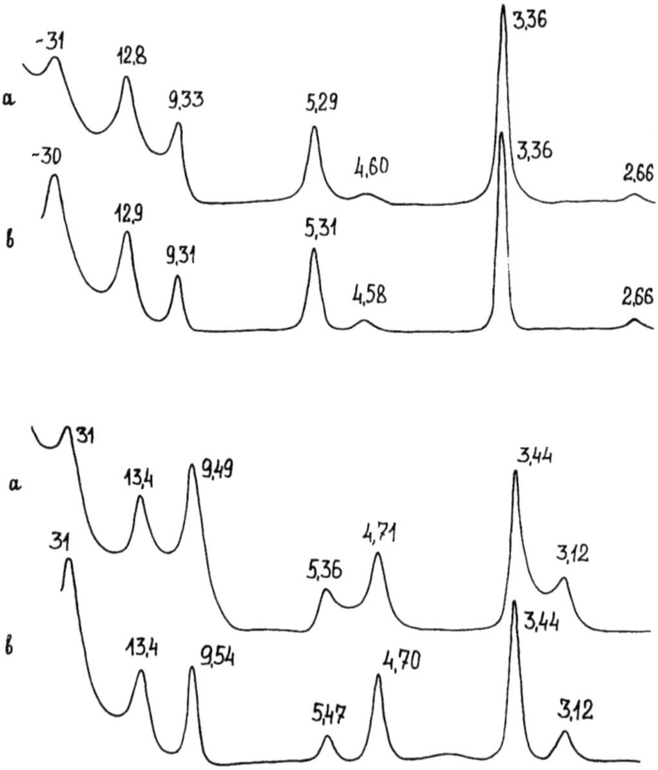

Fig. 9.24. Comparison between experimental (**a**) and calculated (**b**) X-ray diffraction curves for an interstratified illite-montmorillonite (i-m) solvated with ethylene glycol (*upper curve*) or glycerol (*lower curve*). The parameters of the model used for the calculations were $W_i : W_m = 0.65 : 0.35$, $P_{mm} = 0, P_{iii} = 0.4, S = 2, d_i = 10$ Å, $d_m = 16.9$ Å or 17.78 Å, and $10 \leqslant M \leqslant 15$

We shall now examine three examples of semi-quantitative comparison of experimental and calculated spectra of interstratified minerals containing kaolinite layers.

1. Two component (k-m) minerals

Figure 9.26 gives the experimental XRD spectra of an interstratified k-m clay mineral of hydrothermal origin from the southwestern part of the Kambalny mountains (Kamchatka). Curve *a* is the diffractogram given by the Mg^{2+}-exchanged mineral, curve *b* is that of a glycerol-solvated sample, and curve *c* was obtained from the mineral after dehydration at 350 °C. A model with $W_k : W_m = 0.75 : 0.25$ and $S = 0$ yields calculated spectra with the main features of the experimental results, as seen in the Fig. 9.26 d and e respectively. These values of the parameters are a good representation of the true structural characteristics of the mineral (Sakharov and Drits 1973).

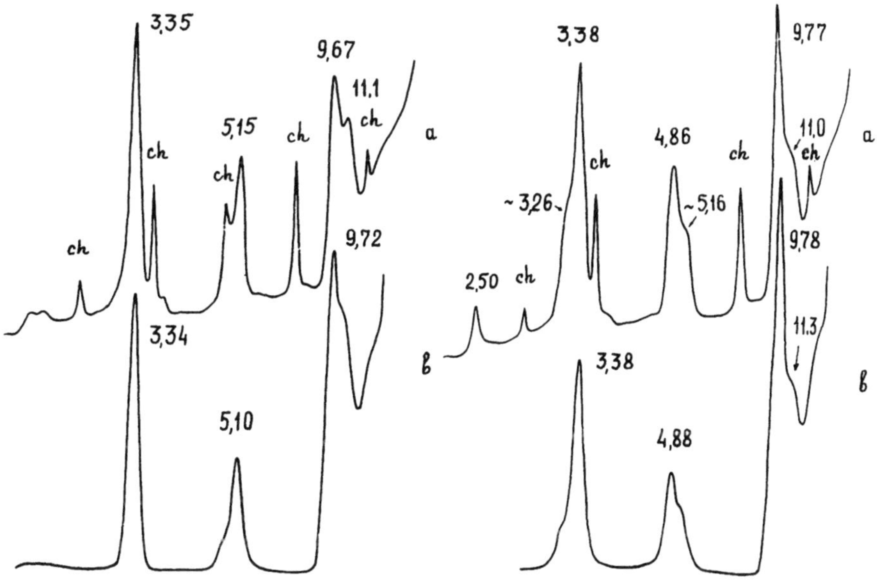

Fig. 9.25. Comparison between experimental (**a**) and calculated (**b**) X-ray diffraction curves for an interstratified illite-montmorillonite (i-m) solvated with ethylene glycol (*left-hand curves*) or glycerol (*right-hand curves*). The parameters of the model used for the calculations were $W_i : W_m = 0.84 : 0.16$, $P_{mm} = P_{mim} = P_{miim} = 0, S = 3, d_i = 10$ Å, $d_m = 16.86$ Å or 17.78 Å, and $15 \leqslant M \leqslant 20$

2. **Three-component (k-m-i) minerals**

This sample, from the Tepacan region (Yucatan), was kindly provided by L. Schultz, who carried out a detailed study of the clays from this part of Mexico (Schultz et al. 1971). The experimental curves presented in Fig. 9.27 correspond to different treatments to which the mineral was submitted: solvation by ethylene glycol (curve *a*), by glycerol (curve *b*), cationic exchange with Mg^{2+} (curve *c*) and dehydration at 350 °C (curve *d*). The comparison between these diffractograms and a series of reference diagrams calculated for an interstratified two-component k-m system with different values of S and of the $W_k : W_m$ ratio (Drits and Sakharov 1976) showed that it was not possible to obtain satisfactory even mere semi-quantitative agreement between the two sets of spectra. More specifically, the second reflection in the diffractogram of the glycol-treated sample is at $d < 3.57$ Å and, furthermore, the basal reflections from the Mg-exchanged sample form a nonrational series. Both observations are in contradiction with the criteria given above and the presence of a third component must be envisaged.

Since the mineral contains nonexchangeable K^+ cations, a 10 Å-thick mica-type layer was considered. The diagrams calculated for a three-component k-m-i system with $S = 0$ and $W_k : W_m : W_i = 0.70 : 0.10 : 0.20$ for the four different states of the natural samples are shown as curves *e*, *f*, *g*, and *h* of Fig. 9.27. In spite of

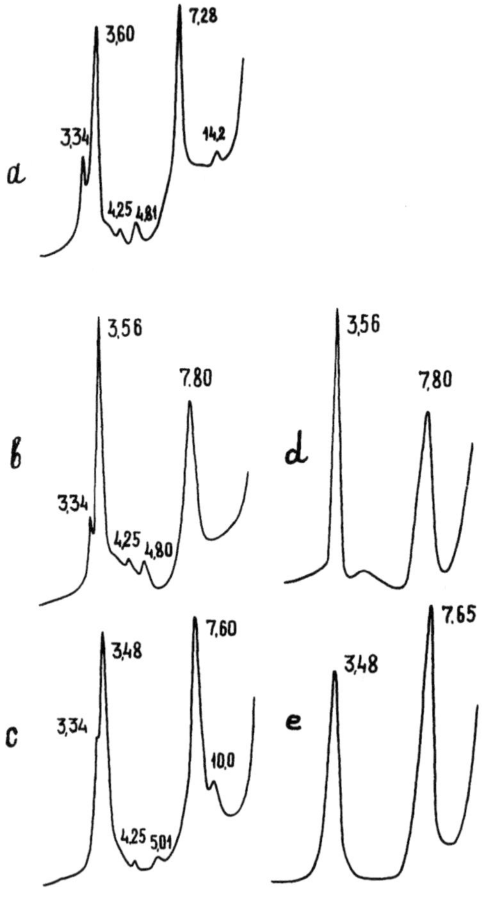

Fig. 9.26. Experimental X-ray diffraction curves: **a** for an interstratified Mg-exchanged kaolinite-montmorillonite (k-m); **b** after saturation with glycerol; **c** after dehydration at 350 °C. The parameters of the models used for the calculations relative to the curves **d** and **e** were: $W_k : W_n = 0.75 : 0.25, S = 0, d_k = 7.15$ Å, $d_m = 17.78$ Å (curve **d**) or 9.60 Å (curve **e**), and $10 \leqslant M \leqslant 20$

some differences between the profiles or intensities of the calculated and experimental curves, there is a fair overall agreement of the two sets of curves and the parameters chosen must therefore be quite close to those of the structure of the mineral under study.

3. **Three-component kaolinite-montmorillonite-swelling chlorite mineral**

From the methodological point of view, the kaolinite-montmorillonite-swelling chlorite (k-m-ch) to be described is a good illustration of the fact that a purely qualitative approach based solely on the analysis of the positions of the reflections may lead to an erroneous interpretation of the actual structure of the mineral (Sakharov et al. 1981).

The sample (from Moldavia) was studied in its natural state, after solvation by glycerine, and after various heat treatments. The diffractograms show the basal reflections from an interstratified mineral and additional reflections from

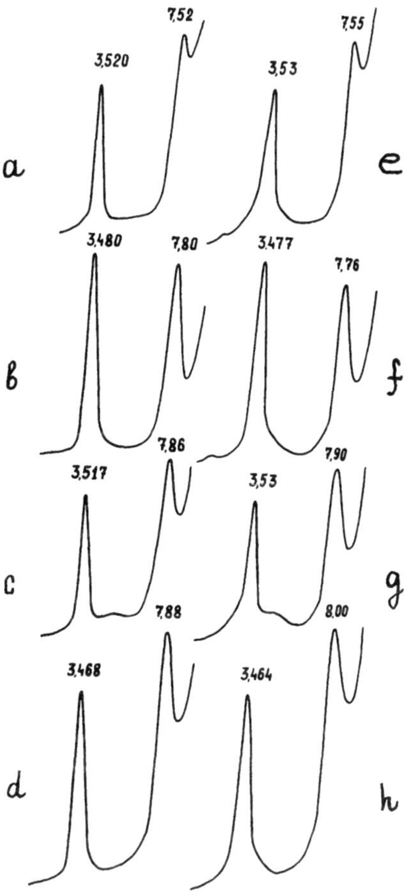

Fig. 9.27. Comparison between experimental and calculated X-ray diffraction curves for an interstratified kaolinite-montmorillonite-illite (k-m-i) either Mg-exchanged (**a**), or ethylene glycol-saturated (**b**) or glycerol-saturated (**c**), or dehydrated at 350 °C (**d**). The parameters of the models used to calculate curves **e, f, g** or **h** were $W_k : W_m : W_i = 0.70 : 0.10 : 0.20$, $S = 0$, $d_k = 7.15$ Å, $d_m = 14.4$ Å, 17.78 Å, 16.68 Å or 9.6 Å, $d_i = 10$ Å, and $5 \leqslant M \leqslant 15$

kaolinite, quartz and goethite present as impurities (curves *a*, *b*, and *c* in Fig. 9.28). There is only one reflection (at $d = 1.498$ Å) between 1.4 Å and 1.5 Å, implying that the silicate 2:1 layers are dioctahedral. Curve *a* shows three basal reflections from the interstratified phase at $d = 10.3$ Å, $d = 4.95$ Å, and $d = 3.326$ Å. The diagram given by the glycerol-solvated sample (curve *b*) is quite different, with three basal reflections at $d = 25.5$ Å, 8.42 Å and 3.56 Å forming a nearly rational series. Finally, the sample heated to various temperatures between 100 °C and 500 °C yields nearly identical diffractograms with basal reflections at $d = 9.8$ Å, 4.95 Å and 3.35 Å characteristic of an interstratified phase (curve *c*).

An interpretation of these diagrams based solely on the analysis of the positions of these maxima could give the impression that the mineral contains only two components, namely illite and montmorillonite: curves *a* and *c* resemble the ones given by Fe-illite, which gives two intense reflections at $d \simeq 10$ Å and $d \simeq 3.3$ Å and a weak reflection at $d \approx 4.9$ Å. Curve *b* could be interpreted as due to a regular

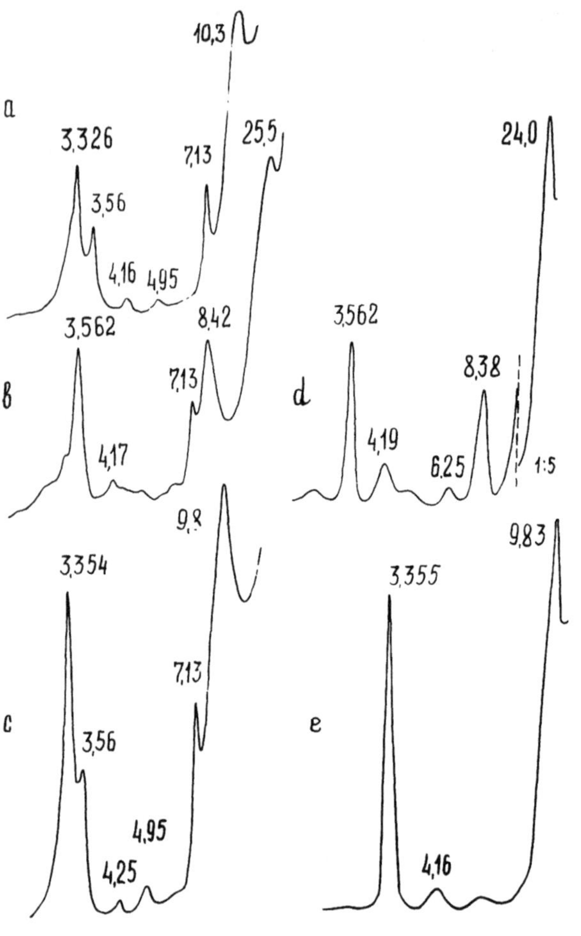

Fig. 9.28. Comparison between experimental and calculated X-ray diffraction curves for an interstratified kaolinite-montmorillonite-swelling chlorite (k-m-ch) in its natural state **(a)**, or glycerol-saturated **(b)**, or dehydrated at 350 °C **(c)**. Diagram **(d)** corresponds to a k-m model with $W_k:W_m = 0.48:0.52$, $P_{kk} = 0.2$, $S = 1$, $d_k = 7.15$ Å and $d_m = 17.78$ Å; diagram **(e)** was calculated using a k-m-ch model with $W_k:W_n:W_{ch} = 0.48:0.26:0.26$, $P_{kk} = 0.2$, $P_{km} = 0.4$, $P_{mk} = 0.74$, $P_{mm} = 0.13$, $S = 1$, $d_k = 7.15$ Å, $d_m = 17.78$ Å or 9.60 Å, $d_{ch} = 13.1$ Å, and $5 \leqslant M \leqslant 15$

ABAB... succession of 10 Å-thick mica layers and 14.4 Å-thick layers of montmorillonite with a single layer of glycerol in the scattering unit. Heating the sample would then have led to a collapse of the montmorillonite interlamellar spaces, resulting in the appearance of a nearly rational series of basal reflections at $d \approx \frac{10}{n}$ Å.

However, if we suppose that the interlayer spaces of the glycerol-solvated montmorillonite scattering units with thickness 17.78 Å contain two layers of organic molecules, the reflections at 25.5 Å can only be explained by admitting the presence of kaolinite layers 7.15 Å-thick alternating regularly with the smectite layers. The reflection at $d = 3.562$ Å on curve *b* seems to confirm this hypothesis.

In order to chose between these possibilities, attempts were made to account for the relative intensities of the reflections, but it was found that:

- on one hand, all the models with two-component glycerol-solvated i-m yielded calculated diagrams with an intense reflection at $d \approx 12$ Å, with no corresponding peak in the experimental diffractogram (cf. curve *b*);
- on the other hand, a two-component k-m model with $W_k : W_m = 0.48 : 0.52$, $P_{kk} = 0.2$ and $S = 1$, and a double layer of glycerol of solvation gave a calculated diagram (curve *d*) in fair agreement with the experimental result (curve *b*), but the same model used with the appropriate values to calculate the diffractogram of the dehydrated mineral did not agree either in positions or in intensities with the characteristics of the peaks in the experimental curve (*c*).

It was therefore necessary to admit that the thickness 17.8 Å found in the glycerol-solvated mineral corresponds to two types of scattering units; one of them is indeed montmorillonite, but the other could be a swelling chlorite. This hypothesis was confirmed by comparison of the experimental diffractogram with the diagram calculated with a model containing kaolinite (k), montmorillonite (m) and chlorite (ch) scattering units having, when anhydrous, thicknesses 7.15 Å, 9.6 Å and 13.1 Å, respectively. A satisfactory semi-quantitative agreement was then obtained with the following values of the characteristic parameters: $W_k : W_m : W_{ch} = 0.48 : 0.26 : 0.26$, $S = 1$, $P_{kk} = 0.2$, $P_{kch} = 0.4$, $P_{mk} = P_{chk} = 0.74$, $P_{mm} = P_{mch} = P_{chm} = 0.13$. For the glycerol-solvated state, the kaolinite scattering units (A), which are 7.15 Å thick, tend to form ABAB... sequences with B diffracting units 17.78 Å thick which can be, with the same probability, either montmorillonite or swelling chlorite.

9.6.3 Study of Hydrated Talcs

Hydrated talc-like minerals, variously known as β-kerolite, 10 Å-garnierite, or hydrated talc, have been described in a number of papers (Maksimovic 1966; Brindley and Hang 1973; Imai et al. 1973; Rosanova 1976; Suzuki et al. 1976).

Brindley and Hang (1973) described these minerals as characterized by the presence, in their XRD patterns, of a basal reflection at $d \approx 10$ Å which remains practically at the same position even if the mineral is dehydrated or solvated, from which it was deduced that the minerals could not contain a high proportion of swelling layers. Moreover, chemical analysis of these minerals showed that the ratio of cations in octahedral sites to those in tetrahedral sites was substantially different from 3/4, the value usually found in talcs.

Justification of the 10 Å thickness for an electrically neutral 2:1 layer was sought in the possible presence of water molecules in the interlamellar spaces and confirmed by differential thermal analysis which showed the presence of relatively important quantities of "structural" water, while three possibilities were explored to explain the 3/4 cationic ratio: a mixture of two phases, namely 2:1 layers 10 Å thick with a small quantity of 1:1 layers, or an interstratification of these two types of layers, or a stacking of 2:1 layers only, with structural defects. For the latter, two possible types were suggested:

- *discontinuities in one of the two tetrahedral sheets*; structurally, this would create local portions of 1 : 1 layers of serpentine in the zones where these discontinuities occurred, but a stack containing this type of structural defect would still give a reflection at 10 Å;
- or, in zones with defective 2 : 1 layers, *absence of two tetrahedral sheets,* leading to the formation of islands of brucite-type layers in the midst of the 2 : 1 layers.

Both the model with two phases and that with defective 2 : 1 layers implicitly imply that the basal reflections must form a rational series but, in actual fact, there is a certain nonrationality which was not detected by Brindley and Hang: the reflection taken to represent the third order reflection corresponding to the 10 Å phase is at 3.15 Å to 3.20 Å, a distance incompatible with true periodicity of layers at 10 Å. As for the hypothesis of an interstratification of 2 : 1 and 1 : 1 layers, this could in no way account for the peak positions observed in the experimental diagrams.

At about the same time, Imai et al. (1973) studied a hydrated talc which yielded reflections at $d \approx 9.7$ Å, 4.7 Å and 3.15 Å. Again, these positions were nearly unaffected by the treatments to which the mineral was submitted. The authors attributed the lack of rationality in the series of basal reflections to the small thicknesses of the interferential coherence domains (cf. Sect. 2.2.2). From the chemical analysis, the structural formula:

$$Na_{0.02}K_{0.02}(Si_{7.73}Al_{0.15})(Mg_{5.81}Mn_{0.4}Fe^{3+}_{0.06})O_{20}(OH)_4, n\,H_2O$$

which was adopted points to an excess of cations in the octahedral sites but a deficit in the tetrahedral sites, as compared to the normal formula of a talc mineral.

Similarly, the study of another hydrated talc from Chichijima (Japan) by Suzuki et al. (1976) gave experimental data analogous to that obtained by Imai et al., but these results were not exploited to yield conclusions concerning the structure of the minerals.

In 1985, Gilan et al. (unpublished) undertook a new analysis of these last two samples kindly provided by Professor Otsuka of Tokyo University. Figure 9.29 gives the experimental spectra of the samples studied by Imai et al. in the air-dry state (curve *a*), or solvated by ethylene glycol (curve *b*), or after heating to 350 °C (curve *c*). The small width of the basal reflections points to fairly large thicknesses of interferential coherence domains. Furthermore, as one goes from the XRD curve of the air-dried sample to that of the ethylene glycol-solvated mineral, then to that of the dehydrated sample, the weak shift in the positions of the basal reflections occurs in a way which is perfectly compatible with the characteristics of an interstratified structure.

Working on these bases, Gilan et al. calculated the theoretical spectra corresponding to structural models with a random sequence of talc-like layers 9.38 Å thick and smectite (s) layers. According to this model, the concentration of s layers varied between 5% and 10%, and their thickness was taken to be 16.86 Å, when

Fig. 9.29. Comparison between experimental and calculated X-ray diagrams for a hydrated talc in its natural state (**a**), or solvated by ethylene glycol (**b**), or dehydrated at 350 °C (**c**). The calculated diagrams correspond to a model of interstratified talc-saponite with defective 2:1 layers either solvated by ethylene glycol (**d**) or dehydrated at 350 °C (**e**), with parameters: $W_t : W_s = 0.9 : 0.1$, $d_t = 9.3$ Å, $d_s = 9.60$ Å or 16.68 Å, S = 0, and $5 \leqslant M \leqslant 15$

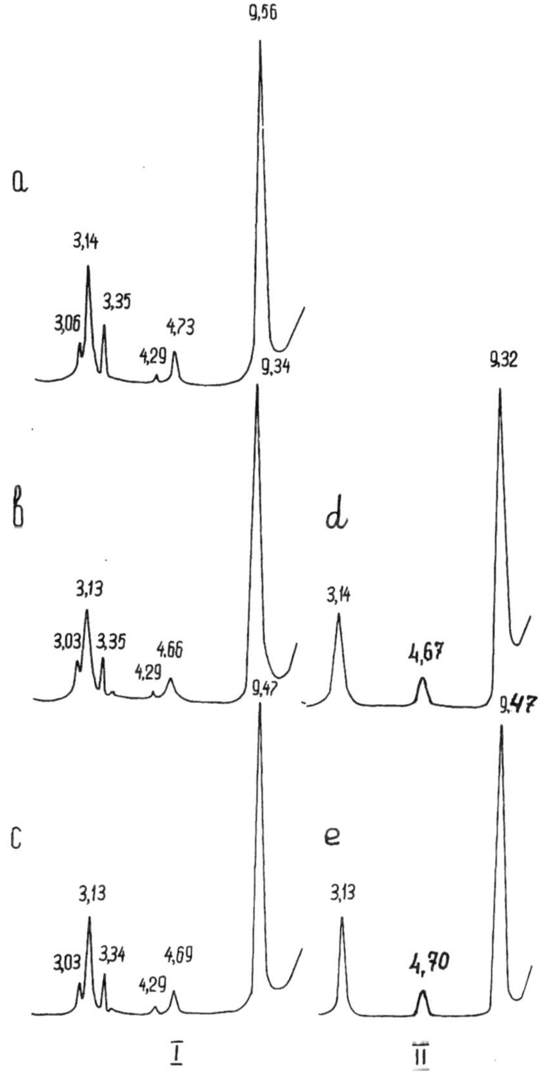

solvated by ethylene glycol, or 9.6 Å after dehydration at 350 °C. In order to find the effect of coherence domain size on the spectra, the values $\overline{M} = 10, 15$, and 20 were chosen for the mean number of layers. From the set of diagrams calculated, the following conclusions were drawn:

- in all the calculated diagrams corresponding to small thicknesses of coherence domains, the reflections are broader than in the experimental XRD diagram,

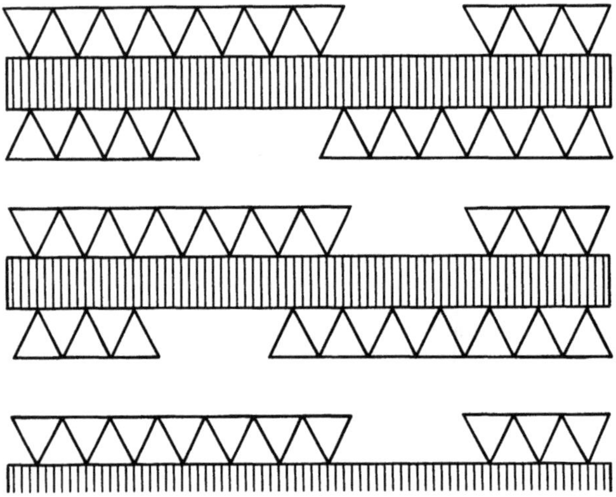

Fig. 9.30. Schematic representation of defective 2:1 layers in a hydrated talc structure

- the model with 10% content of s layer gives diagrams in which the **positions** of the reflections are in good agreement with the corresponding experimental values;
- in the calculated diagram there is not even qualitative agreement with the experimental **intensities** of the basal reflections: in all the calculated spectra, the intensities of the first and third reflections are about equal instead of being in the ratio 3/1, as in the experimental diffractograms (cf. curves *b* and *c* in Fig. 9.29).

As shown in Fig. 9.30, Gilan et al. were therefore led to admit, in the talc-like layers, the presence of defects similar to those described by Brindley and Hang: the tetrahedral sheets forming a 2:1 layer are discontinuous, with consequent modifications of the structure factor, but with no change of the thickness of the scattering units. Curves *d* and *e* of Fig. 9.29 show that a semi-quantitative agreement between experimental and calculated XRD patterns can then be obtained when the ratio of the concentration of defective (serpentine-like) to nondefective (talc-like) layers is taken to be 1/9.

References

Brindley GW, Hang Pham Thi (1973) The nature of garnierites. I. Structures, chemical compositions and color characteristics. Clay Mineral 21:27–40

Drits VA (1987) Mixed layer minerals: Diffraction methods and structural features. Proc Int Clay Conf, Denver, vol 1, pp 33–45

Drits VA, Sakharov BA (1976) X-ray analysis of mixed-layer minerals. Nauka, Moscow, 256 pp (in Russian)

Drits VA, Sakharov BA, Plançon A, Ben Brahim J (1984) The distribution of layers in mixed-layer crystals of identical composition. Kristallografia 29:350–355 (in Russian)

D'yakonov Yu (1961) About the application of Fourier analysis method to the interpretation of X-ray patterns of layer silicates with mixed-layer structure. Kristallografia 6-4:624–625 (in Russian)

D'yakonov Yu (1978) Interpretation of layer distribution curves in the direct method of interpretation of mixed-layer minerals. In: Sidorenko GA (ed) X-ray study of raw materials. Nedra, Moscow, pp 7–14 (in Russian)

D'yakonov Yu (1981) New data about the variety and identification of hydrobiotites. In: Frank-Kamenetsky VA (ed) Crystallochemistry of minerals. Nauka, Leningrad, pp 39–46 (in Russian)

D'yakonov Yu (1983) Mixed-layer minerals. In: Frank-Kamenetsky VA (ed) X-ray study of the basic types of rock-forming minerals. Nedra, Leningrad, pp 177–244 (in Russian)

D'yakonov Yu, Volosnykh GT (1979) Structural characteristics of mixed-layer illite-montmorillonites from the argillaceous rocks close to ores. In: Frank-Kamenetsky VA (ed) Crystallochemistry and structural mineralogy. Nauka, Leningrad, pp 69–82 (in Russian)

Gradusov BP (1971) X-ray patterns and direct Fourier transformation of diffraction patterns of mixtures of illite, mixed-layer mica-montmorillonite and montmorillonite. Izv Akad Nauk SSSR Geol Ser 12:86–92 (in Russian)

Imai N, Otsuka R, Nakamura T, Tsunashima A, Sakamoto T (1973) Hydrated talc: alternation product of wollastonite by reaction with magnesium-bearing hydrothermal solution. Clay Sci 4:175–191

Lipkina MI, Drits VA, Tsipursky SI, Ustinov VI, Srijov VP, Lakusheva IN, Tcherkashin VI (1987) Dioctahedral layer silicates having high content of iron from hydrothermal rocks and sediments of volcagenic mountains of Japan Sea. Izv Akad Nauk SSSR Geol Ser 10:92–111

Maksimovic Z (1966) β-Kerolite-pimelite series from Coles Mountain, Yugoslavia. Proc Int Clay Conf, Jerusalem, vol 1, pp 97–105

Plançon A, Drits VA, Sakharov BA, Gilan ZI, Ben Brahim J (1983) Powder diffraction by layered minerals containing layers and/or stacking defects. Comparison between Markovian and non-Markovian models. J Appl Cryst 16:62–69

Reynolds RC (1980) Interstratified clay systems. In: Brindley GW, Brown G (eds) Crystal structure of clay minerals and their X-ray identification. Mineral Soc, London, pp 249–303

Reynolds, RC, Hower J (1970) The nature of interlayering in mixed-layer illite-montmorillonite. Clays Clay Mineral 18:25–36

Rosanova TV (1976) The series kerolite-pirotin-troïlite in deposits of Hessa. Lithol Polezn Iskop 2:22–34 (in Russian)

Sakharov BA, Drits VA (1973) Mixed-layer kaolinite-montmorillonite. Clays Clay Mineral 21:15–17

Sakharov BA, Gilan ZI, Drits VA (1981) Structure characteristics of three component mixed-layer minerals. Lithol Polezn Iskop 3:136–142 (in Russian)

Sakharov BA, Salyn AL, Drits VA (1989) Possibilities and limitations of various methods for interpretation of mixed-layer mineral X-ray diffraction patterns. In: Frank-Kamenetskii VA (ed) X-ray study of crystalline materials. Nauka, Novosibirsk, pp 38–53

Schultz LG, Shepard AO, Blackmon PD, Starkey HC (1971) Mixed-layer kaolinite-montmorillonite from the Yucatan Peninsula, Mexico. Clays Clay Mineral 19:137–150

Sokolova TN, Drits VA, Sokolova AL (1976) Structural mineralogical characteristics and conditions of formation of leucophyllite from salt-bearing deposits of Inder dome. Lithol Polezn Iskop 6:80–92 (in Russian)

Sokolova TN, Sakharov BA, Drits VA (1978) Mixed-layer leucophyllite-montmorillonite minerals. Lithol Polezn Iskop 6:87–101 (in Russian)

Środon (1980) Precise identification of illite-smectite interstratifications by X-ray powder diffraction. Clays Clay Mineral 28:401–411

Środon J (1984) X-ray powder diffraction identification of illitic materials. Clays Clay Mineral 32:337–349
Środon J, Eberl D (1984) Illite. In: Bailey SW (ed) Micas. Mineral Soc Am, Washington, DC. Rev Mineral 13:495–544
Suzuki S, Nishido N, Otsuka R (1976) Hydrated talc from Chichijima, the Ogasaware Islands, Japan. J Jpn Assoc Mineral Petrol Econ Geol 71:238–247
Watanabe T (1981) Identification of illite/montmorillonite interstratifications by X-ray powder diffraction. J Mineral Soc Jpn Spec Issue 15:32–41 (in Japanese)

Author Index

Adams, J.M. 237
Alcover, J.F. 277
Alexander, I.E. 168, 170, 175
Allegra, G. 71, 113, 155
Amouric, M. 13
Appelo, C.A.Y. 5
Armagan, N. 20, 182, 272, 273
Averbach, B.L. 213
Bailey, S.W. 5, 9, 27, 234, 235, 237, 238, 243, 277
Bak, P. 30
Bansal, R.S. 199
Baronnet, A. 13
Barshad, I. 277
Ben Brahim, J. 20, 25, 127, 130, 182, 196, 234, 242, 243, 253, 272, 273, 278, 280, 281, 284, 341
Berkhin, S.I. 28
Bernard, R. 99
Besson, G. 20, 41–43, 78–80, 182, 254, 255, 260, 262, 264, 267, 269, 272, 273, 278, 280, 281, 285, 286, 289
Bhattacherjee, S. 234
Blackmon, P.D. 351
Blatter, C.L. 28
Blount, A.M. 238
Bokros, J.C. 199
Bookin, A.S. 5, 236, 238, 242, 244, 245, 247–250, 255, 262, 264
Bouraoui, A. 21, 199, 201, 208
Boutouzova, G.Yu. 260
Bradley, W.F. 277
Brooks, J.D. 200
Burger, J. 137
Burnham, C.W. 22, 290
Busek, P. 22
Calle (de la), C. 260, 277
Cardile, C.M. 267
Carmona, F. 201
Cauchois, Y. 168
Cesari, M. 103, 110, 111, 113, 118
Chekin, S.S. 246, 260
Chukhrov, F.V. 28
Clinard, C. 30, 234, 242, 248, 253
Conard, J. 201
Cornuault, P. 208, 212, 218, 222, 227, 228
Courville (de), J. 55, 57, 192, 208, 212
Courville-Brenasin (de), see Courville (de)
Cradwick, P.D. 135
Cvikevich, S. 54
Dainyak, L.G. 255, 262, 264
Debye, P. 166
De Courville, see Courville (de)
De la Calle, see Calle (de la)
Delhaes, P. 201
Deluca, S. 253
Diamond 200
Dickson, D.P.E. 267
Digonskii, V.V. 96
Dmitrik, A.L. 27, 29
Donnay, G. 289
Donnay, J.D.H. 289
Donnet, J.B. 199
Drits, V.A. 5, 13, 22, 25, 27–29, 43, 78–80, 83, 103, 119, 121, 125–127, 130, 135, 150, 155, 159, 236–238, 242, 245, 247–249, 251, 255, 257, 259, 260, 262, 264, 266, 285, 286, 288–290, 311, 313, 314, 321, 322, 331, 332, 341, 343, 348, 350–352
Duchaffault, F. 208, 212, 218, 222, 227, 228
Duffin, M.A. 251
D'yakonov, Yu.S. 135–139, 142, 306, 307, 309–317, 332
Ergun, S. 219
Estrade, H. 201
Ewald, P.P. 34
Favretto, L. 103, 110, 111
Fedotov, A.F. 257, 290
Finck, C. 13
Finko, V.J. 246
Fischbach, D.B. 200
Fitzer, E. 199, 212, 213
Fourme, R. 172
Franklin, R.E. 222
Fraser, A.R. 254, 264
Fuselier, H. 30

Gard, J.A. 22
Gasparoux, H. 9, 200
Gatineau, L. 20, 208, 228, 277
Gaultier, J.P. 255, 260
Gay, R. 9
Giese, R.F. 242, 249, 250, 252, 253
Gilan, Z.I. 127, 341, 352, 358
Gingrich, N.S. 194, 202
Glaeser, R. 254, 255, 267, 275
Goodman, B.A. 254, 264
Goodyear, B. 251
Gorshkov, A.I. 28
Gradusov, B.P. 341, 347
Grim, R.E. 6, 135, 234
Guerard, D. 10
Guet, J.M. 201, 208
Guinier, A. 18–20, 89, 137, 168, 170, 174, 175, 181, 194, 204
Guntert, D.J. 54
Guven, N. 254, 290
Hang, P.T. 355, 358
Harrison, J.L. 249, 252, 253
Heller-Kallai, L. 254, 264
Hendricks, S.B. 71, 103, 135
Herold, A. 10, 22, 28
Hewat, A.W. 237
Hirsch, P.B. 22
Houska, C.R. 93, 96, 222
Hower, J. 321
Howie, A. 22
Hull, A.W. 166
Iakusheva, I.N. 343
Imai, N. 355, 356
Jagodzinski, H. 24, 103
James, R.W. 137
Jefferson, D.A. 200
Johann, H.H. 168, 172
Johnson, J.H. 267
Joswig, W. 237–238
Joyez, P. 208
Kakinoki, J. 71, 76, 103, 135, 150, 155
Kao, C.C. 249, 252, 253
Kameneva, M.Yu. 285, 286, 289, 291
Kashaev, A. 237, 238, 251
Khan, R. 172
Klug, H.P. 168, 170, 175
Kodama, H. 20
Komura, Y. 71, 76, 103, 135, 150, 155
Kossovskaya, A.G. 260
Krylov, V.N. 96
La Calle (de), see Calle (de la)
Lagrange, P. 10
Lahaye, J. 199
Laue, M. 43
Lemonier, M. 172
Lewis, I.C. 200
Lipkina, M.I. 343
Lipsicas, M. 249, 252, 253
Lisitsina, N.A. 260
Longuet-Escard, J. 22
Macewan, D.M.C. 103, 135, 136, 139–142
Maire, J. 20, 43, 93, 201, 221–223, 225, 227, 228, 230
Maksimovic, Z. 355
Mamy, J. 255, 260
Marsh, H. 199
Mathieson, A.M. 277
Melin, J. 28
Mering, J. 20–22, 38, 39, 43, 45–47, 49–51, 61, 78, 82, 93, 96, 103, 199, 201, 208, 212, 218, 221–223, 225–230, 254, 275
Millward, G.R. 200
Mitra, G.B. 234
Moore, A.W. 199
Morelli, G.L. 103, 110, 111
Mueller, K. 199, 212, 213
Murr, L.E. 254
Murray, H.H. 234, 242
Nakamura, T. 355, 356
Naumov, A.S. 135, 150, 155, 159
Newnham, R.E. 238, 242
Nicholson, R.B. 22
Nicolaeva, I.V. 285
Nishido, N. 355, 356
Noble, F.R. 54, 234
Norrish, K. 289
Novikov, V.M. 28
Oberlin, A. 254
Oberlin, M. 229
Odom, I.E. 284, 285
Organova, N.I. 27–29
Osthaus, B.B. 266
Otsuka, R. 355, 356
Pacault, A. 200
Pashley, D.W. 22
Pease, R.W. 254
Pernoux, E. 99
Plançon, A. 22, 25, 28, 43, 55, 57, 72, 73, 78–80, 99, 127, 130, 150, 234, 236, 238, 241–243, 245, 248–250, 252, 253, 341
Pons, C.H. 43, 196, 277
Prado, G. 199
Pynn, R. 30
Radoslovich, E.W. 289, 290
Rang, O. 99
Rautureau, M. 260, 262, 264
Rayner, J.H. 135, 140
Raythatha, R. 249, 252, 253

Reynolds, R.C. 23, 42, 54, 103, 135, 136, 140, 144, 188, 321, 331
Robertson, H.E. 28
Robinson, K. 234, 241, 245
Rosanova, T.V. 355
Ross, M. 42
Rothbauer, R. 3
Rouchy, J.P. 208, 228
Rousseaux, F. 28, 30, 43, 50, 52, 54, 172, 208, 212
Rozenson, I. 254, 264
Rudnitskaya, E.S. 28
Ruland, W. 54, 201
Russell, J.D. 254, 264
Sakamoto, T. 355, 356
Sakharov, B.A. 22, 25, 28, 43, 78–80, 103, 119, 121, 127, 130, 135, 150, 155, 159, 285, 286, 289, 311, 313, 314, 321, 322, 331, 332, 341, 348, 350–352
Salyn, A.L. 238, 247, 311, 313, 314
Samatoin, N.D. 245, 246
Sanz, J. 3
Sato, M. 103, 109, 110, 118
Schaeffer, W. 199, 212, 213
Scherrer, P. 166
Schiller, C. 208, 212, 218, 222, 227, 228
Schultz, L.G. 351
Serratosa, J.M. 3, 277
Shepard, A.O. 351
Shirozu, H. 277
Shterenberg, B.E. 28
Sidorenko, S.V. 257, 290
Simon, C. 201
Singer, L.S. 200
Slaughter, M. 253
Smith, J.V. 9
Smolyar, B.B. 5, 285, 286, 289
Snyder, R. 249, 250, 253
Sokolova, A.L. 332
Sokolova, T.N. 332
Solin, S.A. 22
Sow., C. 234, 233, 242, 248
Srodon, J. 331, 332
Starkey, H.C. 351
Strijov, V.P. 343
Suitch, P.R. 237, 238
Suzuki, S. 355, 356
Takeda, H. 289
Taylor, G.H. 200
Taylor, R.M. 54
Teller, E. 71, 103, 135
Terner, S. 22
Tettenhorst, R. 135
Tcherkashin, V.I. 238, 247, 343
Tchoubar, C. 17, 20, 22, 30, 55, 57, 72, 73, 78–80, 99, 182, 192, 196, 234, 236, 238, 241–243, 245, 248, 249, 253–255, 260, 262, 264, 267, 272, 273, 277, 278, 280, 281
Tchoubar, D. 28, 30, 43, 50, 52, 54, 55, 57, 192, 196, 201, 208, 212, 226, 277
Thomson, G.R. 28
Threadgold, I.M. 238
Tompa, H. 54
Trunz, V. 43
Tsipurski, S.I. 78–80, 259, 260, 262, 264, 286, 288, 290, 343
Tsunashima, A. 355, 356
Ustinov, V.I. 343
Vangelisti, R. 28
Veblen, D.R. 22
Vitovskaya, E.S. 28
Volosnykh, G.T. 310–312, 332
Vrublevskaya, Z.V. 257, 290
Walker, G.F. 277
Walker, P.L. Jr. 199
Warren, B.E. 33, 43, 46, 81, 93, 96, 194, 202, 213, 222
Watanabe, T. 331, 332
Weir, A.H. 135, 140
Whelan, M.J. 22
Wilson, A.J.C. 43, 50, 103
Woodhams, F.W.D. 254, 264
Yoder, H.S. 9
Young, R.A. 237, 238
Yucel, A. 196
Zhoukhlistov, A.P. 257, 290
Zvyagin, B.B. 9, 237, 238, 256, 257, 290

Subject Index

Absolute scale of the intensity measurements 194, 195, 202
– for kaolinite 195, 196
– for Na-montmorillonite 197
Absorption coefficient 170, 179, 182, 183
– correction 183
Anthracene 207
– coke 208, 209–211, 217, 229
AO: see azimutal orientation of a layer
Apical oxygen 6
Apparent interplayer spacing 23
Arrangements of layers with maximum possible degree of order for a given S factor 109, 314, 315
Asbolane 28
Atomic coordinates 5
– for B-layer of kaolinite 251; see also Refinement of kaolinite structure
– of glauconites 291
– of montmorillonite 41
Atomic motif 1, 33
Atomic pattern: see atomic motif
Atomic scattering factors 2
Azimuthal orientation of a layer 8

Basal reflections 23, 135
– d-spacings of 135
– integrated intensity of 138, 139, 306
– irrational series of 23, 135
Beidellite 20, 254, 257, 258, 267–277
– Cs 269–272
– one-water layer 281–284
– two-water layer 273–277
Binary axis 8
Biotite 43
Biotite-vermiculite 317
Biperiodicity 4, 13, 27
– quasi ideal 4
Bragg angle 21

Calculated XRD diagrams for illite-mont morillonite with
– S = 0 323
– S = 1 324, 325, 334, 335
– S = 2 327, 329
– S = 3 330
Carbon 199
– atomic scattering factor 202
– black 199, 200
– coherence domain 208, 212, 216, 222, 224
– defects of the first type 205
– degree of order 221
– disorder of the second type 221
– diffraction bands 204, 206, 207, 216, 219, 224, 226, 229, 230
– fiber 199
– glassy 199, 203, 207–219
– graphitizable 199, 200, 208, 210, 212, 215, 218, 223, 224, 228
– interference function 202, 203, 207
– interlayer distances 229
– interstratification of α and β layers 221
– – with R = 0 227
– layer structure of 202
– – distribution δ^2: see mean square deviation
– – mean square deviation δ^2 of the atoms of 207, 213, 215, 222–226
– modulation function 221, 224, 230
– nongraptizable 200, 208, 212, 217, 224
– position fluctuation δ_c function 221, 223
– random defects 221, 230
– soft 199, 200
– structural and diffraction characteristics of 199
– T defects 223, 225, 226, 230
– well-defined translation defects 221, 224, 229
– 2D hexagonal lattice 202
– α β transition 216, 222
– α(M) function 221
– β-layer 201, 207, 221, 227
– β-state 216, 229
Cation exchange 255
Celadonite 184, 254, 344

Celadonite-nontronite 343
Chlorite 10, 27, 305, 307, 317
Chlorite-montmorillonite 306, 307, 309
Chlorite-smectite 316, 317, 321
Chlorite-swelling chlorite 317
Ch-s: see chlorite-smectite
Cis-octahedra 7
C-n: see celadonite-nontronite
Coherence domains 21, 40, 41
– apparent size of 177
– shape of
– – circle 49
– – rectangle 52, 298
– – regular polygons 49
– – ribbons 53
– thickness of 333, 347
– – distribution function α(M) of 333
Collimation of the X-ray beam 167
Composition: see also Crystallochemical formula
– beidellite 272
– glauconites 284, 285
Compton effect 203
Conditional probailities 25, 103–105, 149, 150–152
Convolution 20
– effects due to X-ray beam width 175
Crystal 2
– actual 2
– lamellar 33
– perfect ordered 2
Crystallite 21
Crystallographic parameters of glauconites 286, 288, 294
Crystal structure 1, 2
– actual 3
– chain 33
– channel 22
– commensurate 27–30
– imperfect 3
– incommensurate 27–30
– interstratified 23
– – irregular 23
– – regular 23
– lamellar 22
– pseudo-lamellar 4
Crystallogchemical formula of: see also Composition
– microcrystalline mica 78
– montmorillonite 41
– nontronite 262, 266
– smectites 260

Deconvolution method 178
Decoration electron microscopy pattern 246, 247
Defects: see also Faults and Stacking faults
– correlated 69
– due to enantiomorphism 69
Frenkel 207
– point 2
– random rotation 69
– random translation 77
– Schottky 207
– structural 33
– uncorrelated 69
– well defined translation 69, 81
– well defined rotation 69, 83
Degree of graphitization 222, 223, 225, 228, 229
Depolarization factor 181
Dickite 233, 235, 236
Diffraction from
– crystal powders 33
– randomly oriented stacks 43
– partionally oriented particles 54–66
Direct methods of structural analysis 2, 3
Disordered kaolinite 22
Disorder of the
– first type 18–20
– second type 18–20
Distribution of
– cations and water molecules in smectites 272–284
– cations in octahedral sites
– – in dioctahedral smectites 254–272
– – in kaolinite: see Vacant octahedral sites in kaolinite
– coherence domain dimensions 50
– Dirac 34, 35
– particles in the powder spatial 55
D'yakonov method 136–139, 305, 311

Electron microscopy 22
– transmission (TEM) 22
– – dark field 22
– – high resolution (HREM) 4, 212
– – scanning (STEM) 22
Enantiomorphous layers 7, 13
Evald sphere 44, 45

Factor
– structure 35, 37
– Lorentz for powder 46
– scattering atomic 36
Faltung: see Convolution
Faults: see also Defects and Stacking faults
– position 70
– random 18

– rotation 13, 70
– three categories of 69
– translation 70
– – well defined 17
Fluctuations in the position of the layers 69
Fluctuations of the translation vectors 87
Fourier
– series 142
– transformation 34, 35, 135, 137
Frustrations 29
Function
– Bessel 50
– Dirac 315
– interference 38, 39, 135, 136, 139–142
– – principal maxima of 39
– modulation 38
– of Dirac distribution 34
– of electron density 34
– Patterson 48
– shape 34
– $\varrho(z)$ 138, 305, 306, 310, 314, 315
– – principal maxima 306
– – secondary peaks 306
– $\omega(z)$ 139
Fundamental
– lattice translations 28
– vectors 1

Garnierite: see hydrated talc
Glauconite 254, 284–300, 344
Goniometer 174
Graphite 9, 27, 199
– defects of 1st type in 205
– hexagonal: see polytype 2H
– highly oriented pyrolytic 199
– HOPG: see highly oriented pyrolytic
– intercalation compound of 30
– layer type of 201
– polytype
– – 2H 199, 220, 223, 229, 230
– – 3R 220, 223, 230
– rhombohedral: see polytype 3R
– shape factor of 204
– structural factor of the layer of 202, 206, 207, 219
– β-layer of 201, 215, 221, 222, 227
Graphitization process 200, 201, 207, 216, 221, 222, 227

Heterogeneity of composition of the interstratified particles 127
Highest treatment temperature of carbones: see HTT of carbon
High-resolution NMR 3
Homogeneous interstratified system 127
– two-component system 129–131
– – with $S = 0$ 129
– – with $S > 0$ 131
Homogeneous mixed-layer models 341
HTT of carbon 200, 201, 207, 212, 215–217, 223, 226, 227, 230
HREM: see Electron microscopy
Hydrated talc 355–358

Illite 305, 317, 332, 335, 346
Illite-montmorillonite 310–313, 316, 323–330, 332–335, 347–359
Illite-smectite 137, 310, 321, 336, 338, 339, 340, 342
I-m: see illite-montmorillonite
Incommensurability: see crystal structure, incommensurate
Indirect method of analysis in layer structures 5
Intensity of diffracted beam
– around the hkl node 45
– from isolated layer 35
– – within a hk rod 36
– from ordered stack of layers 37, 38
– – alongthe Z axis 41
– – within th (hk) rod 45
– from powder with position defects 70

Interaction between
– layers 24, 26
– translations 11, 26
Intercalation compound 4
Interference function 38, 39, 135, 139–142
Interlamelar space 6
Interlayer space: see Interlamelar space
Interstratification: see Interstratified system
– in Cs-beidellite 269
– in glauconites 284
Interstratified crystal
– Patterson function of 137
– – principal maxima of 137
– – secondary maxima of 137
– projection of the electron density of 136
– – self-convolution of 137
– characteristics along 00Z axis of 136
– ampitude of the waves diffracted by 136
– distribution of the intensity diffracted by 137
Interstratified mineral 22–27
Interstratified structures
– disordered 112

Interstratified structures (cont.)
– symmetrical 113
– three component 112–114
Irrationality
– apparent 42
– two sources of 42
I-s: see illite-smectite
Isomorphic substitution 2
– i tetrahedra 266, 267
– in octahedral sites 267
Isotropic powder 45, 165, 166, 187, 192

Kaolin minerals 233, 234, 235
Kaolinite 233–253, 317
– refinement of structure 237
Kaolinite-montmorillonite-swelling chlorite 352–355
Kaolinite-smectite 310, 316, 321, 348–350, 352
Kaolinite-smectite-illite 321, 348–353
Kerolite: see hydrated talc
K-m-ch: see kaolinite-montmorillonite-swelling chlorite
K-s: see kaolinite-smectite
K-s-i: see kaolinite-smectite-illite
K-unit: see scattering unit, kaolin-type 319

Lattice 1
– biperiodic 15, 16
– mean idealized 2
– nodes of reciprocal 2
– one dimensional irregular 143
– pseudo-hexagonal 13
– quasi-ideal biperiodic 4
– reciprocal 2, 35
– – three-dimensional 2
– triperiodic 1
– two-dimensional 35
Layer 4
– dioctahedral 7
– intrinsic structure of 5
– n 60° rotations of 17
– of brucite 10, 23, 26–28
of C_1-type or C_2-type 7, 24
– of chlorite 24
– of kaolinite 23
– of M-type 7
– of mica 23
– of mackinovite 26
– of montmorillonite 23
– of serpentine 24
– of "sulfide" 27, 28
– of T-type 7, 24
– sequences in smectite 267–272
– silicates 233
– stacking 6
– structural characteristics of 5
LC – lower case digits 160
Leucophyllite-montmorillonite 346
Leucophyllite-montmorillonite-"vermiculite" 344, 345
Linear detector: see localization detector
L-m: see leucophyllite-montmorillonite
Localization detector 169, 175, 176, 177, 180, 181, 186
Long-range order 2, 19, 20
Lorentz factor 46, 181, 187–189
Lorentz-polarization factor 138, 181
Lorentz-polarization effects: see Lorentz polarization factor
Lovenstein's rule 3

Mac-Ewan direct method 139
Mackinovite 27
Manganese oxides 28
Markovian statistics 25, 103, 321, 333
Matrix formalism 135, 149–160
Mesophase 200, 208, 209
Mg-vermiculite-chlorite 317
Mica
– microcrystallized 233
– polytypes
– – 1M 284, 285, 288, 291, 294, 297
– – 1Md 285
– – $2M_1$ 294
– – $2M_2$ 297
Mica-chlorite 144
Mica-smectite 144, 146, 147, 341
Microcrystallized lamellar systems 33
Mixed-layer mineral: see Interstratified mineral
Mixed-layer system: see interstratified system
Mixture of two kinds of kaolinites 253
MLM: see mixed layer mineral
Mode of layer stacking 5
Modulation wave 28
Monochromaticity 165
Monochromator 168, 171, 173, 174, 176, 181
– downstream 168
– Johann-Guinier 168, 172
– upstream 169
– with variable radius of curvature 172, 173
Montmorillonite 42, 254, 257, 258, 307, 317, 340, 345, 346
– Fe-containing 258
Montmorillonite-chlorite 317

Mosaic crystal 21
Mosaicity 21
Mounting
- Debeye-Scherrer-Hull 166, 167
- reflection 57, 169, 170
-- asymmetrical 62
- transmission 62, 169, 170, 174, 184, 190
-- non symmetrical 55, 57, 184, 185, 190
-- symmetrical 55, 57, 59, 170, 185, 190
MPDO: see Arrangements of layers with maximum possible degree of order
M-unit: see scattering unit, mica type 319
Muscovite 3, 10, 20, 43

Nacrite 233, 235, 236
Nitric acid 30
Node of reciprocal lattice
- two-dimensional (hk) 35, 39
- three-dimensional (hkl) 39
Noncommensurability: see crystal structure, incommensurate
Nontronite 254, 257, 258, 262-267, 344

Oblique texture electron diffraction (OTED) 256-262
OTED: see Oblique texture electron diffraction
Octahedral sheet 6
- cavities in 7
Order parameter: see interaction between layer or translation
Orientation of particles in a powder 165, 170, 174, 187, 190-193
- random 43
- partial privileged 54

Particle orientation function 175, 187, 190-193
Patterson series 2
Petroleum coke 201
Phengite 10
Phlogopite 305
Phlogopite-biotite 10
Phlogopite-vermiculite 310
Phyllosilicates 233
Pitch coke 201, 211
Planar defects 3, 4
Point defects 2
Polarization of X-ray beam 181, 186
Polyfurfuryl alcohol 212
Polytypes of: see also Mica polytypes
- graphite
-- 2H 9
-- 3R 9
- muscovite
-- 1M 9, 13
-- 2M1 3, 9, 10, 13, 20
-- 3T 9
Polyvinyl coke 230
Preferred orientation of particles: see Orientation of particles
Proportional counter 168, 175
Probabilistic parameters 103
Probabilities of occurrences of the layers 104
Proportion of each type of layer 106, 127
Proportion of layer sequences 106
Pyrocarbons 54, 199
Pyrolysis 199, 212, 216

Quasi-homogeneous interstratified powder 124

Range of interaction: see interaction between layers or translations
Range parameter: see interaction between layers or Reighweite
Rectorite 23
Refinement of the kaolinite structure 237
Reichweite 24
Relative occupancy of translation vectors 11
R factor 26, 270, 296, 298; see also Interaction between translations

SAD: see selected area electron diffraction
Sample thickness
- choice of 179, 180
- convolution effects due to 171, 179, 180
- optimal 179
Saponite 317
Scattering powder: see Structural factor
Scattering unit of 318
- kaolin-type 319
- smectite-type 319
- mica-type 319
- thickness of 332
Selected area electron diffraction 28
S factor: see Interaction between layers
Sherrer relation 21
Short-range order 2, 11, 19
Slit-width choice 177
Smectite 41, 233, 254-284, 305, 317, 319, 335, 340
- Cs 267-272
- dioctahedral 262

Smectite (cont.)
– Fe 254, 257
– Mg-Al 257
Soller slits 188, 189
Space
– object 33
– reciprocal 35
Stacking faults models due to
– additional translation by ±b/3
– – in kaolinites 241, 242, 245–249
– – in nontronites 265, 268–270
– – in glauconites 291
– – in Cs-beidellites 268–271
– – in Na-beidellite 280, 282
– presence of enantiomorphs 69
– – in kaolinite 245–249
– – in glauconite 292, 294–296
– pyrophyllite-like translations 280
– random translations
– – in kaolinites 252
– – in nontronite 265
– – in Na-beidellite 280, 281, 283
– – in Cs-beidellite 269, 271
– – in glauconites 292
– rotation by ±120°
– – in kaolinite 242
– – in nontronite 265, 268, 271
– – in glauconite 284, 292–294
– – in Cs-beidellites 268
– rotation by n 60°
– – in glauconites 296–298
– vacancy displacement in kaolinite 242–245
– vermiculite-like translations 280
Stackings of layers in two-water-layer Na-beidellite 280
STEN: see Electron microscopy
Structural analysis of 1
– interstratified systems by 4
– – direct methods 136
– – indirect methods 5, 136
– partially disordered systems 4
– single crystal 2, 3
Structural factor 136, 140
– of the layer 139
Structural defects in
– kaolinites: see also Mixture of two kinds of kaolinites
– Cs-smectite: see Layer sequences in smectite
– beidellite: see Stackings of layers in two-water-layer Na-beidellite
– glauconite 284–300
Substacks of layers 127, 128
S-unit: see scattering unit, smectite type 319
Symmetrical interstratified structure 113
Symmetrical 0–20 reflection conditions: see Mounting reflection
Synchrotron radiation 172, 173
Synthetic XRD pattern: see calculated XRD diagram

Talc 11, 13, 317, 319, 356
Tangent cylinder approximation 46
Tarasovite 335
TEM: see Electron microscopy
Tetrahedral sheet 6
– hexagonal cavities in 6
Tochilinite 27
Transmission conditions: see Mounting transmission
Trans-octahedra 7
Two-component interstratified structure 106–121
– with range order factor S = 0 106
– – independent parameters 107
– with S = 1 107–112
– – classification 109–111
– – conditional probabilities 107
– – degree of disorder 111
– – degree of segregation 111
– – independent parameters 107, 108
– – MPDO 109
– with S = 2 114–122
– – classification 118–121
– – coefficient of segregation 121
– – conditional probabilities 114
– – degree of disorder 120
– – independent parameters 115, 116, 117
– – MPDO 119
– with S = 3 122, 123
– – conditional probabilities 123, 124

UC – upper case digits 160
Umbrella effect 178
Unit cell 1, 33
Unit cell parameters of: see also Crystallographic parameters of glauconites 286, 288, 294
– dickites 238
– kaolinites 238, 240
– – with different vacancy position 243
– smectites 259

Vacant octahedral sites in kaolinite 235, 237, 242–244, 248–249
Valerite 28

Vermiculite 305, 317, 345

WD: see Wetting and drying cycles
Wetting and drying cycles (WD) 255
Width at half maximum intensity 21

X-ray diffractogram 2
XRD: see X-ray diffraction 1, 22
XRD curve: see X-ray diffractogram
XRD patterens from: see X-ray diffractogram
- powder 43
-- isotropic 43

1:1 layer 233, 234, 237
1D-lattice 18
2D-lattice: see Crystal lattice, biperiodic

2:1 layer
- dioctahedral 254
- cis-octahedral sites in 254
-- vacant 254, 259–262, 272
- swelling 291
- trans-octahedral sites in 254
-- vacant 254, 259–262, 272, 288, 289, 291
- trioctahedral 254

MIX
Papier aus verantwortungsvollen Quellen
Paper from responsible sources
FSC® C105338

If you have any concerns about our products,
you can contact us on
ProductSafety@springernature.com

In case Publisher is established outside the EU,
the EU authorized representative is:
Springer Nature Customer Service Center GmbH
Europaplatz 3, 69115 Heidelberg, Germany

Printed by Libri Plureos GmbH
in Hamburg, Germany